HEATING HANDBOOK

Other Books in McGraw-Hill's Complete Construction Series

Bianchina ▪ *Room Additions*
Carrow ▪ *Energy Systems for Residential Buildings*
Gerhart ▪ *Home Automation and Wiring*
Vizi ▪ *Forced Hot Air Furnaces: Troubleshooting and Repair*
Woodson ▪ *Radiant Heating Systems: Retrofit and Installation*

Dodge Cost Guides ++ Series

All from McGraw-Hill and Marshall & Swift

Unit Cost Book
Repair and Remodel Cost Book
Electrical Cost Book

HEATING HANDBOOK

Chase Powers

McGraw-Hill

New York San Francisco Washington, D.C. Auckland Bogotá
Caracas Lisbon London Madrid Mexico City Milan
Montreal New Delhi San Juan Singapore
Sydney Tokyo Toronto

Library of Congress Cataloging-in-Publication Data

Powers, Chase.
Heating handbook / Chase Powers.
p. cm.
ISBN 0-07-050719-8
1. Heating Handbooks, manuals, etc. I. Title.
TH7225.P63 1999
697—dc21 98-53828
CIP

McGraw-Hill

A Division of The ***McGraw-Hill*** *Companies*

1 2 3 4 5 6 7 8 9 0 DOC/DOC 9 0 4 3 2 1 0 9

ISBN 0-07-050719-8

The sponsoring editor for this book was Zoe G. Foundotos, the editing supervisor was Stephen M. Smith, and the production supervisor was Sherri Souffrance. It was set in Melior by Constance M. Tucker of Lone Wolf Enterprises, Ltd.

Printed and bound by R. R. Donnelley & Sons Company.

This book was printed on recycled, acid-free paper containing a minimum of 50% recycled de-inked fiber.

This book is dedicated to Afton and Adam,
the best children a father could ever hope for.

ABOUT THE AUTHOR

Chase Powers is one of America's favorite construction authorities. The author of the *Builder's Guide to Change-of-Use Properties* and *Kitchens: A Professional's Illustrated Design and Remodeling Guide*, Powers has been a contractor for more than 20 years. He runs a plumbing, heating, and home construction business that builds approximately 60 single-family residences a year.

Contents

Introduction

Heating systems can be complicated to work with. Even trained technicians run into situations that give them trouble. Someone who is highly skilled at working with heat pumps may be at a loss for what to do when faced with an oil-fired boiler. Bring a coal-fired heating system into the mix and you might have mass confusion. There are many types of heating systems in use, and keeping up-to-date on all of them is not a simple task. However, this book will make the job faster, easier, and even enjoyable.

Heat pumps, gas-fired boilers, oil-fired boilers, wood-fired heating systems, coal-fired heating systems, and even solar heating systems are all discussed in the following pages. In this single resource you will find answers to most heating questions, as well as installation tips, general descriptions and advice, and a lot of troubleshooting data that will save you both time and money.

Written by an experienced tradesman and author for people in the trades, this handbook is both comprehensive and easy to understand. You will find the text to be informative, reader-friendly, accurate, and presented in a real-world style. The illustrations placed liberally throughout the pages make it easy to visualize what is being described. Don't let this book's size intimidate you. The chapters are broken down by specific topics on particular types of heating systems. A quick perusal of the Contents or Index will put you right where you want to be, in a hurry.

The troubleshooting sections are true time-savers. Whether you are in business for yourself, working for an employer, or a homeowner with a problem, you will appreciate the speed with which you can pinpoint potential problems in most any type of heating system. Saving time often means making more money.

Do customers sometimes ask for your advice about which type of heating system to install, for example, if an air-source heat pump is better than a water-source heat pump? Sometimes the answers can be difficult to come by. Not anymore. This valuable book includes plenty of suggestions that will help contractors meet their customers' needs. Being knowledgeable about several types of heating systems pays off whether you are a heating contractor, a builder, or a remodeling contractor. If your work involves heating systems in any way, you need the *Heating Handbook*.

A Basic Overview of Heat Pumps

Heat pumps have been in existence for decades, but still a large number of people doubt the efficiency so often raved about when describing the operation of heat pumps. Are these people just die-hard skeptics, or are they correct in their lack of enthusiasm over new, energy-saving heat pumps? Heat pumps can be extremely efficient, both as a heating unit and as an air-conditioning unit. The cost of operating a heat pump can be considerably less than the expense of producing equivalent heat with electric heat or oil heat. Given the many types, models, and configurations of heat pumps available, a heating and cooling system can be designed to give maximum comfort in climate control and cost-effectiveness.

The key to having a favorable experience with a heat pump is knowledge. A heat-pump system that works wonderfully in North Carolina may not fare so well in Maine. It is not practical to assume one type of heat-pump system will work in all parts of the country. There are, however, enough options available when designing a heat-pump system that it is possible to build a suitable system for almost any need.

If you have customers who are thinking of having a heat pump installed, you have a lot to consider. Not taking the time to explain options to your customers will most likely result in disappointment. The subject of heat pumps does have many facets, but these many angles are to your advantage. Having so many ways to build a heating-and-cooling system allows you to customize a system to meet your customer's needs in the most economical and efficient manner.

I've worked in the construction trades for more than 20 years. My career started as a plumber and evolved into much more. My first business was a plumbing and heating business. That business grew to include full-scale remodeling jobs, and from there, the business went on to include building up to 60 single-family homes a year, most of them equipped with heat pumps. As a plumbing and heating contractor, I've installed and worked on many types of heating equipment. Hot-water baseboard heat, radiators, forced-hot-air systems, wall-mount heaters, in-floor radiant heat systems, and heat pumps have all been a part of my working environment.

With the experience and knowledge that I have gained in the construction field, why would I equip so many of the houses I've built, including homes for my personal use, with heat pumps? The reason is simple; heat pumps, in my opinion, are one of the most effective means of providing heating-and-cooling services for a home.

As much as I like heat pumps, I must admit that they are not always the best choice when debating a new heating or cooling system. States with extremely cold climates, like Maine, are not ideal locations for heat pumps to do their best work. This is not to say that heat pumps can't be effective in extremely cold temperatures, but there are several factors to consider in making the right choice in such a climate. I don't want to get too technical yet—this chapter is meant to give you an overview of heat pumps—but let me give you an example to consider.

Assume that it is a winter day. The temperature in Virginia is 20°F. The temperature in Maine is also 20°F. If identical houses, both with identical heat pumps, existed in Maine and Virginia, would both of the heat pumps be equally effective on this winter day? They might, but there is a good chance that the heat pump in Maine would be forced to work much harder than the heat pump in Virginia on this particular day. How could this be when the temperature in both locations is the same?

Maine typically receives much more snow than Virginia does, and falling snow and moisture in the air can make a big difference in the performance of a heat pump. If it were snowing in Maine, the heat pump might have to defrost its coil every hour. The same heat pump in Virginia, where it is not snowing, might not have to defrost its coil but once a week. As we progress into future chapters, I'll give you more of these examples and complete, technical details for working with heat pumps, but for now, let's return to our overview of the systems.

New Construction

As we move on, I want you to put yourself in the position of a customer. In other words, I want you to think of yourself as the potential homeowner. By doing this, you may be able to get a better feel for how your customers feel. New construction of a home is an ideal opportunity for taking advantage of the benefits a heat pump can offer. If you are planning to build a new home, the options for using a heat pump are numerous. When building a new home, you can install insulation easily. Having energy-efficient windows and doors installed is not difficult, and you can plan provisions for the duct work needed with a heat-pump system.

Before your new home is built, a heat gain and heat loss schedule can be worked up to show exactly what your heating and cooling needs are expected to be. With the results of such a test, you can modify your construction plans to maximize the efficiency of your heat pump. New construction certainly provides ample opportunity to make the most of a heat pump (Fig. 1-1).

FIGURE 1-1

Modern heat pumps offer outstanding heating and cooling. (*Courtesy of Trane.*)

Remodeling

Remodeling a home does not offer the same freedom for capitalizing on the use of a heat pump that building a new home does, but extensive remodeling may provide the chance to install a heat pump. Many old heating systems work at an efficiency rating of only 60 to 70 percent. The desire for a more cost-effective heating and cooling system often points remodelers in the direction of heat pumps.

Old houses are frequently poorly insulated and have drafty windows and doors. The glass in old windows is not of the same insulating quality that is available in new windows. A number of older homes are not equipped with duct work, and the ones that are often have small ducts.

Heat pumps require fairly large ducts to be effective. A house with undersized duct work, or no duct work at all, can make the installation of a heat pump a sizable job. Air infiltration around windows and doors hurts the performance of any heating system, and inadequate insulation doesn't allow for maximum heating and cooling comfort at affordable prices.

Simply installing a new heat pump in place of your old furnace could result in frustration and dissatisfaction. If you are planning to replace an old system with a new heat pump, or even if you are planning only to add a heat pump to your existing system, there will be many thoughts to consider first. We will talk more about this later.

Terminology

Let's take some time to talk about the terminology used with heat pumps. It is always helpful to understand the jargon used in the trade when you are dealing with contractors and suppliers.

One-Piece Units

One-piece units are heat pumps that are self-contained. You've probably seen this type of unit in motels that you've stayed in. To make it easy to visualize, think of a one-piece heat pump as a window air conditioner that is also capable of blowing heat.

One-piece units are normally installed in a roof or a wall. This type of heat pump is typically used to heat and cool specific areas, rather than whole houses. If, for example, you were adding a family room to your home, a one-piece heat pump might be a viable choice for heating and cooling the space.

Two-piece heat pumps are the most common in residential applications. (*Courtesy of Trane.*)

Two-Piece Systems

Two-piece systems are made up of two primary pieces of equipment. One piece of the heat pump is installed inside the home, usually in a closet or similar space. The other piece of the heat pump is installed outside of the home, usually on a concrete pad that sits on the ground. The two pieces are connected with pipes that transfer the heat. This is the type of heat pump most often used in residential applications (Fig. 1-2).

Heat Sources

The heat sources for heat pumps are varied and numerous. Let's look at each source of heat you may want to consider.

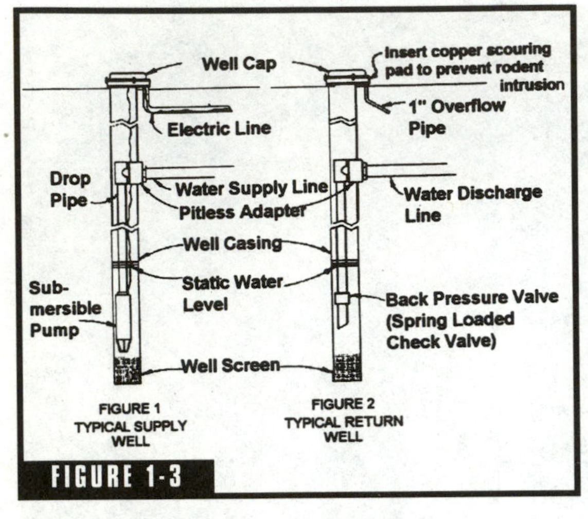

FIGURE 1-3

Typical well setup for a heat pump. (*Courtesy of Bard.*)

Air

Air is a good source of heat, and it is the most common source used for heat pumps in the United States. In states that maintain moderate winter climates, air-source heat pumps work well. If, however, the outside temperature dips into single digits, these systems lose their effectiveness, and a backup heat source (usually electric heat strips) is required.

Well Water

Well water can be an excellent heat source. I know it may seem strange to think of getting your home's heat from a water well, but well water is a very good source for heat pumps to work with (Fig. 1-3).

Surface Water

Surface water can be used as a heat source, but it ranks low on the list of possible choices. We will talk more about surface water and all the other heat sources when we get into the technical part of the book.

Earth

The earth is a good source of heat for heat pumps, but there are some disadvantages to earth systems, too. As you will find later in the book, there is no single choice in heat pumps and heat sources that will serve all needs.

Solar

Solar heat sources are also good for heat pumps. The installation costs associated with solar heat can be extreme, and there are other considerations to be examined before a solar source is chosen.

The Advantages of a Heat Pump

Let's look now at some of the advantages of a heat pump. Assuming that the proper research has gone into the design and installation of a heat pump, the system should be less expensive to operate than other types of heating systems. In mild to moderate climates, heat pumps are at their best. Extreme climates push heat pumps to their extremes, and the units are not always up to the challenge. However, water-source heat pumps can function well in almost any climate (Figs. 1-4 to 1-10).

Sources of heat for heat pumps to pull from are numerous, as you have just seen. Since heat pumps can be configured to use any of the various heat sources, they can be used under varied conditions.

Safety is another advantage gained when heating and cooling with a heat pump. There is no fuel oil to burn, no fumes that must be vented from the home, and no risk of a heating fire escaping its chamber.

A big advantage of a heat pump is its ability to double as both a heating and a cooling unit. There is no need for a furnace and an air conditioner when you install a heat pump; the heat pump is both.

TYPICAL INSTALLATION
Unit installs upright and hooks up conveniently to supply water system. Ideal for closet, basement, or utility room installation. Requires very little floor space. Can be installed with or without return air duct.

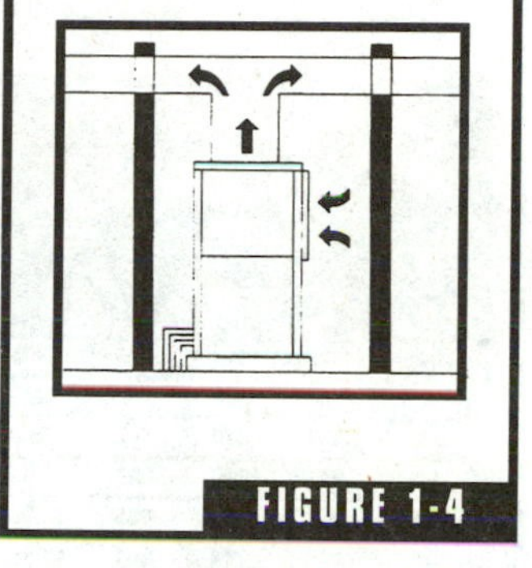

FIGURE 1-4

Water-source packaged heat pump. (*Courtesy of Bard.*)

The space-saving quality of a heat pump is an advantage over some types of heating systems. Since the only portion of the heat pump that takes up space in your home can be concealed in a small closet, it requires much less space than heating systems that demand the storage of an oil tank and considerable floor space for the furnace.

Specifications

MODEL	WPV24C	WPV30C	WPV36C	WPV42C	WPV48C	WPV60C
Electrical Rating (60HZ/V/PH)	230/208-1	230/208-1	230/208-1	230/208-1	230/208-1	230/208-1
Operating Voltage Range	253-197	253-197	253-197	253-197	253-197	253-197
Minimum Circuit Ampacity	15.0	14.4	20.5	27.5	34.5	36.0
+ Field Wire Size	#14	#14	#12	#10	#8	#8
++ Delay Fuse Max. or Ckt. Bkr.	20	25	35	45	50	60
Total Unit Amps 230/208	8.9/9.4	10.3/11.3	14.8/16.3	19.5/21.2	23.7/27.0	25.9/28.8
COMPRESSOR						
Volts	230/208	230/208	230/208	230/208	230/208	230/208
Rated Load Amps 230/208	7/7.5	8.2/9.2	12.7/14.2	14.8/16.5	19.0/22.3	21.2/24.1
Branch Ckt. Selection Current	9.7	12.2	14.7	18.3	23.7	25.0
Lock Rotor Amps 230/208	50/50	61.7/61.7	82/82	109/109	129/129	169/169
BLOWER MOTOR AND EVAPORATOR						
Blower Motor - HP/Spd	1/4 3-spd	1/3 2-spd	1/3 2-spd	1/2 3-spd	1/2 3-spd	1/2 3-spd
Blower Motor - Amps	1.9	2.1	2.1	4.7	4.7	4.7
Face Area Sq. Ft./Row/Fins Per Inch	3.16/3/14	3.16/3/14	3.16/3/14	4.6/3/13	4.6/3/13	4.6/3/13
SHIPPING WEIGHT LBS.	240	240	255	350	370	375

+ 75°C copper wire ++ HACR type circuit breaker

FIGURE 1-5A

Specifications for a water-source heat pump. (*Courtesy of Bard.*)

Dimensions

Model Number	Width A	Depth B	Heigth C	Supply Duct D	Supply Flange E	Return Air Filter Grill Width F	Return Air Filter Grill Heigth G	H	I
WPV24C WPV30C WPV36C	23"	23"	47"	13 7/8"	13 7/8"	22 1/2"	22 1/4"	4"	4 1/4"
WPV42C WPV48C WPV60C	28"	27"	52"	18"	18'	26 7/8"	25 1/2"	5"	4 1/4"

Indoor Blower Performance (CFM - Dry Coil with Filter)①

Model	WPV24C			WPV30C WPV36C		WPV42C, WPV48C, WPV60C Without Optional CW45 Installed			WPV42C, WPV48C, WPV60C With Optional CW45 Installed	
ESP In. WC	Motor Speed			Motor Speed		Motor Speed			Motor Speed	
	High	Med	Low	High	Low	High	Med	Low	High	Med
.00	1,033	946	774	1,300	1,190	1,740	1,650	1,530	1,740	1,600
.10	983	904	757	1,275	1,150	1,695	1,607	1,510	1,695	1,550
.20	942	870	742	1,210	1,110	1,650	1,570	1,480	1,650	1,520
.30	903	836	720	1,150	1,060	1,602	1,532	1,443	1,625	1,500
.40	857	794	688	1,080	1,000	1,550	1,490	1,400	1,500	1,460
.50	799	742	648	1,010	930	1,490	1,435	1,348	1,440	1,380
.60	740	681	603	920	875	1,420	1,365	1,290	1,390	1,310

① For wet coil CFM multiply by .96
ESP = External Static Pressure (inches of water)

Water Coil Pressure Drop

Model	WPV24C		WPV30C		WPV36C		WPV42C		WPV48C, WPV60C	
GPM	PSIG	Ft Hd	PSIG	Ft Hd	PSIG	Ft Hd	PSIG	Ft Hd	PSIG	Ft Hd
4	3.00	6.93	2.50	5.78	—	—	—	—	—	—
5	3.50	8.08	3.20	7.39	2.20	5.08	—	—	—	—
6	4.10	9.50	5.30	12.24	2.75	6.36	1.00	2.31	1.65	3.82
7	4.70	10.85	6.40	14.78	3.40	7.86	1.49	3.44	2.35	5.43
8	—	—	9.60	22.18	4.15	9.59	2.02	4.67	3.10	7.16
9	—	—	—	—	5.00	11.56	2.60	6.01	3.86	8.92
10	—	—	—	—	5.95	13.75	3.22	7.44	4.65	10.75
11	—	—	—	—	—	—	3.90	9.01	5.50	12.71
12	—	—	—	—	—	—	4.60	10.63	6.40	14.79
13	—	—	—	—	—	—	—	—	7.45	17.22
14	—	—	—	—	—	—	—	—	8.60	19.88
15	—	—	—	—	—	—	—	—	9.90	22.89

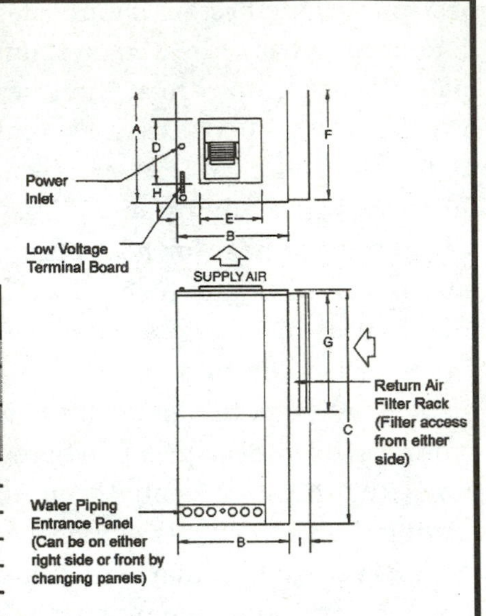

Piping Connections from Three Sides

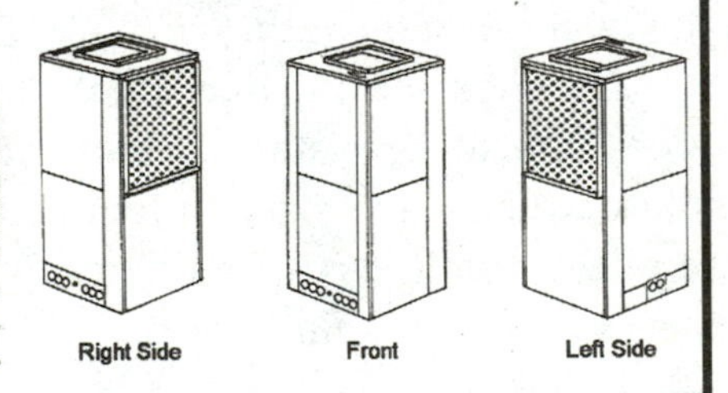

As a special design feature, Bard Water Source Heat Pumps include three interchangeable access panels. One model meets most installation requirements.

NOTE: For 208V operation deduct 600 BTU's from ARI certified performance rating chart for Models WPV24C, WPV30C and WPV36C. Deduct 1000 BTU's for Models WPV42C, WPV48C and WPV60C.

FIGURE 1-5B

Dimensions, indoor blower performance, and water coil pressure drop charts for water-source heat pump. (*Courtesy of Bard.*)

Capacity and Efficiency Ratings (Closed Loop)

Model	CFM / ESP		Recommend Airflow Range	GPM	Cooling 77° EWT		Heating 32° EWT	
					BTU/HR	EER	BTU/HR	COP
WPV24C	800	.40	720 - 880	5	21,600	13.0	15,600	3.0
WPV30C	1,000	.22	900 - 1,090	6	27,000	13.0	19,000	2.8
WPV36C	1,150	.27	1,070 - 1,345	7	32,600	12.5	26,000	3.0
WPV42C	1,550	.25	1,400 - 1,700	9	42,000	11.5	34,000	2.8
WPV48C	1,550	.25	1,400 - 1,700	9	47,500	11.5	36,000	2.7
WPV60C	1,570	.20	1,400 - 1,700	11	54,000	11.5	40,000	2.5

Certified in accordance with ARI Standard 330-90 "Ground Source Closed-Loop Heat Pumps" which includes watt allowance for water pumping.
Cooling capacity based on 80° F DB 67° WB entering air temperature.
Heating capacity based on 70° DB entering air temperature.

Capacity and Efficiency Ratings (Open-Loop)

Model	CFM / ESP	GPM	Cooling ① 70° F EWT		Cooling ① 50° F EWT		Heating ① 70° F EWT		Heating ① 50° F EWT	
			BTU/HR	EER	BTU/HR	EER	BTU/HRR	COP	BTU/HR	COP
WPV24C	800 / .41	4	22,400	12.10	23,400	14.30	24,200	3.8	19,400	3.1
WPV30C	1,000 / .22	4	28,400	13.80	29,800	16.30	29,200	4.0	23,200	3.3
WPV36C	1,150 / .27	5	34,300	13.30	36,000	15.70	39,500	3.8	31,700	3.1
WPV42C	1,550 / .25	6	46,200	12.80	48,400	15.30	54,300	3.7	43,600	3.0
WPV48C	1,550 / .25	6	48,000	11.70	50,700	13.90	57,400	3.7	45,900	3.0
WPV60C	1,570 / .20	8	54,100	11.70	56,800	13.90	62,600	3.4	50,200	2.8

① Rated in accordance with ARI standard for 325, "Standard for Ground Water Source Heat Pumps," which includes Watt allowance for water pumping.
Cooling capacity based on 80° F DB 67° WB entering air temperature.
Heating capacity based on 70° DB entering air temperature.

Flow Rates for Various Fluids

	WPV24C	WPV30C	WPV36C	WPV42C	WPV48C	WPV60C
Flow rate required GPM fresh water	4	4	5	6	6	8
Flow rate required GPM 15% Sodium Chloride	5	6	7	9	9	11
Flow rate required GPM 25% GS4	5	6	7	9	9	11

FIGURE 1-6

Capacity and efficiency ratings for a heat pump. (*Courtesy of Bard.*)

Capacity and Efficiency Application Ratings - (Based on 15% Sodium Chloride)

WPV24C - 800 CFM 5 GPM

Dry Bulb/ Wet Bulb	Cooling Capacity	Fluid Temperature Entering Water Coil °F 30° ②	40° ②	50°	60°	70°	80°	90°	100°	110°
75/62	Total Cooling	23,400	22,750	22,090	21,428	20,767	20,100	19,444	18,782	18,121
	Sensible Cooling	17,700	17,100	16,700	16,200	15,700	15,200	14,600	14,100	13,600
	Total Heat of Rejection	27,500	26,600	25,700	24,800	23,900	23,000	22,100	21,200	20,300
	EER ①	23.50	21.31	19.09	16.86	14.60	12.40	10.15	7.90	5.75
80/67	Total Cooling	24,900	24,203	23,500	22,796	22,000	21,380	20,685	19,981	19,277
	Sensible Cooling	18,200	17,700	17,200	16,600	16,200	15,600	15,100	14,600	14,000
	Total Heat of Rejection	29,300	28,300	27,400	26,400	25,400	24,500	23,500	22,500	21,600
	EER ①	24.30	21.99	19.70	17.40	15.10	12.80	10.50	8.20	5.90
85/72	Total Cooling	27,400	26,624	25,850	25,075	24,300	23,527	22,753	21,979	21,205
	Sensible Cooling	19,100	18,600	18,000	17,500	16,900	16,400	15,800	15,300	14,800
	Total Heat of Rejection	32,200	31,200	30,100	29,000	28,000	26,900	25,900	24,800	23,700
	EER ①	25.90	23.50	21.00	18.60	16.10	13.70	11.20	8.70	6.30

Dry Bulb	Heating Capacity	Fluid Temperature Entering Water Coil °F 25° ②	30° ②	40° ②	50°	60°	70°	80°
70	Total Heating	14,400	15,100	17,200	19,400	21,500	23,600	25,700
	Total Heat of Absorption	10,900	11,700	13,500	15,300	17,000	18,800	20,600
	COP ①	3.20	3.26	3.53	3.80	4.06	4.33	4.60

① Unit Only ② Requires anti-freeze solution

WPV30C - 1,000 CFM 6 GPM

Dry Bulb/ Wet Bulb	Cooling Capacity	Fluid Temperature Entering Water Coil °F 30° ②	40° ②	50°	60°	70°	80°	90°	100°	110°
75/62	Total Cooling	31,600	30,300	28,900	27,600	26,300	25,000	23,600	22,300	21,000
	Sensible Cooling	23,300	22,700	22,000	21,300	20,600	19,900	19,300	18,600	17,900
	Total Heat of Rejection	36,700	35,500	34,300	33,100	32,000	30,700	29,600	28,400	27,200
	EER ①	25.30	22.90	20.50	18.10	15.70	13.30	10.90	8.50	6.10
80/67	Total Cooling	33,600	32,200	30,800	29,400	27,900	26,500	25,100	23,700	22,300
	Sensible Cooling	24,100	23,400	22,700	22,000	21,300	20,600	19,800	19,100	18,400
	Total Heat of Rejection	39,000	37,700	36,500	35,200	34,000	32,700	31,400	30,200	28,900
	EER ①	26.10	23.60	21.20	18.70	16.20	13.70	11.20	8.80	6.30
85/72	Total Cooling	36,900	35,400	33,800	32,300	30,700	29,200	27,600	26,100	24,600
	Sensible Cooling	25,300	24,500	23,800	23,000	22,300	21,600	20,800	20,100	19,400
	Total Heat of Rejection	42,900	41,500	40,100	38,700	37,400	36,000	34,600	33,200	31,800
	EER ①	27.90	25.30	22.60	20.00	17.30	14.60	12.00	9.40	6.70

Dry Bulb	Heating Capacity	Fluid Temperature Entering Water Coil °F 25° ②	30° ②	40° ②	50°	60°	70°	80°
70	Total Heating	16,666	18,333	21,666	25,000	27,700	30,500	33,300
	Total Heat of Absorption	12,600	13,900	16,400	18,900	21,400	23,900	26,400
	COP ①	3.00	3.20	3.50	3.90	4.20	4.50	4.90

(1) Unit Only (2) Requires anti-freeze soul

WPV36C - 1,150 CFM 7 GPM

Dry Bulb/ Wet Bulb	Cooling Capacity	Fluid Temperature Entering Water Coil °F 30° ②	40° ②	50°	60°	70°	80°	90°	100°	110°
75/62	Total Cooling	37,200	35,700	34,200	32,600	31,100	29,600	28,000	26,500	25,000
	Sensible Cooling	26,900	26,200	25,400	24,600	23,800	23,000	22,200	21,400	20,600
	Total Heat of Rejection	41,200	40,400	39,600	38,800	38,000	37,200	36,400	35,600	34,800
	EER ①	24.00	21.60	19.20	16.90	15.45	12.10	9.70	7.30	5.00
80/67	Total Cooling	39,600	38,000	36,400	34,700	33,100	31,500	29,800	28,200	26,600
	Sensible Cooling	27,800	27,000	26,200	25,400	24,500	23,700	22,900	22,100	21,300
	Total Heat of Rejection	43,900	43,000	42,200	41,300	40,500	39,600	38,800	37,900	37,000
	EER ①	24.80	22.30	19.90	17.40	15.00	12.50	10.00	7.60	5.10
85/72	Total Cooling	43,600	41,800	40,000	38,200	36,400	34,600	32,800	31,000	29,200
	Sensible Cooling	29,200	28,300	27,500	26,600	25,800	24,900	24,000	23,200	22,300
	Total Heat of Rejection	48,300	47,300	46,400	45,400	44,500	43,600	42,600	41,700	40,700
	EER ①	26.50	23.90	21.20	18.60	16.00	13.30	10.70	8.10	5.40

Dry Bulb	Heating Capacity	Fluid Temperature Entering Water Coil °F 25° ②	30° ②	40° ②	50°	60°	70°	80°
70	Total Heating	23,400	25,200	29,000	32,700	36,400	40,100	43,800
	Total Heat of Absorption	16,400	18,100	21,500	25,000	28,400	31,900	35,300
	COP ①	2.90	3.00	3.40	3.70	4.00	4.30	4.60

① Unit Only ② Requires anti-freeze solution

FIGURE 1-7

Capacity and efficiency ratings for a heat pump. (*Courtesy of Bard.*)

WPV42C - 1,550 CFM 9 GPM

Dry Bulb/ Wet Bulb	Cooling Capacity	Fluid Temperature Entering Water Coil °F 30° ②	40° ②	50°	60°	70°	80°	90°	100°	110°
75/62	Total Cooling	53,500	50,700	47,900	45,100	42,300	39,500	36,800	34,000	31,200
	Sensible Cooling	38,700	37,100	35,600	34,000	32,400	30,800	29,200	27,700	26,100
	Total Heat of Rejection	61,500	59,500	57,400	55,400	53,300	51,200	49,200	47,100	45,100
	EER①	23.30	21.00	18.60	16.20	13.80	11.50	9.10	6.70	4.40
80/67	Total Cooling	57,000	53,900	51,000	48,000	45,000	42,100	39,100	36,100	33,200
	Sensible Cooling	40,000	38,300	36,700	35,000	33,400	31,800	30,100	28,500	26,900
	Total Heat of Rejection	65,500	63,200	61,100	58,900	56,700	54,500	52,300	50,100	47,900
	EER①	24.00	21.60	19.20	16.70	14.30	11.90	9.40	7.00	4.50
85/72	Total Cooling	62,600	59,300	56,100	52,800	49,500	46,300	43,000	39,800	36,500
	Sensible Cooling	42,000	40,200	38,500	36,800	35,100	33,400	31,600	30,000	28,200
	Total Heat of Rejection	72,000	69,600	67,200	64,800	62,400	60,000	57,600	55,100	52,700
	EER①	25.70	23.10	20.50	17.90	15.30	12.70	10.00	7.50	4.84

Dry Bulb	Heating Capacity	Fluid Temperature Entering Water Coil °F 25° ②	30° ②	40° ②	50°	60°	70°	80°
70	Total Heating	29,800	32,800	38,800	45,000	51,100	57,200	63,300
	Total Heat of Absorption	19,300	22,100	27,800	33,500	39,100	44,800	50,500
	COP①	2.75	3.00	3.30	3.80	4.20	4.60	5.00

① Unit Only ② Requires anti-freeze solution

WPV48C - 1,550 CFM 9 GPM

Dry Bulb/ Wet Bulb	Cooling Capacity	Fluid Temperature Entering Water Coil °F 30° ②	40° ②	50°	60°	70°	80°	90°	100°	110°
75/62	Total Cooling	54,467	52,378	50,290	48,201	46,112	44,023	41,934	39,845	37,756
	Sensible Cooling	37,675	36,346	35,017	33,687	32,358	31,029	29,699	28,370	27,041
	Total Heat of Rejection	62,530	61,486	60,442	59,397	58,363	57,308	56,264	55,219	54,175
	EER①	18.97	17.43	15.89	14.34	12.80	11.26	9.71	8.17	6.63
80/67	Total Cooling	57,944	55,722	53,500	51,277	49,055	46,833	44,611	44,238	40,166
	Sensible Cooling	38,840	37,470	36,100	34,729	33,359	31,988	30,618	29,248	27,877
	Total Heat of Rejection	66,522	65,411	64,300	63,188	62,077	60,966	59.855	58,744	57,633
	EER①	19.58	17.99	16.40	14.80	13.21	11.62	10.02	8.43	6.84
85/72	Total Cooling	63,738	61,294	58,850	56,405	53,961	51,516	49,072	46,627	44,183
	Sensible Cooling	40,782	39,343	37,905	36,466	35,027	33,588	32,149	30,710	29,271
	Total Heat of Rejection	73,174	71,952	70,730	69,507	68,285	67,063	65,841	64,618	63,396
	EER①	20.91	19.21	17.51	15.81	14.11	12.41	10.71	9.01	7.30

Dry Bulb	Heating Capacity	Fluid Temperature Entering Water Coil °F 25° ②	30° ②	40° ②	50°	60°	70°	80°	90°
70	Total Heating	32,500	35,000	40,000	45,000	50,000	55,000	60,000	65,000
	Total Heat of Absorption	21,516	23,933	28,766	33,600	38,433	43,266	48,100	52,933
	COP①	2.79	2.90	3.12	3.35	3.57	3.79	4.01	4.23

① Unit Only ② Requires anti-freeze solution

WPV60C - 1,570 CFM 11 GPM

Dry Bulb/ Wet Bulb	Cooling Capacity	Fluid Temperature Entering Water Coil °F 30° ②	40° ②	50°	60°	70°	80°	90°	100°	110°
75/62	Total Cooling	58,777	57,071	55,366	53,660	51,954	50,248	48,542	46,836	45,130
	Sensible Cooling	44,972	42,759	40,546	38,332	36,119	33,906	31,693	29,480	27,267
	Total Heat of Rejection	65,511	65,232	64,954	64,675	64,396	64,118	63,839	63,561	63,282
	EER ①	19.61	18.04	16.47	14.90	13.33	11.76	10.20	8.63	7.06
80/67	Total Cooling	62,529	60,714	58,900	57,085	55,270	53,455	51,640	49,825	48,011
	Sensible Cooling	46,362	44,081	41,800	39,518	37,237	34,955	32,674	30,392	28,111
	Total Heat of Rejection	69,692	69,396	69,100	68,803	68,507	68,211	67,914	67,618	67,322
	EER ①	20.23	18.61	17.00	15.38	13.76	12.14	10.52	8.90	7.28
85/72	Total Cooling	68,782	66,786	64,790	62,793	60,797	58,801	56,804	54,808	52,812
	Sensible Cooling	48,681	46,285	43,889	41,494	39,098	36,703	34,307	31,912	29,516
	Total Heat of Rejection	76,661	76,335	76,010	75,684	75,358	75,032	74,706	74,380	74,054
	EER ①	21.61	19.88	18.15	16.42	14.69	12.96	11.24	9.51	7.78

Dry Bulb	Heating Capacity	Fluid Temperature Entering Water Coil °F 25° ②	30° ②	40° ②	50°	60°	70°	80°
70	Total Heating	35,400	38,700	45,200	51,800	58,300	64,900	71,400
	Total Heat of Absorption	22,000	24,800	30,300	35,800	41,300	46,800	52,300
	COP ①	2.50	2.70	3.00	3.25	3.50	3.80	4.00

① Unit Only ② Requires anti-freeze solution

FIGURE 1-8A

Capacity and efficiency ratings for a heat pump. (*Courtesy of Bard.*)

Capacity Multiplier Factors

% of Rated Air Flow	-10	Rated	+10
Total BTUH	.975	1	1.02
Sensible BTUH	.95	1	1.05

Correction Factors for Performance at Other Water Flows

Rated Flow Plus - GPM	Heating		Cooling	
	BTUH	Watts	BTUH	Watts
2	1.00	98	1.01	1.00
4	1.01	97	1.03	1.01
6	1.02	96	1.05	1.02
8	1.02	95	1.06	1.02

FIGURE 1-8B

Capacity multiplier and correction factors for a heat pump. (*Courtesy of Bard.*)

Domestic Hot Water Heating Performance

Model	Thermostat Position	Ground Water Temperature	Water Heating Recovery Rate Gal/Hr Water Temp to 135°	HP Ground Water Flow Rate GPM① with Domestic Hot Water Heat Exchanger Operating (GPM)
WPV24C	Heat ②	50 70	3.6 5.0	4 4
	Cool ③	50 70	2.6 3.6	4 4
WPV30C	Heat ②	50 70	4.0 6.0	4 4
	Cool ③	50 70	3.5 4.0	4 4
WPV36C	Heat ②	50 70	4.4 6.5	5 5
	Cool ③	50 70	2.9 4.2	5 5
WPV42C WPV48C	Heat ②	50 70	5.6 8.0	6 6
	Cool ③	50 70	2.5 4.7	6 6
WPV60C	Heat ②	50 70	5.8 10.0	8 8
	Cool ③	50 70	4.4 5.6	8 8

① Automatically regulated with water constant flow valves.
② When heat pump is in heating mode and operating, it heats water at a COP of 3.2 to 3.4 for up to 70% energy savings over conventional electric water heaters.
③ When heat pump is in cooling mode and operating, it heats water for free, offering a 100% energy savings over conventional electric water heaters. Annual energy savings will vary depending on water usage and number of heating and cooling operating hours per season.

Indoor Thermostat Options

Part No.	Model No.	Description
8403-002 8404-014	T87F3111 Q539J1089	For WPV units not using duct heaters
8403-017 8404-009	T874R1129 Q674L1181	For WPV units using duct heaters
8403-027	IF92-1	For all WPV units

FIGURE 1-9

Specifications for domestic hot water heating. (*Courtesy of Bard.*)

Duct Heaters

Part Number	PH	Volt	KW	Minimum Ampacity	Wire Size① Cu	Wire Size① A1	Max. Fuse	Dimensions A	B	C	D	E	F
8604-080	1	240	5.0	27	#10	#8	30	8	10	4	7	7	12
8604-081	1	240	9.8	52	#6	#4	55	8	10	4	7	7	16
8604-082 ②	1	240	14.7	78	#4	#1	80	15	18	5	11	9	18
8604-083 ②	1	240	19.2	100	#2	#0	100	15	18	5	11	9	18

① Use wire suitable for at least 75° C. ② Fused units (over 48 amperes).
NOTE: All duct heaters are supplied with backup protection and internal fusing as required by NEC.

Water Flow Control Accessories - (Ground Water)

Required for All Models				Required for Installation of WPV42C or WPV60C with CW45 Chilled Water Coil		
Model	Part No.	Qty.	Description	Part No.	Qty.	Description
WPV24C WPV30C	8603-010	1	Constant flow valve 4 GPM 3/4" NPT	8603-006	1	Solenoid Valve 1" NPT
WPV36C	8603-011	1	Constant flow valve 5 GPM 3/4" NPT	8603-007	2	Constant flow valve 6 GPM 3/4" NPT
WPV42C WPV48C	8603-007	1	Constant flow valve 6 GPM 3/4" NPT	8603-008	1	Constant flow valve 8 GPM 3/4" NPT
WPV60C	8603-008	1②	Constant flow valve 8 GPM 3/4" NPT	Optional Flow Meter 1 - 10 GPM ③		
All	8603-006	1	Solenoid valve 1" NPT	8603-012	1	3/4" NPT used to measure waste flow through system may be permanently installed.

① Select either (2) 6 GPM valves for 12 GPM flow capacity or (1) 8 GPM valve as required to obtain desired cooling capacity as shown in performance chart. Order individual components by Bard Part Number.
② For units operating below 50° water temperature install (2) 8 GPM constant flow valves in parralel.
③ Can be used with all models with or without chilled water coil.

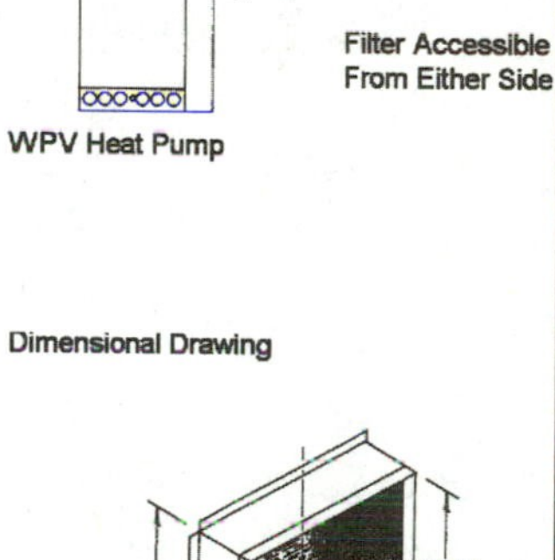

Chilled Water Coil - Cooling Capacity and Efficiency Ratings

1,700 CFM (80°DB/67°WB)							1,550 CFM (80°DB/67°WB)						
Water Temp °F	GPM	Total BTUH	Sensible BTUH	S/T Ratio	Blowe Watts	EER*	Water Temp °F	GPM	Total BTUH	Sensible BTUH	S/T Ratio	Blower Watts	EER*
45	8	37,800	30,000	79%	780	30.0	45	8	35,800	28,100	79%	720	29.8
	12	44,900	32,200	72%	780	29.9		12	42,800	30,000	70%	720	29.7
50	8	29,700	27,600	93%	780	21.9	50	8	28,500	25,300	88%	720	23.8
	12	33,600	29,400	88%	780	22.4		12	32,700	26,500	81%	720	22.7
52	8	25,800	25,000	97%	770	20.6	52	8	24,500	22,400	91%	720	20.4
	12	29,800	26,300	88%	780	19.9		12	28,800	23,600	82%	720	20.0

*Includes well pump watts at a rate of 60 watts per GPM. Ex: 8 GPM x 60 watts = 480 watts.

Dimensional Drawing

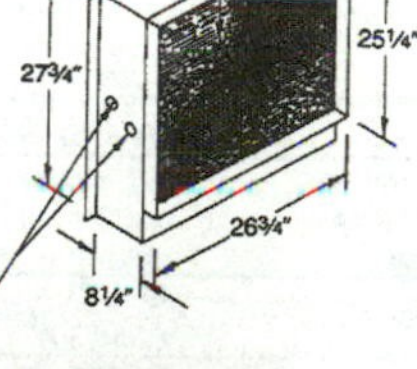

Chilled Water Cooling Up to 30 E.E.R.

Reduces energy costs. For direct ground water cooling without aid of compressor. Used in areas where ground water temperatures are 52°F or less. The only electricity used is for operating the indoor blower and well pump motor. Designed specifically for Models WPV42C, WPV48C and WPV60C.

Chilled Water Coil Pressure Drop

Model CW45		
GPM	PSIG	Ft. Hd.
8	1.5	6.4
8	3.8	16.3
10	6.0	25.8
12	9.2	39.5
14	11.3	48.5

FIGURE 1-10

Options and accessories for a heat pump. (*Courtesy of Bard.*)

The Disadvantages

What are the disadvantages of a heat pump? If you buy a quality system that has been properly designed for your home, there is very little to consider that might be termed a disadvantage.

Since a heat pump becomes less effective as the temperature of its heat source drops, you might make a case for this being a disadvantage. Most heat pumps pull their heat source from air (Fig. 1-11). When the outside air temperature drops, homeowners want the interiors of their homes to remain warm. The heat pump loses some of its heat-producing ability as the temperature gets colder, which is when the heating aspect of the heat pump is needed most, resulting in a possible disadvantage. If, however, the heat-pump system has been designed and installed properly, this so-called disadvantage should not come into play.

People in colder climates may find the initial cost of a heat-pump system to be a disadvantage. It is not unusual for a heat-pump installation to cost more than a standard furnace, but remember, you are getting not only a furnace, but a central air-conditioning system as well. Since many homes in colder climates, such as Maine, don't have or need central air conditioning, the additional cost for a heat pump may not be warranted.

TABLE 2 — RATED CFM AND AIRFLOW DATA (WET COIL — COOLING)

Condensing Unit Model Number	Evaporator Coil Model Number	Rated Airflow CFM	Rated Airflow Pressure Drop H_2O ①	Rated ESP ②	Motor Speed Tap	Recommended Air Flow Range	System Orifice Required
24UHPSC	BC24C	800		.15	High	700 - 910	.063 ◀
	A36AS-A	800	.15			700 - 910	.063 ◀
30UHPSC	BC36C	1,050		.25	Low	900 - 1150	.069 ◀
	A36AS-A	1,050	.25			900 - 1150	.069 ◀
	A37AS-A	1,050	.20			900 - 1150	.069 ◀
36UHPSC	BC36C	1,200		.15	High	1,020 - 1,320	.072
	A37AS-A	1,200	.25			1,020 - 1,320	.072
42UHPSC	BC48C	1,525		.35	Low	1,300 - 1,750	.078
	A61AS-A	1,525	.20			1,300 - 1,750	.078 ◀
48UHPSB	BC48C	1,675		.25	High	1,450 - 1,950	.078
	A61AS-A	1,650	.25			1,400 - 1,900	.078 ◀
48UHPSC	BC48S	1,675		.20	High	1,425 - 1,925	.078
	A61AS-A	1,650	.25			1,400 - 1,900	.078 ◀
60UHPSC	BC60C	1,800		.20	High	1,530 - 2,050	.092
	A61AS-A	1,750	.30			1,530 - 2,050	.092

① Measured across the evaporator coil assembly, including drain pan.

② External static pressure available for the duct system—supply and return. All blower coils have multi-speed motors, and value shown is at the recommended rated speed. Consult specification airflow charts with the blower coil units for complete information at other speeds.

◀ IMPORTANT — Proper sized orifice is not factory installed in indoor section. Proper orifice size is shipped with outdoor unit packaged with its installation instructions for indoor sections listed on this page. The orifice must be replaced with the proper orifice shown above.

FIGURE 1-11

Specifications for an air-source heat pump. (*Courtesy of Bard.*)

So, Is a Heat Pump Right for You?

So, is a heat pump right for you? There is still much to consider before you begin to make that decision. The odds are good, however, that a heat pump is right for you.

I have lived in homes with electric heat, wood heat, hot-water baseboard heat, forced hot-air heat, radiators, and heat pumps. Allow me to share my feelings about each of these types of heat. My personal experiences may help you in making your decision about heat pumps.

Electric Heat

The homes I've lived in that were equipped with electric heat, both in Virginia and Maine, have been very expensive to keep warm. Electric heat has rarely seemed to feel warm, even when the thermostat indicated an interior temperature that should have been more than comfortable. Electric heat produces heat quickly, but it is expensive to operate.

Wood Heat

I love wood heat. The aroma of wood smoke, the warmth, and the sense of going back in time all combine to make wood heat a favorite of mine. However, as I've gotten older, carrying wood in for the stove seems to get more difficult with each trip to the barn. Not only is carrying wood physically demanding, the wood leaves debris throughout the house. It is impossible to clean out the stove without creating a dust layer, and with a small child in the home, a wood stove can be a disaster waiting to happen. There is also the danger of fire and the expense of having the chimney cleaned.

In my younger days, I heated my home exclusively with wood. During those years I didn't mind the mess and the trouble. Now, I use wood only for the sense of pleasure it gives me to see it burning in the fireplace. The warmth is still welcomed, but the wood stove is gone, and I don't depend on wood to heat my home any longer.

Hot-Water Baseboard Heat

Hot-water baseboard heat is the heat of choice in Maine. More homes have this type of heating system than any other. Hot-water baseboard

A typical boiler. (*Courtesy of Burnham.*)

is good for Maine's extreme climate, but the boilers and oil tanks take up considerable space, there are oil deliveries that must be made on snowy roads, and the heating system does nothing to cool the home during those few weeks of the year when it is hot in Maine (Fig. 1-12).

Forced Hot-Air Heat

Forced hot-air heat is common. Some people don't like it because of the dust it moves around, but in general, it is a well-accepted form of heat. My experience with forced hot-air heat in Maine has been acceptable, but somewhat expensive. When you consider that a heat pump can give heat and air conditioning through the same ducts, it makes little sense to use only a forced hot-air furnace when a heat pump could be used.

Heat Pumps

All of the new homes I've built for myself have been equipped with heat pumps. The benefit of having my heating and air-conditioning needs met by a single piece of equipment has always made me happy. The performance of the heat pumps in my homes has always been satisfactory. Maintenance on the heat pumps have been low, and the units have never forced me to live in a cold house.

Of all the heating systems available, I would opt first for a heat pump and second for a hot-water baseboard system. This choice is based on overall average conditions. In extreme conditions, I might lean more toward the hot-water baseboard for one reason: the resale value of the home. In some areas, such as Maine, home buyers have preconceived opinions of value, and many rank hot-water baseboard heat higher than heat pumps. As far as overall cost, efficiency, comfort, and use, I prefer heat pumps.

As you may recall, I told you earlier that I have built as many as 60 single-family homes in a single year. Most of my building activity was carried out in Virginia. The winter temperatures in the part of Virginia where I worked frequently go into the teens and sometimes much lower. Summer temperatures hanging in the upper 80s and 90s are common.

A state where the cold temperature can hit zero and the hot temperature can exceed 100°F should be a good testing ground for heat pumps. I can tell you from experience, not one of my home-buying customers ever complained about the performance of their heat pumps in summer or in winter. I think this says a lot for heat pumps.

In Maine, there are extremely cold days during the winter. The summers are typically cool, so air conditioning in homes is not a requirement. These two factors combine to make heat pumps less prevalent in Maine. However, commercial buildings in Maine, where air conditioning is a needed convenience, do use heat pumps. Do you think savvy investors and real estate developers would invest big bucks in systems that can't handle extreme cold? I don't, and I know, again from first-hand experience, that heat pumps do work in Maine. Heat pumps are a viable consideration in almost all cases, so don't rule them out.

CHAPTER

2

Components of Heat Pumps

When talk turns to equipment controls and components used with heat pumps, the discussion can take many turns. These items can involve everything from the heat pump's compressor to a tiny air valve. There are electrical items, coils, all types of valves, and more to cover when you take on the broad spectrum of controls and components for heat pumps. This chapter is going to give you a solid understanding of what these important parts of a heat-pump system are and of what they do.

As far as installing a heat pump goes, you may never need to understand what a discharge muffler does or why a crankcase heater is needed. You may not care about electric resistance heaters until the day backup heat is needed, your customer is not getting it, and you are being called for answers. While much of the information can be ignored during a standard installation, there may come a time during a troubleshooting phase when you, or your service department, must at least know these items exist. There is much to cover in this chapter, so let's start with one of the most important parts of a heat pump, the compressor (Figs. 2-1 to 2-3).

A Compressor

A compressor is a most important element of a heat pump. It is the piece of equipment that controls the flow of refrigerant through the system. The compressor does not work alone; it depends upon other controls to get its job done, but the compressor is the key player in the game of circulating refrigerant.

Due to the construction of most compressors, there is little that can be done with them in the field. If a problem occurs with the unit, it will normally have to be shipped to an authorized repair facility. Other than understanding the difference between types of compressors when deciding what type of heat pump to buy, there is not much more that a homeowner needs to know. When your customers are shopping for a heat pump, you will probably discover that they may choose between a model with a reciprocating compressor or a rotary compressor. Either type can be used with satisfactory results.

Reciprocating compressors can be used with a variety of refrigerants, and they are noted for their durability. Add to this their simple design and their ability to be efficient at high condensing pressures, and you have a good compressor. Rotary compressors are not as common as reciprocating compressors, but they do offer at least two pos-

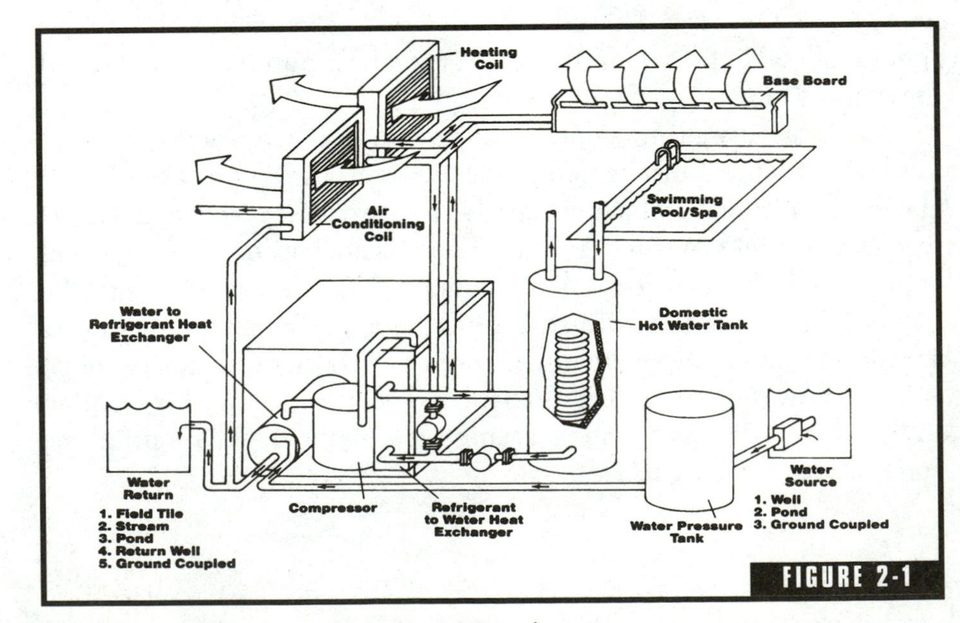

Hydronic heating system. (*Courtesy of Tetco.*)

sible advantages. One of the most noticeable ad-vantages is that rotary compressors can run with less noise than what is created with a reciprocating compressor. A deeper vacuum is also normally possible with a rotary compressor.

Reciprocating compressors are the type you are most likely to encounter, and they work well. If you have an interest in rotary compressors, do some comparison work. Read specifications from various manufacturers and assess the detailed information to determine which type will best suit your needs.

The Compressor's Helpers

I mentioned that a compressor does not work alone, and now I'd like to introduce you to the compressor's helpers. These helpers are controls that are used to direct the flow of refrigerant. The simplest of the group is the check valve.

A check valve is not a complicated device. It is simply an in-line valve that opens and closes to allow or block the flow within a pipe.

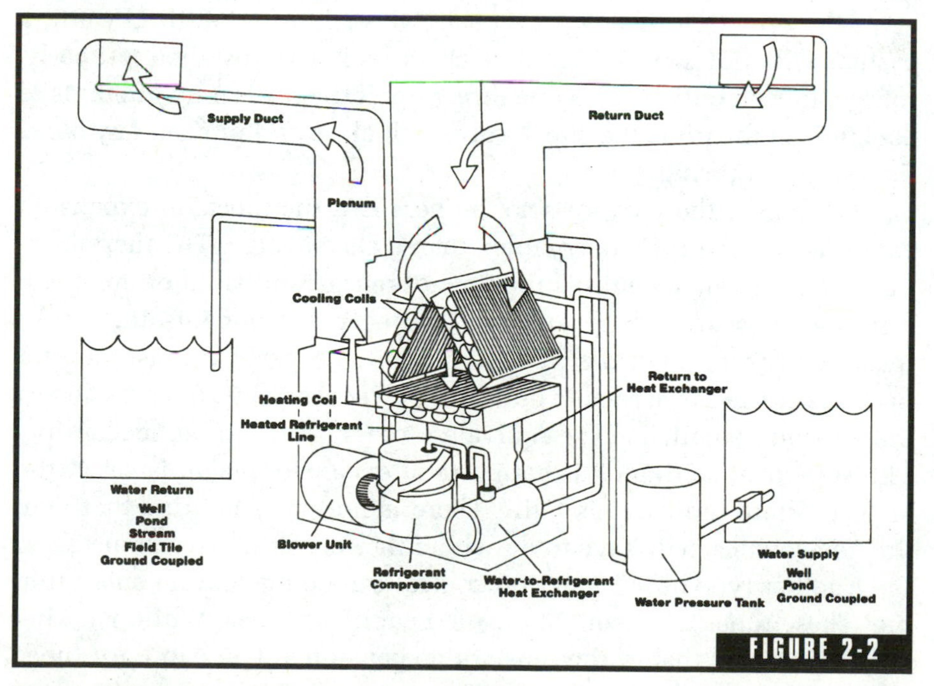

FIGURE 2-2

Forced-air heating system. (*Courtesy of Tetco.*)

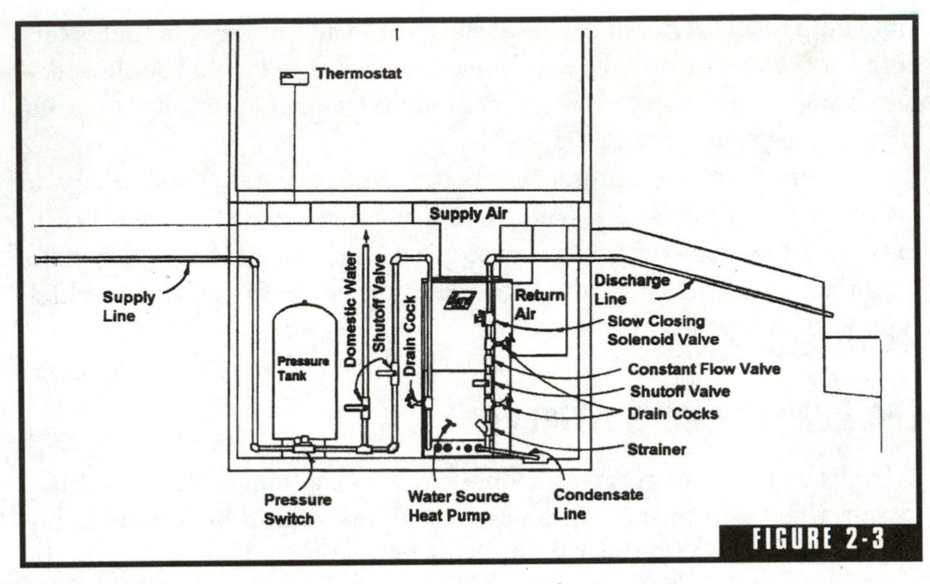

Standard piping arrangement for a water-source heat pump. (*Courtesy of Bard.*)

In the case of heat pumps, check valves are spring-loaded. The valve is marked on the outside of its casing with an arrow. The arrow indicates the direction of flow. If a check valve is installed backward, a system will not work. Once the check valve is installed properly, refrigerant can only flow in one direction. If the refrigerant attempts to backflow in the pipe, the check valve will close and prevent any backflow from occurring.

Another of the compressor's helpers is a thermostatic expansion valve. This valve is more complex than a check valve. The thermostatic expansion valve controls the flow of refrigerant based on temperature and pressure. There are three primary components that make these valves work. For the valve to function properly, it must monitor pressure created by a remote bulb and power assembly, the pressure in the heat-pump coil, and the equivalent pressure of a superheat spring. The superheat setting of a thermostatic expansion valve is set at the factory where the valve is built. There is normally nothing that you, the homeowner, will have to do with a thermostatic expansion valve.

Another type of refrigeration control can be made from small tubing. These tubes are frequently called capillary tubes. While working very differently than a thermostatic expansion valve, capillary tubes perform the same basic function. They regulate the flow of refrigerant based pressure ratings. Capillary tubes are inexpensive, and they work well when all aspects of the system remain constant. If, however, the system becomes unstable with pressure changes, problems can arise. Too much or too little refrigerant can pass through the tubes that are always open and cause the system to fail. There are pros and cons to capillary tubes. While they are the least-expensive method of controlling refrigerant, they are not necessarily the best way to get the job done.

Accessory Valves

There are a number of accessory valves that may be found in a heat-pump system. These valves perform a variety of duties. For example, compressor service valves allow gauges to be attached to the system for tests without shutting down the system. Air valves allow the removal of air and the checking of pressures. Relief valves are safety devices, and there are other types of valves that can play important roles in the successful operation of a heat-pump system. Let's take a look at each of these types of valves on a one-by-one basis.

Relief Valves

Relief valves are safety devices that are typically required by local code enforcement offices. Paperwork supplied by the manufacturer of a heat pump will often recommend a specific type of relief valve that should be used with a given type of heat pump. Basically, a relief valve is intended to open if excessive pressure builds up within a system. This allows the pressure to be vented out of the system without property damage or personal injury.

Air Valves

Air valves can be used to check system pressures. These valves are very similar to the type used in the stems of tires for cars and bicycles. A service technician can install an adapter on test gauges and use the air valve as a means of access to system pressures.

Service Valves

Service valves can be used to test system pressures in place of air valves. The service valves are more expensive and take up more room,

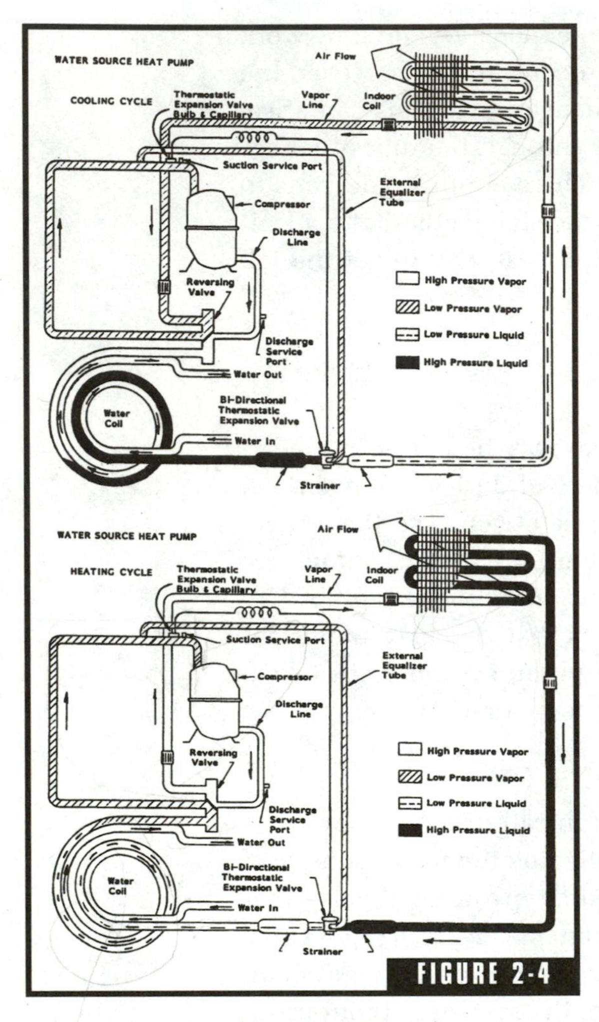

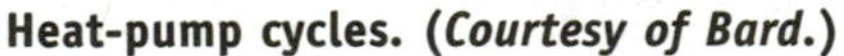
Heat-pump cycles. ***(Courtesy of Bard.)***

but there is less likelihood of a leak developing when service valves are used. These valves have a special port in the side of their body that allows a gauge to be screwed into it. By doing this, a technician can adjust the valve stem and take readings of the system pressure without shutting down the system. Once the testing is complete, the stem can be repositioned to close off the gauge port.

Regulating Water

Regulating the flow of water with water-source systems is done to control pressure and conserve water. This feat is accomplished with the use of water-regulating valves.

The Magic Valve

The magic valve in a heat-pump system is the reversing valve. This valve is sometimes called a four-way valve, because it has four connection ports. What makes the reversing valve magic? This valve is used to control the direction of flow within the heat-pump system when alternating between heating and cooling requirements. Since it can sense its duty for either heating or cooling, it may appear to be magical.

Reversing valves (Fig. 2-4) maintain a regular flow through two of their four ports. The hot gas from the compressor accounts for one of the ports, and the suction line back to the compressor takes up the other. That leaves two ports to be dealt with. These two ports are connected to the outdoor coil and the indoor coil. The direction of flow for refrigerant is controlled by the reversing valve based on the mode that the heat pump is operating in. In the cooling mode, refrigerant moves one way. When in the heating mode, refrigerant flows in the opposite direction. This action is made possible with the use of a solenoid-controlled reversing valve.

Other Refrigerant Controls

Other refrigerant controls can also be found in heat-pump systems (Fig. 2-5). These controls perform various functions. For example, an accumulator is used to contain and control liquid refrigerant and oil that might otherwise make their way into the compressor. Accumulators are typically installed in the suction line at a point between the compressor and the reversing valve.

Accumulators are needed most when a heat pump is in its heating cycle. If the outdoor coil is unable to evaporate refrigerant as it should, interior flooding would be possible without the installation of an accumulator. If an air-source heat pump was in a defrosting cycle, a similar problem could arise. Due to the design of the device, an accumulator will catch excessive liquids and prevent serious damage to equipment.

A Line Drier

A line drier is often installed between the indoor and outdoor units of a heat pump. These driers, or dehydrators as they are also called, are mounted in the refrigerant liquid line. Any refrigerant passing through the pipe must also pass through the drier.

Dehydrators contain drying agents of one sort or another that remove moisture from the system. Many types of driers

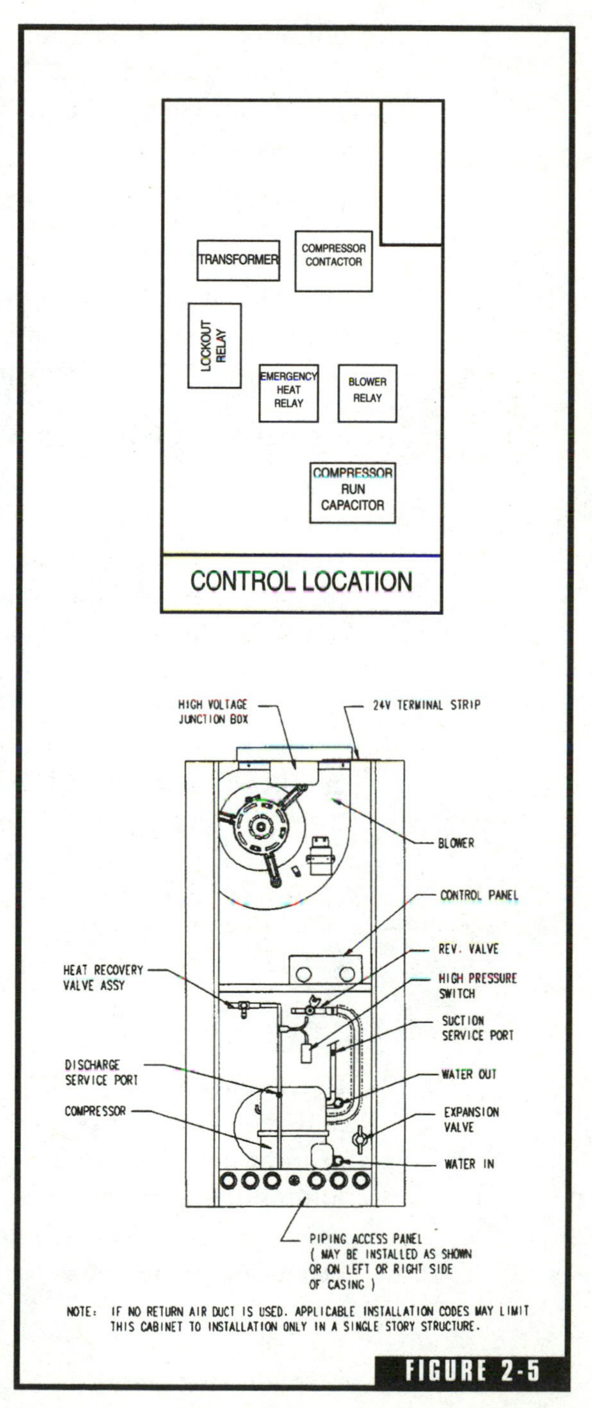

FIGURE 2-5

Heat-pump control locations. (*Courtesy of Bard.*)

double as filters, and a number of them are made with the intent of being discarded after a set period of use. Standard procedure calls for an existing drier element to be replaced anytime the refrigerant system is opened.

Mufflers

Noise can be a problem with some heat-pump installations. When it is an installer's intent to minimize noise, a muffler is installed to reduce the discharge noise from a compressor. This concept is similar to the one used to quiet cars by muffling the exhaust.

Crankcase Heaters

Crankcase heaters are often used to stop refrigerant from being absorbed into the oil contained in the crankcase of a compressor. When a heat pump is not running, refrigerant will make its way into the crankcase. During cold seasons, this can make it very difficult, if not damaging, for the heat pump to start. By keeping the crankcase warm, the refrigerant doesn't mingle with the oil in a way that will lead to problems.

Some heat pumps don't use crankcase heaters. Instead, their crankcase area is kept warm by a trickle charge of power running through windings in the compressor's motor. If you don't know which method is employed by your heat pump, check the specifications that came with your unit.

Coils

Both water-source and air-source heat pumps depend on coils to heat and cool homes. There is one coil in the inside unit and another coil in the outside unit. The coils are sometimes called heat exchangers, and they are usually made of metallic pipe that is covered with aluminum fins. The tubing that makes up the coil could be made from aluminum or copper. The aluminum fins on the coils are easily damaged, and caution should be used to protect them at all times. Damaged fins result in a less-effective heat pump.

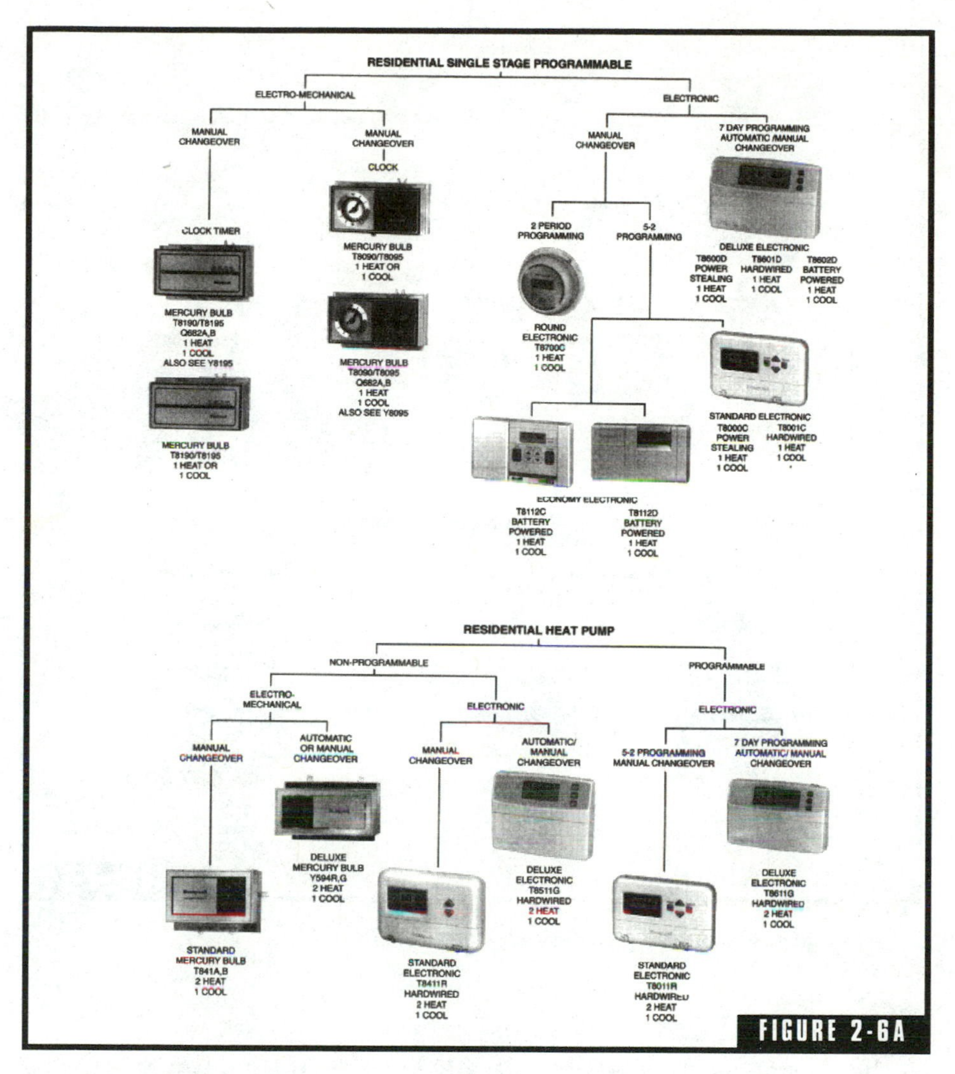

FIGURE 2-6A

A variety of thermostats. (*Courtesy of Honeywell.*)

Electrical Controls

Electrical controls are responsible for the starting, running, and stopping of heat pumps. Without electrical controls, not much would happen with a heat pump. Let's look first at thermostats, since they are the electrical control most people are familiar with.

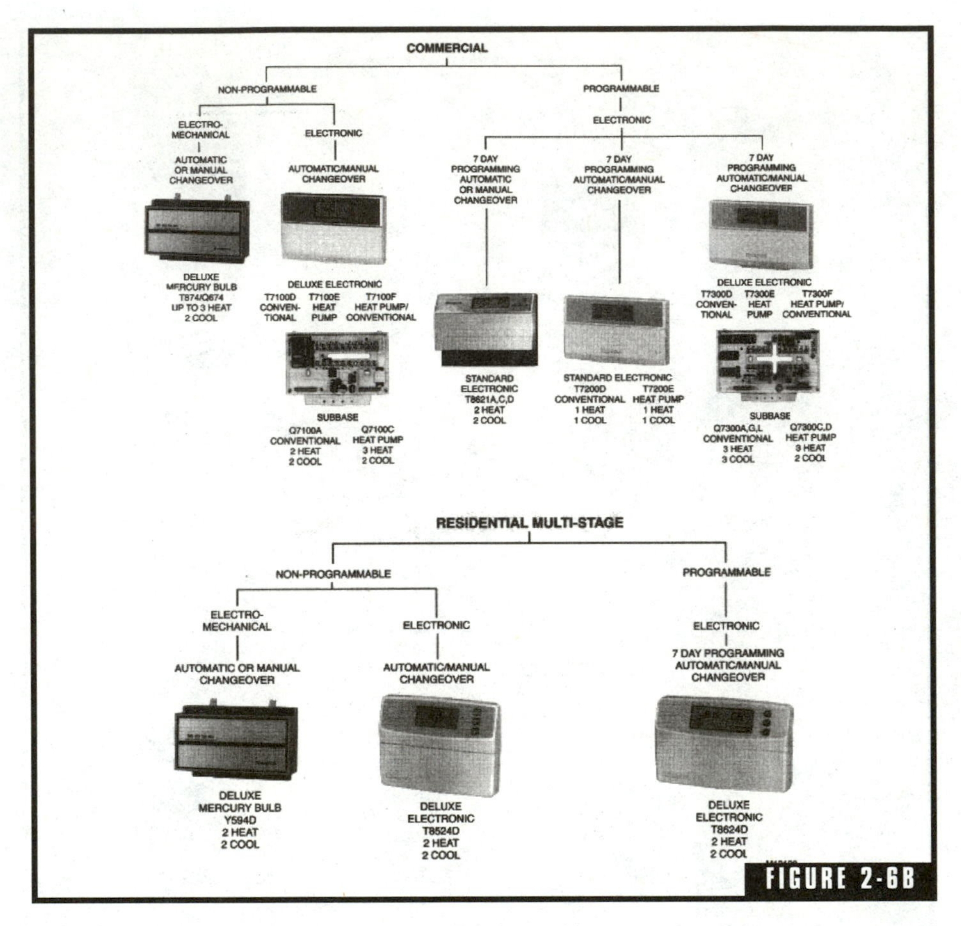

FIGURE 2-6B

A variety of thermostats. (*Courtesy of Honeywell.*)

Thermostats

A thermostat, as you probably already know, is the control that tells a heat pump when to cut on and when to cut off (Fig. 2-6). The options available in thermostats are vast. It is also possible for a heat pump to be controlled by more than one thermostat. For example, you might have one thermostat in a home and another thermostat in the outdoor unit that controls the backup heat. In the case of an air-source heat pump, another thermostat may be used to control the defrosting cycles.

Low-Voltage Controls

Low-voltage controls are frequently used with heat pumps. These controls can work with either 110- or 220-volt current, depending upon the type of control purchased. A transformer is used to convert the standard voltage into low voltage. A thermostat is an example of a low-voltage control.

Pressure Controls

Pressure controls are also used with heat pumps. These controls open and close based on pressure changes. In the case of a residential heat pump, a low-pressure control is the only type normally needed. However, high-pressure controls also exist and may be used on some types of heat pumps.

Motors

The motors used on heat pumps are not normally homeowner-friendly. In other words, there is not much an average person can do to work with the motor of a heat pump. Air-source heat pumps commonly have three motors involved in their operation. One of the motors is responsible for running the compressor, one is required to circulate air over the outside coil, and a third is used to run the blower on the inside unit. A water-source unit will likely have only two motors, one for the compressor and one for the inside blower. All of these motors are induction motors. If you experience problems with the motors in your heat pump, call a qualified professional to service your equipment.

Matching Controls to Your Equipment

Matching controls to your equipment is not a job to be taken lightly. Using an incompatible thermostat can give you hours of trouble before you discover what the real cause of the problem is. Not all heat pumps are the same, and not all controls will work on all heat pumps. Matching controls and heat pumps is best done by following the recommendations made by manufacturers.

If you are buying all of your equipment from a single, reputable supplier, you should not run into problems with mismatched parts. However, if you are trying to salvage existing materials, such as a thermostat, from an old system you are replacing, problems could arise. Buying used equipment or purchasing one piece of equipment from one supplier and another piece of equipment from another supplier can put you up against some difficult situations.

Before you begin mating controls to your major equipment, make sure they are the right controls. For example, one relief valve looks pretty much like the next relief valve. Their physical appearances can be very deceptive. The relief valve for your heat pump will be required to have a specific pressure rating. A relief valve with a rating below the recommended rating will blow off before a safety interruption is required by your system. On the other hand, a relief valve with a rating higher than the specified rating for your system won't release built-up pressure until the pressure has exceeded the safe-operating pressure for your system.

In the case of relief valves, there is a little tag that dictates the pressure rating. Reading the tag will tell you if the relief valve is of the proper rating for your system. Not all parts are so easy to identify. Spend enough time reading the recommendations and installation instructions that will come with your heat pump to know that you are installing suitable controls.

Sensitive Objects

Some controls are very sensitive objects. A marginal mistake in a setting on a control can render a heat pump useless or at the least inefficient. When you are working with controls, make sure you understand them and their operation well.

I mentioned earlier that a check valve installed backward would not allow a normal flow. It may seem unlikely that you could make such a mistake, but I've seen numerous jobs where professional installers have installed the check valves backward. Even though there is a direction-of-flow arrow on the outside of the valve, installing the valves backward is not as uncommon as you might think. The installation of heat pumps is not particularly difficult for trained, experienced people. If you don't fall into this category, make sure that the people performing installations for you and your customers are.

CHAPTER

3

Choosing the Right Heat Pump for Your Job

Shopping for a new heat pump can get confusing. There are so many options available that the average person will have no idea where to begin. You could leave the decision of what type of heat pump to buy up a mechanical contractor who you hire to do the job. However, some contractors deal in better equipment than others do, and not all contractors are equal in their knowledge of heating and cooling systems. Do you really want to trust an unknown contractor to provide you with the best heat pump for your customers? If you are a general contractor, you can rely on your subcontractors, but it is wise to get to know as much as you can about what your subcontractors are doing.

When you begin to consider the purchase of a heat pump seriously, it is easy to see that if you don't know much about heating-and-cooling equipment, you may wind up with a less-than-perfect heat pump. So, what should you do to ensure that your customers get exactly what they want and need? You should read this chapter carefully; it will give you the data needed to make a smart decision about what type of heat pump to suggest.

Before you send your customers shopping for a new heat pump, there are some aspects of the equipment you should already be familiar with. For example, what size heat pump is needed? Is a 1- or a 2-ton unit needed (Fig. 3-1)? Will your customers be looking for an air-to-air system or an air-to-water system? Will the heat pump you install

Bard Ground Coupled Loop Design

1. Return Design To: ATTN:____________ Company: ________________________________
Address: ____________________________ City & State: _______________,___ Zip: ______
Telephone No.:_______________ Fax No. : _______________________ Date: ______________
Dealer/Job Name:__________________ Job Location: _______________________, ______
City State

2. Building Design Load: **HEATING**: ____________ BTUs/h Sensible: ____________BTUs/h
NOTE: Buildings to large for one heat pump, complete a worksheet for each zone. Latent :____________
TOTAL COOLING :____________BTUs/h

3. Weather City used to calculate #2 above. (See Manual "J" table 1): ____, _______________
State City

4. **NEAREST** Earth Temperature City (See back of this form): ____, _______________
State City

5. **TYPE of SOIL** at Trench Depth or Around Bore Hole: Select One

EPRI Classification
[ ] Sand or Gravel
[ ] Loam
[ ] Clay
[ ] Silt, Saturated Clay or Silt
[ ] Saturated Sand

ASHRAE Classification
[ ] Light Soil Dry
[ ] Light Soil Damp
[ ] Heavy Soil Dry
[ ] Heavy Soil Damp
[ ] Heavy Soil Saturated
[ ] Dense Rock
[ ] Average Rock
[ ] Dense Concrete

6. Provide COST OF OPERATION ESTIMATES: [ ] No [ ] Yes Local cost $_____Winter kW
$_____Summer kW

7. **Bard Ground Source Heat Pump** to be used: **Model**:________________

8. **TYPE of LOOP SYSTEM** to be used: Select One

HORIZONTAL SYSTEM Horizontal Pipe Configuration in Trench: SEE back of sheet
[ ] One Pipe: 1x1
[ ] Slinky Pitch ___In.
( ) Flat ( ) Edge
[ ] Pond Loop
( ) Mat ( ) Coil
[ ] 2 Pipes: stacked 1x2
[ ] 3 Pipes: stacked 1x3
[ ] 4 Pipes: stacked 1x4
[ ] 4 Pipes: stacked pairs 2x2
[ ] 6 Pipes: stacked 3 layers 2x3
[ ] 2 Pipes: side/by/side 2x1
[ ] 2 Pipes: horizontal bore
[ ] 3 Pipes: side/by/side 3x1
[ ] 4 Pipes: side/by/side 4x1
[ ] 6 Pipes: stacked 2 layers 3x2
Desired Trench Depth ___________Feet. & Trench Width _______ Inches.

VERTICAL SYSTEM Vertical Pipe Configuration in Bore Holes:
[ ] One U-Bend (2 pipes per bore hole) [ ] Two U-Bends (4 pipes per bore hole)

9. **TYPE of PIPE** to be used: [ ] SDR-11 Polyethylene Pipe [ ] SCH-40 Polyethylene Pipe & **SIZE**: [ ] 3/4 [ ] 1 [ ] 1 1/4 [ ] 1 1/2 [ ] 2

10. **FLUID FLOW** Thru Loop System: [ ] **SERIES**: Typically 1-1/4" to 2" pipe used. [ ] **PARALLEL**: Typically 3/4" to 1" pipe used.
Total header length (unit to loop)________Ft. & Pipe ____dia. In.

11. **ANTI-FREEZE** Solution: [ ] None [ ] Methanol [ ] Potassium Acetate (GS4tm) [ ] Propylene Glycol [ ] Environal 1000 [ ] Environal 2000

Fax: 419-636-2640 or Send To: Ground Coupled Loop System Design
Bard Manufacturing Company
P.O. Box 607
Bryan, Ohio 43506

If you have any questions, see manual 2100-099. F1115-298

FIGURE 3-1

Typical heat-pump sizing chart. (*Courtesy of Bard.*)

or have installed for your customer be a one-piece heat pump or a two-stage heat pump? Are your customers interested in a heat pump with a rectangular outside unit or a circular one? Does the shape of the outside unit make any real difference? These are just some of the questions that will come up when you work with customers in choosing a heat pump.

Until you understand your customers' needs for a heating-and-cooling system, you shouldn't voice opinions. Your customers may just become confused and frustrated. It is best to do you homework first and then send your customers window shopping. Once your customers have narrowed the field of competitive heat pumps, they can begin to talk seriously with sales representatives to wind up at a final decision.

To prepare you for the myriad of choices when offering advice on a heat pump, we are going to spend the time in this chapter to take you on a step-by-step tour of important considerations that may affect your customer's buying decision. Let's begin our trek into the purchasing possibilities for heat pumps by talking about the major types of them that are available.

One-Piece Heat Pumps

One-piece heat pumps (Fig. 3-2) are normally used when only a relatively small area of living space is to be climate controlled. These units are fine for motel rooms and home additions. They are not well suited to heating and cooling whole houses. A one-piece heat pump is fine if you want to beef up the heating and cooling in a specific part of your home, but it will not serve you well as your primary source of heating and cooling for the entire home.

Two-Piece Heat Pumps

Two-piece heat pumps, or two-stage heat pumps as they are often called, are by far the most popular style for most heating and cooling needs. One piece of the system is installed outside the home; this section of the equipment is referred to as the outside unit, or the condenser unit. The second piece of equipment in a two-piece system is installed inside the home. This unit doesn't require a lot of room and

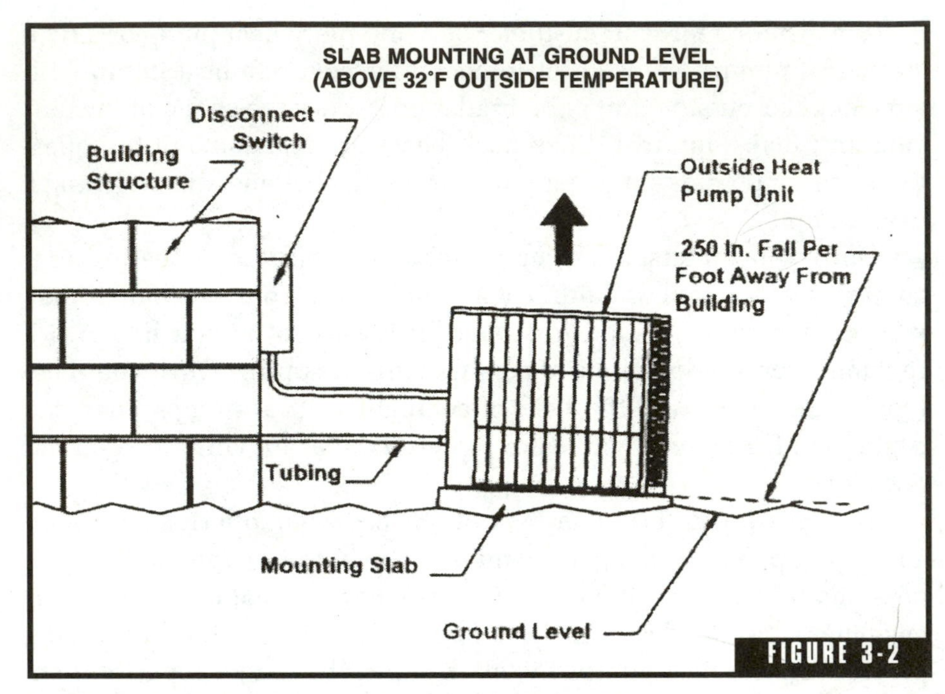

Outside unit of a two-piece heat pump. (*Courtesy of Bard.*)

can be concealed in a closet. For an average home, a two-piece heat pump will be the equipment style of choice.

Air-to-Air Heat Pumps

Air-to-air heat pumps are the most common type of heat pump used in residential applications. This is not to say they are the best type available, but they are the most popular.

Air-to-air heat pumps derive their heat source from the atmosphere. The outside unit pulls in warmth from the outside air and delivers it to the inside unit. Then the inside unit distributes the warmth throughout the home via duct work. This type of heat source, like any type, has its advantages and its disadvantages. Let's take a moment to examine the pros and cons of an air-to-air system.

Advantages

What are the advantages of an air-to-air system? There are several. First, air is available anywhere people live normal lives. This makes the source of heat for an air-to-air heat pump abundant. Also, air is considered to be a good source of heat for a heat pump.

Cost is another advantage. An air-to-air heat pump is the least expensive of all types of heat pumps. When you are evaluating the initial cost of a new heating and cooling system, you will be hard pressed to find a better value.

A final advantage is not one to be overlooked. Air-to-air heat pumps have been around a long time. This has allowed manufacturers to work the glitches out of them; so air-to-air units are typically dependable and long-lasting.

Disadvantages

Are their any disadvantages to air-to-air heat pumps? Yes, there are some disadvantages. The operation costs for an air-to-air heat pump are higher than those for other types of heat pumps. While some of the types of heat pumps we will address later in this chapter cost more to acquire, their operating expenses may be low enough to warrant the higher acquisition cost.

Frosting (Fig. 3-3) can be a problem with an air-to-air heat pump. In moist, cold temperatures, the outside unit will frost. This not only requires the unit to run frequently in its defrost cycle (increasing operating costs), it tends to make the heating unit less efficient in the production of warm air.

People who live in areas where the winter temperatures stay below 20°F for extended periods of time may find that an air-to-air heat pump is not as economical as they thought it would be. If frequent snowfall accompanies the low temperatures, even more cost-effectiveness will be lost (Fig. 3-4).

During extreme cold, heat pumps can be made to switch automatically to a backup heat source, such as electric resistance heating in the heat-pump system. This will bring the inside temperature of a home up to a comfortable level, but the operating costs will be substantial.

For a heat pump to work at its best, the difference in the temperature desired inside a home and the temperature of the heat source should be as minimal as possible. If you want your home to be a warm 72°F on a day when the outside temperature is 3°F, you are asking a lot of your air-to-air heat pump.

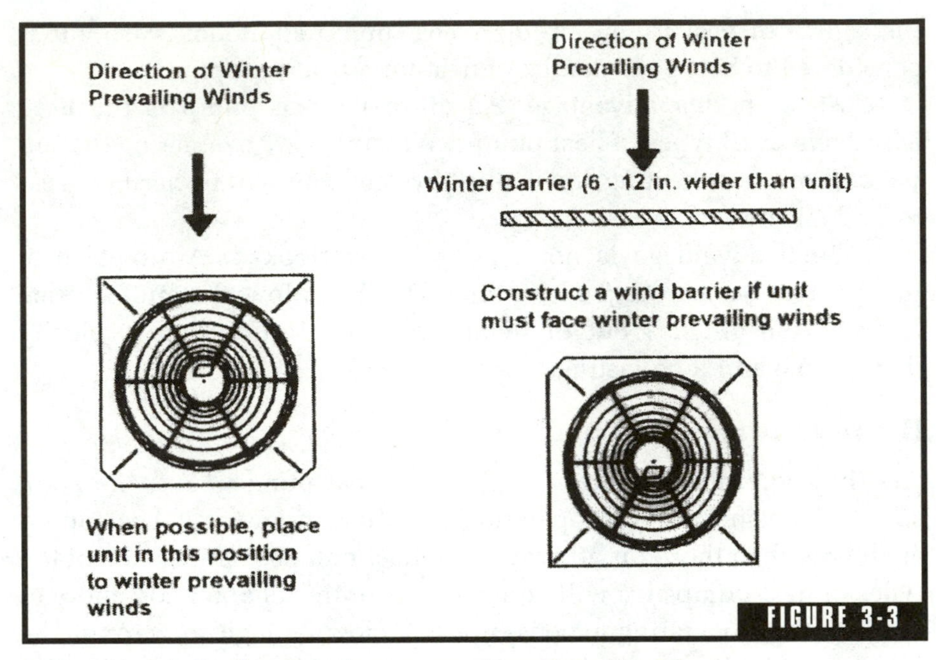

Proper wind deflection for a heat pump. (*Courtesy of Bard.*)

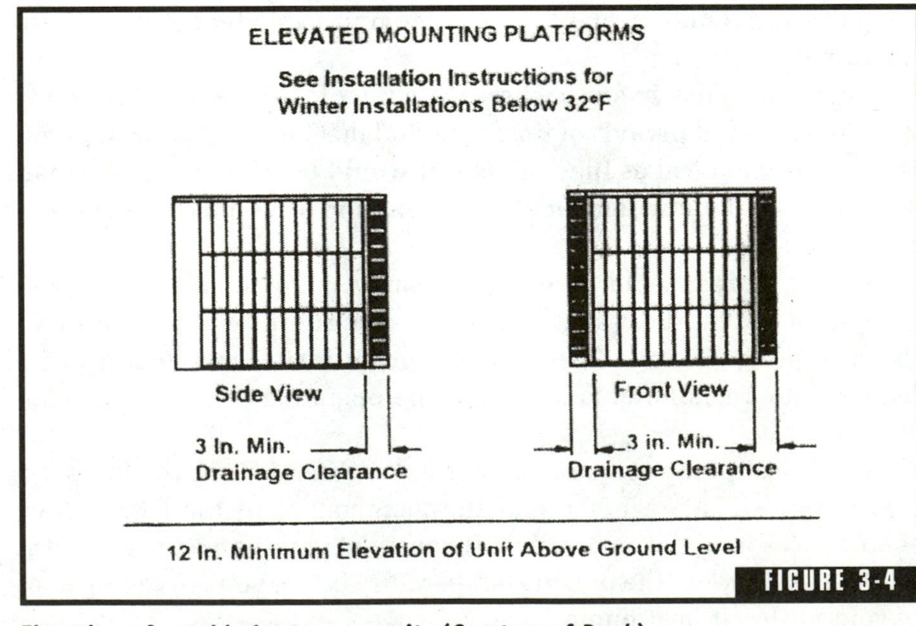

Elevation of outside heat pump unit. (*Courtesy of Bard.*)

Of course, if you live in Florida, you don't have to be concerned with subzero temperatures. In fact, many parts of the United States have moderate enough temperatures for the consideration of backup heat to be nonexistent. You must take your location into consideration when planning the purchase of any type of heat pump.

Air-to-Ground Heat Pumps

Air-to-ground heat pumps are not as well known as air-to-air heat pumps, but there does seem to be a growing interest in them. When you are thinking about the different types of heat pumps, remember that the closer the temperature of the heat source is to the desired inside temperature of the home, the better the heat pump will work. This fact is one of the most desirable aspects of an air-to-ground system.

Earth-source heat pumps (Fig. 3-5) pull their heat from the earth. Piping coils are installed a foot or two below the frost line. The frost line is the depth in the earth that winter frost will penetrate. The depth varies from state to state and region to region. For example, where I used to live in Virginia, the frost line was 18 inches. Here in Maine, the frost line is 4 feet, over twice the depth of that in Virginia.

By installing the coils well below the frost line, it is possible for an air-to-ground heat pump to enjoy a fairly constant temperature from its heat source. Unlike an air-source system, an earth-source system will normally have a heat source with a stable temperature of around 50°F. What does this mean to you? It means that during cold weather, an earth-source heat pump will not have to work as hard as an air-source heat pump. Let me explain (Figs. 3-6 and 3-7).

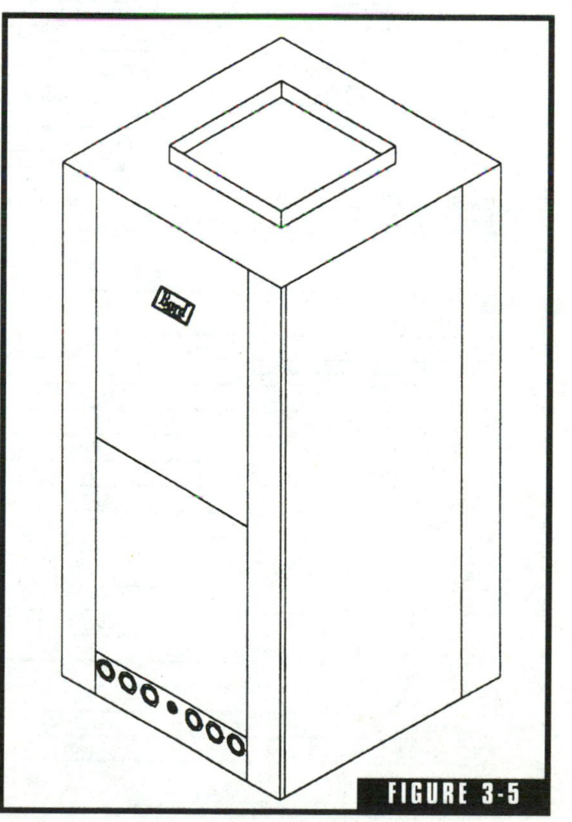

FIGURE 3-5

Earth-loop heat pump (*Courtesy of Bard.*)

TABLE 1
ARI CERTIFIED (1) CAPACITY AND EFFICIENCY RATINGS (Closed-Loop Earth Coupled Ground Loop Applications)

Model	CFM	ESP	Recommended Airflow Range	GPM	Cooling 77°EWT BTU/HR	Cooling 77°EWT EER	Heating 32° EWT BTU/HR	Heating 32° EWT COP
WPV24C	800	0.40	720 - 880	5	21,600	13.0	15,600	3.0
WPV30C	1,000	0.22	900 - 1,090	6	27,000	13.0	19,000	2.8
WPV36C	1,150	0.27	1,070 - 1,345	7	32,600	12.5	26,000	3.0
WPV42C	1,550	0.25	1,400 - 1,700	9	42,000	11.5	34,000	2.8
WPV48C	1,550	0.25	1,400 - 1,700	9	47,500	11.5	36,000	2.7
WPV60C	1,570	0.20	1,400 - 1,700	11	54,000	11.5	40,000	2.5

Certified in accordance with ARI Standard 330-90 "Ground Source Closed -Loop Heat Pumps" which includes watt allowance for water pumping.
Cooling capacity based on 80°F DB 67°WB entering air temperature.
Heating capacity based on 70°DB entering air temperature.

TABLE 2
CAPACITY AND EFFICIENCY RATINGS (Open Loop Well System Applications)

Model	CFM / ESP	GPM	Cooling 70° F. EWT BTU/HR	Cooling 70° F. EWT EER	Cooling 50° F EWT BTU/HR	Cooling 50° F EWT EER	Heating 70° F EWT BTU/HR	Heating 70° F EWT COP	Heating 50° F EWT BTU/HR	Heating 50° F EWT COP
WPV24C	800 / .41	4	22,400	12.1	23,400	14.3	24,200	3.8	19,400	3.1
WPV30C	1,000 / .22	4	28,400	13.8	29,800	16.3	29,200	4.0	23,200	3.3
WPV36C	1,150 / .27	5	34,300	13.3	36,000	15.7	39,500	3.8	31,700	3.1
WPV42C	1,550 / .25	6	46,200	12.9	48,400	15.3	54,300	3.7	43,600	3.0
WPV48C	1,550 / .25	9	48,000	11.7	50,700	13.9	57,400	3.7	45,900	3.0
WPV60C	1,570 / .20	8	54,100	11.7	56,800	13.9	62,600	3.4	50,200	2.8

Rated in accordance with ARI Standard 325 "Standard for Ground Water Source Heat Pumps" which includes watt allowance for water pumping.
Cooling capacity based on 80° F DB 67° WB entering air temperature.
Heating capacity based on 70° DB entering air temperature.

TABLE 3
Specifications

MODEL	WPV24C	WPV30C	WPV36C	WPV42C	WPV48C	WPV60C
Electrical Rating (60HZ/V/PH)	230/208-1	230/208-1	230/208-1	230/208-1	230/208-1	230/208-1
Operating Voltage Range	253-197	253-197	253-197	253-197	253-197	253-197
Minimum Circuit Ampacity	15.0	14.4	20.5	27.5	34.5	36.0
+ Field Wire Size	#14	#14	#12	#10	#8	#8
++ Delay Fuse Max. or Ckt. Bkr.	20	25	35	45	50	60
Total Unit Amps 230/208	8.9/9.4	10.3/11.3	14.8/16.3	19.5/21.2	23.7/27.0	25.9/28.8
COMPRESSOR						
Volts	230/208	230/208	230/208	230/208	230/208	230/208
Rated Load Amps 230/208	7/7.5	8.2/9.2	12.7/14.2	14.8/16.5	19.0/22.3	21.2/24.1
Branch Ckt. Selection Current	9.7	12.2	14.7	18.3	23.7	25.0
Lock Rotor Amps 230/208	50/50	61.7/61.7	82/82	109/109	129/ 129	169 / 169
BLOWER MOTOR AND EVAPORATOR						
Blower Motor - HP/Spd	1/4 3-spd	1/3 2-spd	1/3 2-spd	1/2 3-spd	1/2 3-spd	1/2 3-spd
Blower Motor - Amps	1.9	2.1	2.1	4.7	4.7	4.7
Face Area Sq. Ft./Row/Fins Per Inch	3.16/3/14	3.16/3/14	3.16/3/14	4.6/3/13	4.6/3/13	4.6/3/13
SHIPPING WEIGHT LBS.	240	240	255	350	370	375

FIGURE 3-6

Capacity and efficiency ratings. (*Courtesy of Bard.*)

To illustrate the advantage that an earth-source system holds over an air-source system, we will examine two examples. For convenience, we will look at heat pumps for homes in Virginia and Maine, since those are the areas I have the most experience with.

The Virginia House

The Virginia house we will talk about has an air-to-air heat pump in the first phase of our illustration. During the winter, the inside temperature of this house is kept at 70°F, not an unusually warm temperature. Much of the winter in the part of Virginia where the house is located sees daytime highs in the 30s and occasionally the 40s.

TABLE 4
WATER COIL PRESSURE DROP

Model	WPV24C		WPV30C		WPV36C		WPV42C		WPV48C, WPV60C	
GPM	PSIG	Ft Hd	PSIG	Ft Hd	PSIG	Ft Hd	PSIG	Ft Hd	PSIG	Ft HD
4	3.00	6.93	2.50	5.78	—	—	—	—	—	—
5	3.50	8.08	3.20	7.39	2.20	5.08	—	—	—	—
6	4.10	9.50	5.30	12.24	2.75	6.36	1.00	2.31	1.65	3.82
7	4.70	10.85	6.40	14.78	3.40	7.86	1.49	3.44	2.35	5.43
8	—	—	9.60	22.18	4.15	9.59	2.02	4.67	3.10	7.16
9	—	—	—	—	5.00	11.56	2.60	6.01	3.86	8.92
10	—	—	—	—	5.95	13.75	3.22	7.44	4.65	10.75
11	—	—	—	—	—	—	3.90	9.01	5.50	12.71
12	—	—	—	—	—	—	4.60	10.63	6.40	14.79
13	—	—	—	—	—	—	—	—	7.45	17.22
14	—	—	—	—	—	—	—	—	8.60	19.88
15	—	—	—	—	—	—	—	—	9.90	22.89

TABLE 5
INDOOR BLOWER PERFORMANCE (CFM – Dry Coil with Filter) (1)

Model	WPV24C			WPV30C WPV36C		WPV42C, WPV48C, WPV60C Without Optional CW45 Installed			WPV42C, WPV48C, WPV60C With Optional CW45 Installed	
ESP In. WC	Motor Speed			Motor Speed		Motor Speed			Motor Speed	
	High	Medium	Low	High	Low	High	Medium	Low	High	Medium
.00	1,033	946	774	1,300	1,190	1,740	1,650	1,530	1,740	1,600
.10	983	904	757	1,275	1,150	1,695	1,607	1,510	1,695	1,550
.20	942	870	742	1,210	1,110	1,650	1,570	1,480	1,650	1,520
.30	903	836	720	1,150	1,060	1,602	1,532	1,443	1,625	1,500
.40	857	794	688	1,080	1,000	1,550	1,490	1,400	1,500	1,460
.50	799	742	648	1,010	930	1,490	1,435	1,348	1,440	1,380
.60	740	681	603	920	875	1,420	1,365	1,290	1,390	1,310

(1) For wet coil CFM multiply by .96
ESP = External Static Pressure (inches of water)

FIGURE 3-7

Performance data. ***(Courtesy of Bard.)***

Nighttime lows linger in the 20s and 30s. A week or two of each winter finds the house surrounded by nighttime lows in the single digits, but almost never below zero.

The air-to-air heat pump is equipped with electric backup heat, and it provides the warmth desired in the home. It does, however, have to run its expensive electric heat to keep the house warm on the colder nights. The operation of the electric heat makes the electric bill rise at the same time it is bringing the inside temperature up. Would this home benefit from the installation of an air-to-ground heat pump? Yes, and I'll show you why.

The air-to-air heat pump is maintaining an inside temperature of 70°F. To do this, it is pulling from a daytime heat source that is typically in the 30s. For the sake of our comparison, we will say the air temperature during the day is 35°F. We still also assume the nighttime temperature averages 25°F. This means that the spread between the inside temperature and the heat source during the day is 35°F. At night, the spread is 45°. Now, what would the spread be if an earth-source heat pump was being used?

A properly installed earth-source heat pump pulls from a heat source where the temperature is normally 50°F. The daytime spread for the house in our example would be 20°. This is 15° warmer than the heat source for an air-to-air heat pump. At night, the spread for an earth-source system is still only 20°F. The heat-source temperature for a air-to-ground system doesn't fluctuate with the cycle of days and nights. This amounts to a nighttime heat-source temperature that is 25°F warmer than the air-to-air system in the example.

As you can imagine, it takes much less effort on the part of the heat pump to keep the house warm when it is getting its heat source from the earth. Now to expand on this, let's look at an abbreviated version of the example for my home in Maine.

The daytime highs for the last few weeks have been below 10°F. Nighttime temperatures have been well below zero, with temperatures of 20° below not being unusual. This means that an earth-source system would have a 40° advantage during the day and a 70° advantage at night. It should be obvious why an earth-source system provides more consistent performance and lower operating costs than an air-to-air system does. Now let's run down the same list of advantages and disadvantages that we used for air-to-air systems.

Advantages

One of the best advantages of an earth-source system has just been explained to you, but there are others. For example, just as air is an abundant resource for air-to-air systems, the earth is under the feet of people most of the time. In other words, there is no shortage of availability for earth as a heat source. Additionally, operating costs for an air-to-ground system tend to be low.

Disadvantages

What are the disadvantages of an earth-source system? Money is the major disadvantage. Air-to-ground heat pumps cost more to buy than air-to-air heat pumps do. Not only do they cost more to purchase, they are more expensive to install.

Aside from just the issue of money, a leak in an underground system can be very difficult, and expensive, to locate and correct.

Earth-source systems are not as difficult to install during the construction of a home as they are after a house has been built and lived in. Under these conditions, the excavation work needed to bury the coils can damage existing water pipes, gas pipes, electrical wires, and septic systems, not to mention the landscaping.

A last factor to consider when thinking of earth-source systems is their relative youth in the heat-pump industry. Since these systems are newer, and less tested, than air-to-air systems, there may be more bugs to work out of the equipment before it is as dependable as the old standby of air-to-air.

Well-Water Heat Pumps

Well-water heat pumps are probably the most efficient heat pumps available. If we go back to the fact that the temperature of the heat source is relative to the heating ability of a heat pump, you can hardly beat a well-water system.

Well water, like the earth, maintains a fairly constant temperature throughout the year. In fact, it often mirrors the temperature of the earth in many parts of the country. In such cases, a well-water system will perform with about the same effectiveness as an earth-source system. There are, however, many parts of the United States where the temperature of well water rises well above 50°F. This is when a well-water system leads the pack in efficiency.

Going back to our earlier example of the houses in Maine and Virginia, we can draw a comparison between a well-water system and an earth-source system. Well water in the part of Maine where I live runs at a temperature of approximately 48°F. There is no real temperature advantage between a well-water system and an earth-source system. In Virginia, where I used to live, the well water runs at a temperature of around 60°F. This is a full 10° warmer than the estimated temperature of an earth heat source in the same area. Based on these facts, a well-water system would have a constant 10° advantage over an earth-source system in central Virginia.

The temperature advantage for a well-water system in states where cold winter nights are common will range from 6 to 10°. In extremely cold states, the earth-source heat pump may have a warmth advantage over the well-water system of 2 to 6°. As you can see, it is necessary to refine your personal needs based on where you live (Figs. 3-8 to 3-12).

Advantages

One of the biggest advantages of a well-water system is the low cost of operating the system. As long as the well supplying the system has a constant reserve of water, the stability of the heat source is good. Well-water heat pumps have proven themselves over time, so there are not many kinks to get out of the works (Figs. 3-13 and 3-14).

Disadvantages

There are some disadvantages to a well-water heat pump. Unlike earth and air, well water is not available in all areas of the United States, but it can be found in more than three-fourths of the country. The fact that well-water is not available in 100 percent of the country counts as a minor disadvantage.

The initial cost of a well-source heat pump will be more than that of an air-to-air system and about the same as that of an earth-source system. However, the installation cost for a well-source heat pump may be less than that of most earth-source systems. The cost of drilling a well varies, but it is never cheap.

Some of the problems that may occur with a well-source heat pump are severe. The well could run dry, causing the heat pump to

CAPACITY AND EFFICIENCY APPLICATION RATINGS BASED ON 15% SODIUM CHLORIDE

MODEL WPV24C
800 CFM
5 GPM

Dry Bulb/ Wet Bulb	Cooling Capacity	Fluid Temperature Entering Water Coil °F								
		30° (2)	40° (2)	50°	60°	70°	80°	90°	100°	110°
75/62	Total Cooling	23,400	22,750	22,090	21,428	20,767	20,100	19,444	18,782	18,121
	Sensible Cooling	17,700	17,100	16,700	16,200	15,700	15,200	14,600	14,100	13,600
	Total Heat of Rejection	27,500	26,600	25,700	24,800	23,900	23,000	22,100	21,200	20,300
	EER (1)	23.50	21.31	19.09	16.86	14.60	12.40	10.15	7.90	5.75
80/67	Total Cooling	24,900	24,203	23,500	22,796	22,000	21,380	20,685	19,981	19,277
	Sensible Cooling	18,200	17,700	17,200	16,600	16,200	15,600	15,100	14,600	14,000
	Total Heat of Rejection	29,300	28,300	27,400	26,400	25,400	24,500	23,500	22,500	21,600
	EER (1)	24.30	21.99	19.70	17.40	15.10	12.80	10.50	8.20	5.90
85/72	Total Cooling	27,400	26,624	25,850	25,075	24,300	23,527	22,753	21,979	21,205
	Sensible Cooling	19,100	18,600	18,000	17,500	16,900	16,400	15,800	15,300	14,800
	Total Heat of Rejection	32,200	31,200	30,100	29,000	28,000	26,900	25,900	24,800	23,700
	EER (1)	25.90	23.50	21.00	18.60	16.10	13.70	11.20	8.70	6.30

Dry Bulb	Heating Capacity	Fluid Temperature Entering Water Coi l °F						
		25° (2)	30° (2)	40° (2)	50°	60°	70°	80°
70	Total Heating	14,400	15,100	17,200	19,400	21,500	23,600	25,700
	Total Heat of Absorption	10,900	11,700	13,500	15,300	17,000	18,800	20,600
	COP (1)	3.20	3.26	3.53	3.80	4.06	4.33	4.60

(1) Unit only. (2) Requires anti-freeze solution

MODEL WPV30C
1,000 CFM
6 GPM

Dry Bulb/ Wet Bulb	Cooling Capacity	Fluid Temperature Entering Water Coil °F								
		30° (2)	40° (2)	50°	60°	70°	80°	90°	100°	110°
75/62	Total Cooling	31,600	30,300	28,900	27,600	26,300	25,000	23,600	22,300	21,000
	Sensible Cooling	23,300	22,700	22,000	21,300	20,600	19,900	19,300	18,600	17,900
	Total Heat of Rejection	36,700	35,500	34,300	33,100	32,000	30,700	29,600	28,400	27,200
	EER (1)	25.30	22.90	20.50	18.10	15.70	13.30	10.90	8.50	6.10
80/67	Total Cooling	33,600	32,200	30,800	29,400	27,900	26,500	25,100	23,700	22,300
	Sensible Cooling	24,100	23,400	22,700	22,000	21,300	20,600	19,800	19,100	18,400
	Total Heat of Rejection	39,000	37,700	36,500	35,200	34,000	32,700	31,400	30,200	28,900
	EER (1)	26.10	23.60	21.20	18.70	16.20	13.70	11.20	8.80	6.30
85/72	Total Cooling	36,900	35,400	33,800	32,300	30,700	29,200	27,600	26,100	24,600
	Sensible Cooling	25,300	24,500	23,800	23,000	22,300	21,600	20,800	20,100	19,400
	Total Heat of Rejection	42,900	41,500	40,100	38,700	37,400	36,000	34,600	33,200	31,800
	EER (1)	27.90	25.30	22.60	20.00	17.30	14.60	12.00	9.40	6.70

Dry Bulb	Heating Capacity	Fluid Temperature Entering Water Coil °F						
		25° (2)	30° (2)	40° (2)	50°	60°	70°	80°
70	Total Heating	16,666	18,333	21,666	25,000	27,700	30,500	33,300
	Total Heat of Absorption	12,600	13,900	16,400	18,900	21,400	23,900	26,400
	COP (1)	3.00	3.20	3.50	3.90	4.20	4.50	4.90

(1) Unit only. (2) Requires anti-freeze solution

FIGURE 3-8

Performance rating when 15% sodium chloride is used. (*Courtesy of Bard.*)

MODEL WPV36C
1,150 CFM
7 GPM

Dry Bulb/ Wet Bulb	Cooling Capacity	Fluid Temperature Entering Water Coil °F								
		30° (2)	40° (2)	50°	60°	70°	80°	90°	100°	110°
75/62	Total Cooling	37,200	35,700	34,200	32,600	31,100	29,600	28,000	26,500	25,000
	Sensible Cooling	26,900	26,200	25,400	24,600	23,800	23,000	22,200	21,400	20,600
	Total Heat of Rejection	41,200	40,400	39,600	38,800	38,000	37,200	36,400	35,600	34,800
	EER (1)	24.00	21.60	19.20	16.90	14.50	12.10	9.70	7.30	5.00
80/67	Total Cooling	39,600	38,000	36,400	34,700	33,100	31,500	29,800	28,200	26,600
	Sensible Cooling	27,800	27,000	26,200	25,400	24,500	23,700	22,900	22,100	21,300
	Total Heat of Rejection	43,900	43,000	42,200	41,300	40,500	39,600	38,800	37,900	37,000
	EER (1)	24.80	22.30	19.90	17.40	15.00	12.50	10.00	7.60	5.10
85/72	Total Cooling	43,600	41,800	40,000	38,200	36,400	34,600	32,800	31,000	29,200
	Sensible Cooling	29,200	28,300	27,500	26,600	25,800	24,900	24,000	23,200	22,300
	Total Heat of Rejection	48,300	47,300	46,400	45,400	44,500	43,600	42,600	41,700	40,700
	EER (1)	26.50	23.90	21.20	18.60	16.00	13.30	10.70	8.10	5.40

Dry Bulb	Heating Capacity	Fluid Temperature Entering Water Coil °F						
		25° (2)	30° (2)	40° (2)	50°	60°	70°	80°
70	Total Heating	23,400	25,200	29,000	32,700	36,400	40,100	43,800
	Total Heat of Absorption	16,400	18,100	21,500	25,000	28,400	31,900	35,300
	COP (1)	2.90	3.00	3.40	3.70	4.00	4.30	4.60

(1) Unit only. (2) Requires anti-freeze solution

MODEL WPV42C
1,550 CFM
9 GPM

Dry Bulb/ Wet Bulb	Cooling Capacity	Fluid Temperature Entering Water Coil °F								
		30° (2)	40° (2)	50°	60°	70°	80°	90°	100°	110°
75/62	Total Cooling	53,500	50,700	47,900	45,100	42,300	39,500	36,800	34,000	31,200
	Sensible Cooling	38,700	37,100	35,600	34,000	32,400	30,800	29,200	27,700	26,100
	Total Heat of Rejection	61,500	59,500	57,400	55,400	53,300	51,200	49,200	47,100	45,100
	EER (1)	23.30	21.00	18.60	16.20	13.80	11.50	9.10	6.70	4.40
80/67	Total Cooling	57,000	53,900	51,000	48,000	45,000	41,500	39,100	36,100	33,200
	Sensible Cooling	40,000	38,300	36,700	35,000	33,400	31,800	30,100	28,500	26,900
	Total Heat of Rejection	65,500	63,200	61,100	58,900	56,700	54,500	52,300	50,100	47,900
	EER (1)	24.00	21.60	19.20	16.70	14.30	11.90	9.40	7.00	4.50
85/72	Total Cooling	62,600	59,3000	56,100	52,800	49,500	46,300	43,000	39,800	36,500
	Sensible Cooling	42,000	40,200	38,500	36,800	35,100	33,400	31,600	30,000	28,200
	Total Heat of Rejection	72,000	69,600	67,200	64,800	62,400	60,000	57,600	55,100	52,700
	EER (1)	25.70	23.10	20.50	17.90	15.30	12.70	10.00	7.50	4.84

Dry Bulb	Heating Capacity	Fluid Temperature Entering Water Coil °F						
		25° (2)	30° (2)	40° (2)	50°	60°	70°	80°
70	Total Heating	29,800	32,800	38,800	45,000	51,100	57,200	63,300
	Total Heat of Absorption	19,300	22,100	27,800	33,500	39,100	44,800	50,500
	COP (1)	2.75	3.00	3.30	3.80	4.20	4.60	5.00

(1) Unit only. (2) Requires anti-freeze solution

FIGURE 3-9

Performance ratings. (*Courtesy of Bard.*)

MODEL WPV48C
1,550 CFM
9 GPM

Dry Bulb/ Wet Bulb	Cooling Capacity	Fluid Temperature Entering Water Coil °F								
		30° (2)	40° (2)	50°	60°	70°	80°	90°	100°	110°
75/62	Total Cooling	54,467	52,378	50,290	48,201	46,112	44,023	41,934	39,845	37,756
	Sensible Cooling	37,675	36,346	35,017	33,687	32,358	31,029	29,699	28,370	27,041
	Total Heat of Rejection	62,530	61,486	60,442	59,397	58,353	57,308	56,264	55,219	54,175
	EER (1)	18.97	17.43	15.89	14.34	12.80	11.26	9.71	8.17	6.63
80/67	Total Cooling	57,944	55,722	53,500	51,277	49,055	46,833	44,611	42,388	40,166
	Sensible Cooling	38,840	37,470	36,100	34,729	33,359	31,988	30,618	29,248	27,877
	Total Heat of Rejection	66,522	65,411	64,300	63,188	62,077	60,966	59,855	58,744	57,633
	EER (1)	19.58	17.99	16.40	14.80	13.21	11.62	10.02	8.43	6.84
85/72	Total Cooling	63,738	61,294	58,850	56,405	53,961	51,516	49,072	46,627	44,183
	Sensible Cooling	40,782	39,343	37,905	36,466	35,027	33,588	32,149	30,710	29,271
	Total Heat of Rejection	73,174	71,952	70,730	69,507	68,285	67,063	65,841	64,618	63,396
	EER ()	20.91	19.21	17.51	15.81	14.11	12.41	10.71	9.01	7.30

Dry Bulb	Heating Capacity	Fluid Temperature Entering Water Coil °F							
		25° (2)	30° (2)	40° (2	50°	60°	70°	80°	90°
70	Total Heating	32,500	35,000	40,000	45,000	50,000	55,000	60,000	65,000
	Total Heat of Absorption	21,516	23,933	28,766	33,600	38,433	43,266	48,100	52,933
	COP (1)	2.79	2.90	3.12	3.35	3.57	3.79	4.01	4.23

(1) Unit only. (2) Requires anti-freeze solution

MODEL WPV60C
1,570 CFM
11 GPM

Dry Bulb/ Wet Bulb	Cooling Capacity	Fluid Temperature Entering Water Coil °F								
		30° (2)	40° (2)	50°	60°	70°	80°	90°	100°	110°
75/62	Total Cooling	58,777	57,071	55,366	53,660	51,954	50,248	48,542	46,836	45,130
	Sensible Cooling	44,972	42,759	40,546	38,332	36,119	33,906	31,693	29,480	27,267
	Total Heat of Rejection	65,511	65,232	64,954	64,675	64,396	64,118	63,839	63,561	63,282
	EER (1)	19.61	18.04	16.47	14.90	13.33	11.76	10.20	8.63	7.06
80/67	Total Cooling	62,529	60,714	58,900	57,085	55,270	53,455	51,640	49,825	48,011
	Sensible Cooling	46,362	44,081	41,800	39,518	37,237	34,935	32,674	30,392	28,111
	Total Heat of Rejection	69,692	69,396	69,100	68,803	68,507	68,211	67,914	67,618	67,322
	EER (1)	20.23	18.61	17.00	15.38	13.76	12.14	10.52	8.90	7.28
85/72	Total Cooling	68,782	66,786	64,790	62,793	60,797	58,801	56,804	54,808	52,812
	Sensible Cooling	48,681	46,285	43,889	41,494	39,098	36,703	34,307	31,912	29,516
	Total Heat of Rejection	76,661	76,335	76,010	75,684	75,358	75,032	74,706	74,380	74,054
	EER ()	21.61	19.88	18.15	16.42	14.69	12.96	11.24	9.51	7.78

Dry Bulb	Heating Capacity	Fluid Temperature Entering Water Coil °F						
		25° (2)	30° (2)	40° (2	50°	60°	70°	80°
70	Total Heating	35,400	38,700	45,200	51,800	58,300	64,900	71,400
	Total Heat of Absorption	22,000	24,800	30,300	35,800	41,300	46,800	52,300
	COP (1)	2.50	2.70	3.00	3.25	3.50	3.80	4.00

(1) Unit only. (2) Requires anti-freeze solution

FIGURE 3-10

Performance ratings. ***(Courtesy of Bard.)***

fail. Mineral deposits left in the equipment from the well water could gum up the works. These are factors that don't come into play with an air-to-air or air-to-ground system.

An Alternative

There is an alternative to well-water systems that still use water as a heat source. These systems pull their water from ponds, lakes, and rivers. This type of system certainly isn't for everyone, but if you have access to suitable surface water, you may want to consider such a system.

Operating costs for a surface-water system are usually low, and the total cost of installation is usually second only to air-to-air systems. The stability of the heat source is better than that of an air-source system but not as good as a well-water source or an earth-source.

CAPACITY MULTIPLIER FACTORS

% of Rated Air Flow	-10	Rated	10
Total Btuh	0.975	1.00	1.02
Sensible Btuh	0.95	1.00	1.05

CORRECTION FACTORS FOR PERFORMANCE AT OTHER WATER FLOWS

Rated Flow Plus – GPM	Heating		Cooling	
	Btuh	Watts	Btuh	Watts
2	1.00	98	1.01	1.00
4	1.01	97	1.03	1.01
6	1.02	96	1.05	1.02
8	1.02	95	1.06	1.02

FIGURE 3-11

Rating data. (*Courtesy of Bard.*)

COOLING

Model	Return Air Temperature	Pressure	30°	35°	40°	45°	50°	55°	60°	65°	70°	75°	80°	85°	90°	95°	100°	105°	110°
			Fluid Temperature Entering Water Coil °F																
	75° DB	Low Side	70	71	72	73	74	75	76	77	78	79	81	82	83	84	85	86	87
WPV24C	62° WB	High Side	142	149	156	162	169	176	183	189	196	203	209	216	223	229	236	243	249
Rated Flow	80° DB	Low Side	75	76	77	78	80	81	82	83	84	85	86	87	88	90	91	92	93
Rated GPM *	67° WB	High Side	146	153	160	167	174	180	187	194	201	208	215	221	228	235	242	249	256
Rated CFM 800	85° DB	Low Side	81	82	83	84	86	87	88	89	90	91	93	94	95	96	97	99	100
	72° WB	High Side	151	158	165	172	180	187	194	201	208	215	222	229	236	243	251	258	265
	75° DB	Low Side	66	67	68	69	70	70	71	72	73	74	75	76	77	77	78	79	80
WPV30C	62° WB	High Side	96	107	118	129	140	151	162	173	184	195	206	217	228	239	250	261	272
Rated Flow	80° DB	Low Side	71	72	73	74	75	75	76	77	78	79	80	81	82	83	84	85	86
Rate GPM *	67° WB	High Side	98	110	121	132	144	155	166	177	189	200	211	223	234	245	256	268	279
Rated CFM 1000	85° DB	Low Side	76	77	78	79	80	81	82	83	84	85	86	87	88	89	90	91	92
	72° WB	High Side	102	113	125	137	149	160	172	184	195	207	219	230	242	254	265	277	289
	75° DB	Low Side	57	58	59	60	61	62	63	64	65	66	67	68	70	71	72	73	74
WPV36C	62° WB	High Side	91	102	113	124	135	146	157	168	179	190	201	212	223	234	245	256	267
Rated Flow	80° DB	Low Side	61	62	63	64	66	67	68	69	70	71	72	73	74	76	77	78	79
Rate GPM *	67° WB	High Side	93	105	116	127	139	150	161	172	184	195	206	218	229	240	251	263	274
Rated CFM 1150	85° DB	Low Side	66	67	68	69	70	72	73	74	75	76	78	79	80	81	82	84	85
	72° WB	High Side	97	108	120	132	143	155	167	178	190	202	214	225	237	249	260	272	284
	75° DB	Low Side	59	60	61	62	63	64	65	66	67	68	69	70	71	72	73	75	76
WPV42C	62° WB	High Side	101	112	123	134	145	156	167	178	189	200	211	222	233	244	255	266	277
Rated Flow	80° DB	Low Side	63	64	65	66	68	69	70	71	72	73	74	75	76	78	79	80	81
Rate GPM *	67° WB	High Side	103	115	126	137	149	160	171	182	194	205	216	[illegible]	239	[illegible]	261	[illegible]	284
Rated CFM 1550	85° DB	Low Side	[illegible]	[illegible]	70	71	73	74	75	76	77	79	80	81	82	83	85	86	87
	72° WB	High Side	107	119	130	142	154	165	177	189	200	212	224	236	247	259	271	282	294
	75° DB	Low Side	68	67	67	66	66	65	65	65	64	64	63	63	63	62	62	61	61
WPV48C	62° WB	High Side	111	121	132	142	153	163	174	184	195	205	215	225	235	246	256	267	277
Rated Flow	80° DB	Low Side	73	72	72	71	71	70	70	70	69	69	68	68	68	67	67	66	66
Rate GPM *	67° WB	High Side	114	125	136	146	157	167	178	189	200	211	221	232	243	254	264	275	286
Rated CFM	85° DB	Low Side	79	78	78	77	77	76	76	75	75	74	73	73	73	72	72	71	71
	72° WB	High Side	118	129	140	151	162	173	184	195	206	217	229	240	250	262	273	284	295
	75° DB	Low Side	59	60	61	62	63	64	65	66	67	68	69	70	71	72	73	75	76
WPV62C	62° WB	High Side	90	101	112	123	134	145	156	167	178	189	200	211	222	233	244	255	266
Rated Flow	80° DB	Low Side	63	64	65	66	68	69	70	71	72	73	74	75	76	78	79	80	81
Rate GPM *	67° WB	High Side	92	104	115	126	138	149	160	171	183	194	205	217	228	239	250	262	273
Rated CFM 1570	85° DB	Low Side	68	69	70	71	73	74	75	76	77	79	80	81	82	83	85	86	87
	72° WB	High Side	96	107	119	131	142	154	166	177	189	201	212	224	236	248	259	271	283

HEATING

Model	Return Air Temperature	Pressure	25°	30°	35°	40°	45°	50°	55°	60°	65°	70°	75°	80°
			Fluid Temperature Entering Water Coil °F											
WPV24C Rated Flow Rate GPM * Rated CFM 800	70° DB	Low Side	35	40	44	49	54	59	63	68	73	77	82	87
		High Side	148	177	181	186	190	195	199	203	208	212	217	221
WPV30C Rated Flow Rate GPM * Rated CFM 1000	70° DB	Low Side	30	34	38	42	46	51	55	59	63	67	71	76
		High Side	179	183	187	191	195	200	204	208	212	216	220	225
WPV36C Rated Flow Rate GPM * Rated CFM 1150	70° DB	Low Side	33	37	41	45	49	54	58	62	66	70	74	79
		High Side	190	195	201	206	212	218	223	229	234	240	245	251
WPV42C Rated Flow Rate GPM * Rated CFM 1550	70° DB	Low Side	32	36	40	44	48	53	57	61	65	69	73	78
		High Side	173	178	184	189	195	201	206	212	217	223	228	234
WPV48C Rated Flow Rate GPM * Rated CFM	70° DB	Low Side	28	32	36	40	43	47	51	54	57	61	65	69
		High Side	184	189	195	200	206	212	217	223	228	234	239	245
WPV60C Rated Flow Rate GPM* Rated CFM 1570	70° DB	Low Side	31	35	39	43	47	52	56	60	64	68	72	77
		High Side	214	219	225	230	236	242	247	253	258	264	269	275

Low side pressure ± 2 PSIG
High side pressure ± 5 PSIG
Tables are based upon rated CFM (airflow) across the evaporator coil and rated fluid flow rate through the water coil. If there is any doubt as to correct operating charge being in the system, the charge should be removed, system evacuated and recharged to serial plate specifications.

*Flow Rates for Various Fluids	WPV24C	WPV30C	WPV36C	WPV48C WPV42C	WPV60C
Flow rate required GPM fresh water	4	4	5	6	8
Flow rate required GPM 15% sodium chloride	5	6	7	9	11
Flow rate required GPM 25% GS4	5	6	7	9	11

FIGURE 3-12

Performance ratings. (*Courtesy of Bard.*)

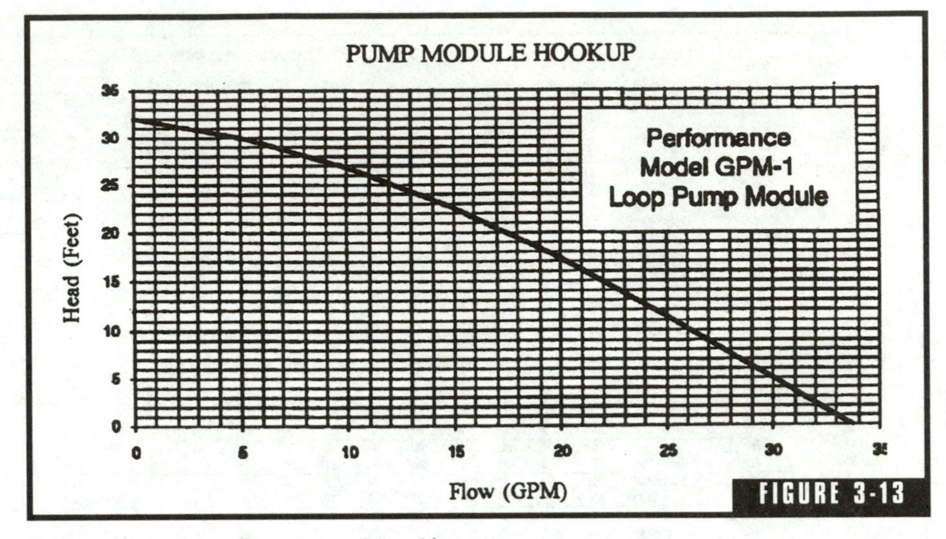

FIGURE 3-13

Performance data. (*Courtesy of Bard.*)

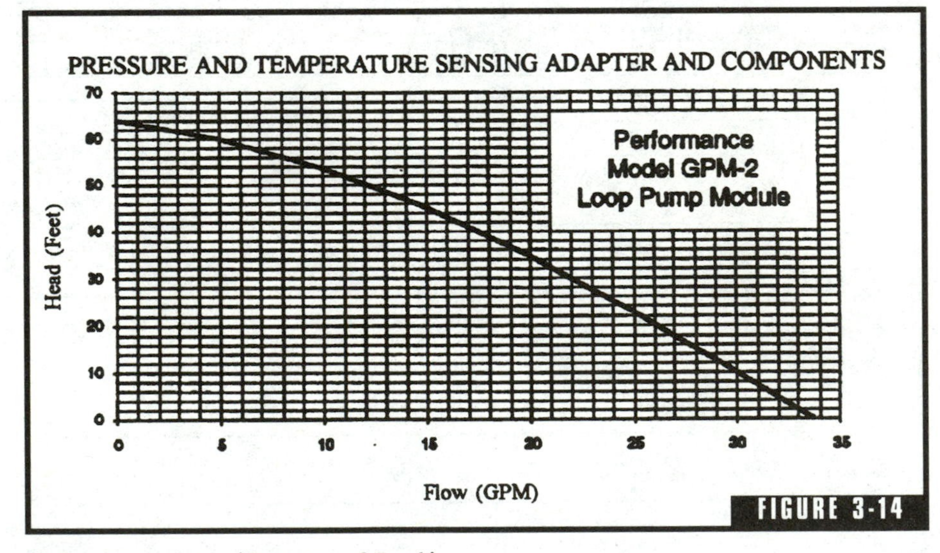

FIGURE 3-14

Performance data. (*Courtesy of Bard.*)

Solar Heat Source

Have you considered a solar heat source for your new heat pump? Solar heat-pump systems are continuing to be perfected, and they show a lot of promise for the future. They are, however, expensive, perhaps too expensive to be feasible in many homes.

Most people have some idea how solar collection and storage works for heating water and homes. If you've driven down a neighborhood street and seen homes with large, glass-looking panels on their roofs, you can bet the home is using solar power. Other, less noticeable, uses of solar energy can be seen in homes equipped with sunrooms, large areas of glass in the main walls of the home, and so forth. Tile on the floors of homes with a lot of glass in the walls helps to catch and store the warmth from the sun. With these principles accepted as commonplace, it stands to reason that solar-powered heat pumps are not too far away from becoming popular. The catch is in figuring out how to make the installation of such systems affordable.

Advantages

Let's talk about the advantages of a solar-source heat pump. First of all, the availability of solar power is very good. So is its suitability for use with heat pumps. With a properly installed solar-source heat pump, operating costs can be kept very low. These are all certainly good reasons to look to the sun for low-cost energy.

Disadvantages

Now for the disadvantages. While it is true that the sun will provide plenty of low-cost energy as a heat source, the expense of developing a system to harness that energy is not so agreeable. Since solar power is not always dependable, such as on cloudy days and at night, storage of the energy collected is a must. The price of solar panels and storage requirements needed to use the sun as a heat source for a heat pump can be a real eye-opener.

There are many factors that come into play when considering a solar-source heat pump. You must take into consideration the orientation of your home to the heat source. Obviously, an unobstructed southerly exposure is best. If your home sits in a valley, surrounded by mountains and tall trees, solar power is not going to be as effective as it could be for homes in other locations.

A house that is being built can be adapted to use solar power with less trouble than would be encountered if doing the same job on an older home. Remodeling an existing home to capitalize on solar power can become quite expensive, indeed.

Before you get to excited about cashing in on the warmth of the sun, talk to some qualified contractors in your area. They can help you determine the financial feasibility of a solar conversion or installation.

Coil Material

The coil material in a heat pump can affect your operating costs. Coils in heat pumps look like a large mass of metal fins. These fins are usually made of aluminum, and they will bend easily. Since the fins play a crucial role in the effective operation of a heat pump, it is important that they be protected from abuse.

Coil fins attach to tubing in order to effectively expose the tubing to more air temperature. Copper tubing provides the best efficiency as a coil material, so look for a unit that is fitted with a copper coil.

Shape versus Efficiency

When it comes to shape versus efficiency, would you rather have a heat pump that looks good or one that works well? There are a lot of round outside units installed as part of two-stage heat pump systems. Seeing so many round units might give you the idea that a round unit is the best unit. This, however, is not necessarily true. In fact, round units are generally not the most efficient design.

Americans often buy products for reasons other than reliability and performance. A practical car may sit on a sales lot until it becomes an antique, but a pretty car will be sold before the first layer of dust settles on it. Since it is unusual to get something that is both pretty and practical, it seems most people opt for the pretty. This is the case, to some extent, with heat pumps.

Outside units that are round normally appeal to more people than rectangular units do. Does the fact that a round outside unit has more curb appeal make it a wiser choice? No, it doesn't. Can it be said then that rectangular units are better than round units? No, rectangular units may not be the most efficient. There is, however, a way to tell what units are the most efficient.

There are two ratings that give performance information on heat pumps. One is the coefficient of performance (COP), and the other is an energy efficiency ratio (EER). Check these two ratings to determine the effectiveness of prospective heat pumps. With both the COP and the EER, the higher the ratio rating is, the better the heat pump is. Research is the key to success in many aspects of life, including the purchase of a heat pump. By spending adequate time researching your options, you will gain the ability to help your customers in their decision-making duties.

Installing an Air-Source Heat Pump

Air-source heat pumps are, by far, the most common type of heat pumps used today. The systems work extremely well in moderate climates. There are a number of factors associated with the installation of these systems that affect their effectiveness. Key issues range from unit location, pad placement, duct sizing, and so forth. One of the often overlooked aspects is the pad that outdoor units sit on. Because of this, we will start with the pad requirements. Then, we will go over the basic steps involved with the remaining installation requirements of an air-source heat pump.

Rooftop Locations

Rooftop locations can be used for the installation of heat pumps, but a roof is not the best place to put a heat pump. You will be hard pressed to find a residential heat pump installed on a roof, but roofs are often used for commercial installations. Heat pumps that are roof mounted do not function at their best efficiency levels. They are exposed to direct, hot sunshine in the summer. This makes cooling more difficult for the heat pump. In winter, the rooftop unit is exposed to brisk winds that handicap the unit's ability to produce comfortable warmth. While you may see heat pumps set atop the roofs of garages, avoid these locations. Setting the unit near the ground will produce more favorable results all through the year.

Ground-Level Locations

Ground-level locations are the most desirable sites for heat-pump pads. A unit that is mounted close to the ground will be affected less by winter winds than a unit mounted at a higher elevation. With careful placement, a ground-level unit can be shaded by deciduous trees in the summer to improve air-conditioning efficiency. In winter, the trees will lose their leaves and allow the sun to help warm the outside unit. An ideal location will find the heating unit placed with a southerly exposure that is not close to a bedroom (units close to bedrooms can create objectable noise) and is surrounded by deciduous trees and shrubs.

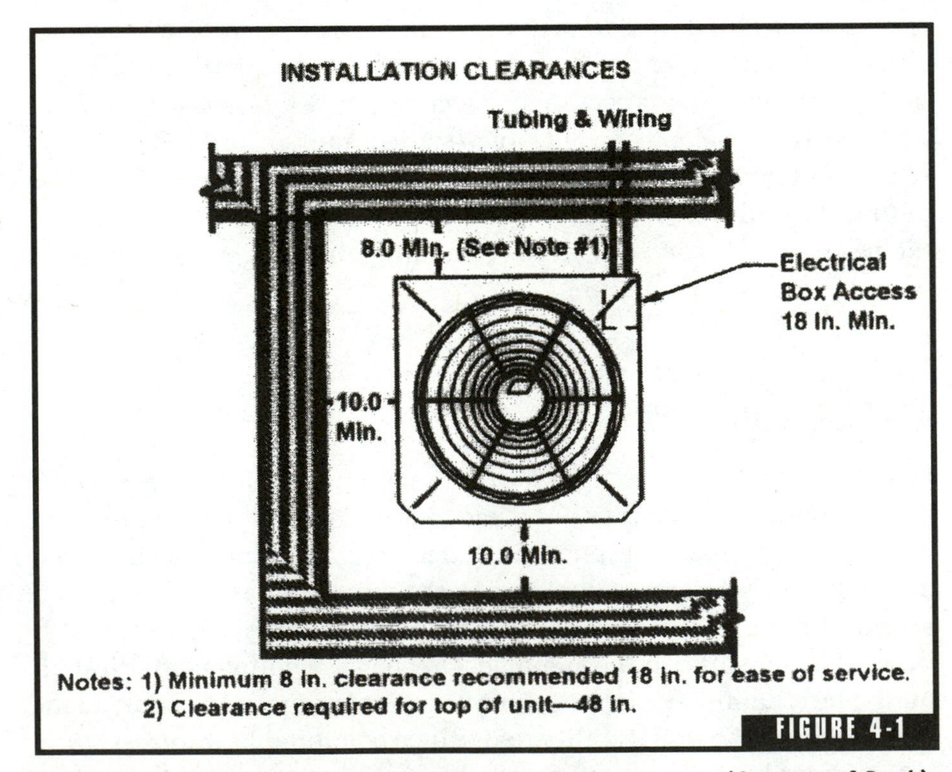

Recommended clearance for the outside unit of a heat pump. (*Courtesy of Bard.*)

Distances

The distances for clearance on a heat pump's outside unit must always be taken into consideration when planning the placement of the equipment pad (Fig. 4-1). Always refer to the manufacturer's specific recommendations for the unit you will be using. Normally, the outside compressor unit will sit at least 30 inches away from the side of the house. The distance from the air intake on the outside unit to any obstruction should ideally be about 3 feet. This is not a mandatory measurement, but it is a good practice to observe.

If you are adding landscaping to hide the outside unit from view, keep in mind the growth pattern of selected shrubs. Many people plant little bushes close to the outside unit only to find that in a few years the shrubs are blocking the air intake. Keep in mind that the outside unit should be located so that the distance between the inside unit and the outside unit is not more than about 60 feet when using an air-to-air heat pump.

Snow

Snow is another consideration when thinking in terms of distance. People with houses in southern states don't have to worry much about deep snow, but residents of northern states do. High accumulations of snow can disrupt the performance of a heat pump that is set too close to the ground. There are many ways to raise the level of an outside platform. Most contractors have a stand made out of angle iron for the unit to sit on. There are some factory-made stands (Fig. 4-2) available as well. Standard cinder blocks can even be used to increase the distance between the ground and the bottom of the outside unit. The key is to keep the unit above any anticipated snow buildup.

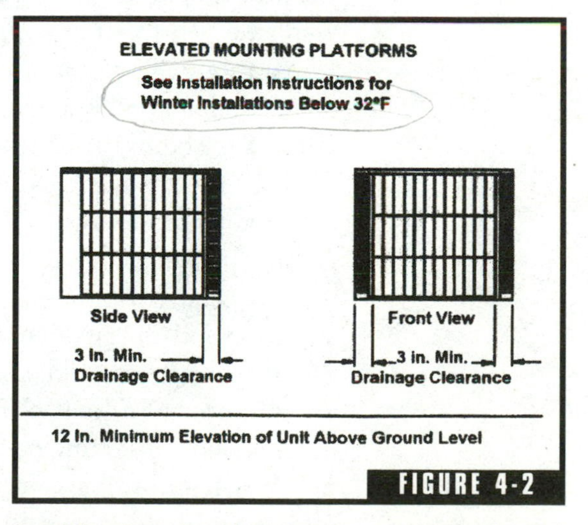

FIGURE 4-2

Specifications for elevating an outside unit of a heat pump. (*Courtesy of Bard.*)

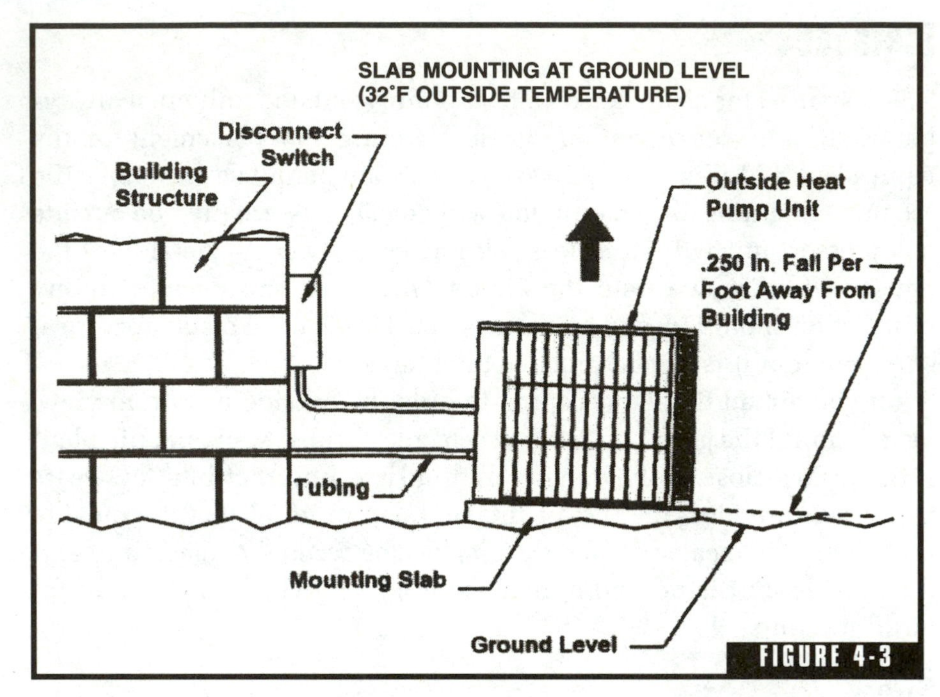

Proper setting of an outside unit for a heat pump. (*Courtesy of Bard.*)

Ground Preparation

Ground preparation for a heat-pump pad is very important. You don't want the pad to sink or tip. This can adversely affect the operation of a heat pump. How you prepare the ground to accept a pad will depend on several factors. For example, preparation techniques used when the ground is stable differ from those used when the earth has been disturbed recently, such as in the case of new construction. Another factor in your prep work will be the type of soils that must be prepared. A hard-packed clay will not require as much work as soggy soil will.

New construction creates a need for ground preparation. Installing a pad on soil that has been recently filled in around a house, without the proper preparation, can result in a sinking and tipping base for the outside unit of a heat pump. To overcome weak soil conditions, you must spend some time compressing the ground with some type of tamper. When ground is compacted, it should be done in thin layers. This may mean digging a hole in the new dirt at the location for the pad. The hole should be at least 12 to 18 inches deep when you begin the compaction. It is best if you go all the way down to undisturbed ground to begin your work, but this is not always feasible or needed.

Types of Pads

There are many types of pads that can be used to support the outside unit of a heat pump. Concrete pads are the most common. Many contractors do little more than place cinder blocks under the outside units they install. This is a cheap and effective method, but it is not nearly as desirable as a good concrete pad. If you opt for a block foundation to support your equipment, make sure you install the blocks in a way that will give uniform, level support (Fig. 4-3).

Foam-filled plastic is another option to consider using as a support for your condenser units. These pads are lightweight and easy to handle. There are, however, a couple of drawbacks to the foam pads. First, they are expensive. Second, the equipment may vibrate and slide on the pad unless an antiskid material (such as a sheet of rubber) is placed between the equipment and the pad.

Why Bother

Some people wonder why they should bother going to all the trouble to prepare solid ground and a good base for outside units. The tilting of an outside condenser can damage heating equipment. Condenser units use oil for lubrication. If a support pad shifts and allows the condenser to tilt, the lubricating oil may not be able to maintain proper lubrication of the equipment.

Sinking can occur if the ground under an outside unit is not firm. The settling can be a subtle process that doesn't catch a homeowner's attention, especially if the heating unit is camouflaged by shrubbery. As the pad and outside unit sinks, stress will be put on the refrigeration pipes and electrical wires. With enough settling, a broken wire or pipe could result.

Water can build up under a heat-pump pad if the ground around it is not suitable for allowing water to perk down into the earth. Therefore, you should put stone and sand under the pad. When an outside unit defrosts itself, water runs out of it. Without adequate drainage, the runoff water can cause problems. In winter, snow can cause problems. Snow that builds up around an outside unit blocks the air intake openings. This causes the heating system to malfunction. Failure to plan for snow by building an elevated pad or using a snow stand can result in a cold house when your customers need warmth the most.

Air-Source Heat Pumps

Air-source heat pumps are normally the least expensive type of heat pump to install. They are also the most prolific of the three major types of heat pumps. Air-source heat pumps have limitations on their efficiency. If you live in a state that experiences extremely cold temperatures during the winter months, an air-source heat pump probably isn't the best heating solution for heating homes in your region.

Most air-source heat pumps begin to lose their efficiency when the outside temperature drops below 28°F. Once the outside temperature dips to 10°F, air-source heat pumps are normally straining to maintain a comfortable temperature in the homes they serve. Most of these heat pumps have backup heating elements to assist them when the outside temperatures are extremely cold, but the operation of the backup heat drastically affects the efficiency of the heat pump.

Assuming that you have decided to install an air-source heat pump, you will need to take several factors into consideration. Since almost all residential applications of heat pumps involve split systems, those are the ones we will cover. A split system is one where one part of the heat pump is installed outside the home, and the other part of the heat pump is installed inside the home.

Before you can get too far into your design stage, you must know what size heat pump will be required for the home you are working on (Fig. 4-4).

Rooftop Installations

Rooftop installations are common on commercial buildings, but rare in residential applications. There are several reasons why residential dwellings don't utilize rooftop installations.

TABLE 1

Model	BC36C
Electrical Rating – 60HZ – 50HZ	240/208V 1PH 240/220V 1PH
Operating Voltage Range	197-253
Fusing and Ampacity	See Electric Heat Table
Blower and Motor	10x9 Direct
Motor – RPM / Speed	1075 / 2 Speed
Motor – HP / Amps	1/3 / 2.1
Evaporator Face Area Sq. Ft. / Rows/ Fins Per Inch	3.12 / 3 / 10
Filter – Throwaway	16x20x1 T
Refrigerant Cont./ R22 Charge	Orifice
Maximum Electric Heat	18 KW

TABLE 2 – MAXIMUM E.S.P. OPERATION BC36C (1)

Type of Application	Upflow Position		Counterflow Position (2)		Horizontal Position	
	Low Speed	High Speed	Low Speed	High Speed	Low Speed	High Speed
Heat Pump w/18KW	.30	.45	(3)	(3)	.20	.40
Heat Pump w/14KW	.40	.55	(3)	(3)	.35	.50
Heat Pump w/ 9KW	.60	.60	.40	.50	.50	.55
Heat Pump w/ 5KW	.60	.60	.60	.60	.55	.60
Heat Pump Only	.60	.60	.60	.60	.60	.60
18KW Only	.60	.60	.60	.60	.60	.60
14KW Only	.60	.60	.60	.60	.60	.60
9KW Only	.60	.60	.60	.60	.60	.60
5KW Only	.60	.60	.60	.60	.60	.60

(1) Values shown are for bottom and side return air opening.
(2) Side inlet not available on counterflow applications.
(3) 18KW and 14KW not approved in counterflow position when used with heat pump heating.

NOTE: 14KW is the maximum electric heat approved for 50HZ applications.

TABLE 3 – OPTIONAL FIELD INSTALLED ELECTRIC HEATER TABLE

Heater Package Model No.	Heater Package Volts/Phase	Heater KW and Capacity @ 240 Volts		Heater KW and Capacity @ 208 Volts		Heater Amps @240/208 Volts	Max. Fuse Size (3)	Maximum Circuit Breaker (3)	Minimum Circuit Ampacity (3)	Field Wire Size (4)	Ground Wire Size (5)
		KW	BTU	KW	BTU						
None	—	—	—	—	—	—	15	HACR Type 15	15	14	14
EH3BA-A05N,C	240/208–1	4.5	15345	3.38	11525	18.8/16.3	30	HACR Type30	28.1	10	10
EH3BA-A09N,C	240/208–1	9	30690	6.75	23018	37.5/32.5	50	HACR Type 50	50	6	10
EH3BB-A14N,C	240/208–1	13.5	46035	10.13	34543	56.3/48.7	80	80	74.9	3	8
EH3BA-A18N,C	240/208–1	18	61380	13.5	46035	75/64.9	100	100	98.3	1	8

(3) Includes blower motor
(4) Suggested size based on use of 60 degree C wiring material for ampacities less than 100A.
(5) Based upon table 250-95 degree F 1987 N.E.C.

NOTE: 9KW is the maximum electric heat approved for 50 HZ applications.

FIGURE 4-4

Specifications for a heat pump. (*Courtesy of Bard.*)

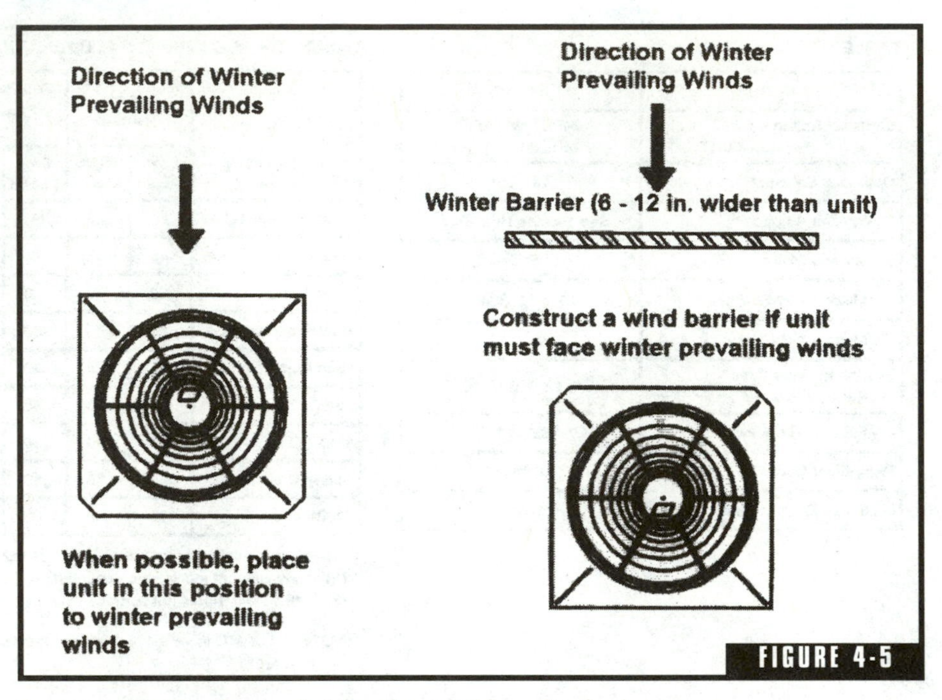

FIGURE 4-5

Proper protection from wind for the outside unit of a heat pump. (*Courtesy of Bard.*)

Since most homes don't have flat roofs, setting the outside unit of a heat pump on the roof is rarely practical. Even if you have a flat roof that has the structural capacity to house an outside unit, the roof is still not the best place to put the equipment.

Outside units that are installed on roofs are subjected to blazing heat in the summer and cold winds in the winter. These two factors play a large role in the efficiency of a heat pump. To be at its best, the outside unit of a heat pump should be shaded in the summer and protected from wind in the winter. This goal is much easier to accomplish when the equipment is installed at ground level (Fig. 4-5).

Ground-Level Installations

Ground-level installations of outside units for heat pumps far exceed rooftop installations when residential properties are involved. Not only are ground-level installations more convenient to work with, they also allow the heat pump to be more energy efficient.

There are several rules that should be obeyed when designing the location for an outside unit. Let's look at them on an individual basis.

Clearance

When you are deciding where to set the outside unit, you must keep the clearance requirements for the unit in mind. A good rule-of-thumb is to allow 30 inches of clearance, horizontally, in all directions from the coil of the outside unit. Vertically, a clearance of 5 feet will generally keep you out of trouble.

Wind Protection

Wind protection should be provided, in some form, for the outside unit. The house will protect the coil of the unit on at least one side. Many people plant low-growing bushes around their outside units to both improve the appearance of their home and to protect the coil from wind. If you decide to plant shrubbery as a wind screen, remember that many types of low-growing bushes grow out in width. Plant the bushes far enough away from the outside unit to maintain a clear distance of at least 2 feet, and preferably 30 inches.

Some soil conditions make planting natural wind screens difficult. If you can't accomplish the protection with shrubbery, you can build a decorative wind screen out of some other type of material, such as wood.

Overhangs

The overhangs of roofs can cause problems for heat pumps that are installed under them. Avoid a location that will put the outside unit in the path of falling ice, snow, and rain that may come off the roof.

Snow

Deep snow can bury the outside unit of a heat pump if the equipment is not elevated on a stable platform. While homeowners in southern states can set their outside units on concrete pads, close to the ground, residents in snow country must take accumulated snow buildup into consideration.

To keep snow from interfering with the clearance requirements of an outside unit, the unit must be elevated to a point above an anticipated snow accumulation. There are special stands available for setting outside units on, or a homemade stand can be used. Options for elevating the outside unit include the use of cinder blocks, concrete, angle iron, commercial stands, and others. The important thing is to make sure the stand is level and stable.

INDOOR BLOWER COIL PERFORMANCE [DRY COIL @ 230 VOLTS 60 HZ (3)] (1)

Model	KW	Speed	Position	IN H_2O						
				.00	.10	.20	.30	.40	.50	.60
BC36C	0	Hi	Upflow/Horizontal	1348	1297	1230	1174	1110	1009	883
	0	Low	Upflow/Horizontal	1169	1135	1090	1042	987	853	707
BC36C	5	Hi	Upflow/Horizontal	1333	1282	1215	1159	1095	994	868
	5	Low	Upflow/Horizontal	1154	1120	1075	1027	972	838	692
BC36C	9	Hi	Upflow/Horizontal	1275	1219	1143	1090	956	832	672
	9	Low	Upflow/Horizontal	1173	1105	1055	974	867	734	619
BC36C	14	Hi	Upflow/Horizontal	1260	1198	1128	1075	941	817	657
	14	Low	Upflow/Horizontal	1158	1090	1040	959	852	719	604
BC36C	18	Hi	Upflow/Horizontal	1245	1183	1113	1060	926	802	642
	18	Low	Upflow/Horizontal	1143	1075	1025	944	837	704	589

TABLE 5 INDOOR BLOWER COIL PERFORMANCE [DRY COIL @ 230 VOLTS 60 HZ (3)] (1)

Model	KW	Speed	Position	IN H_2O						
				.00	.10	.20	.30	.40	.50	.60
BC36C	0	Hi	Counterflow	1305	1243	1173	1120	986	862	702
	0	Low	Counterflow	1203	1135	1085	1004	897	764	649
BC36C	5	Hi	Counterflow	1290	1228	1158	1105	971	847	687
	5	Low	Counterflow	1188	1120	1070	989	882	749	634
BC36C	9	Hi	Counterflow	1275	1219	1143	1090	956	832	672
	9	Low	Counterflow	1173	1105	1055	974	867	734	619
BC36C	14	Hi	Counterflow	1260	1198	1128	1075	941	817	657
	14	Low	Counterflow	1158	1090	1040	959	852	719	604
BC36C	18	Hi	Counterflow	1245	1183	1113	1060	926	802	642
	18	Low	Counterflow	1143	1075	1025	944	837	704	589

(1) Values shown are standard for both bottom and side return air opening.

(2) Values shown are standard for bottom return air opening, side return air opening nor available for counterflow.

(3) Reduce airflow values shown by 130 CFM for 208 volt operation.

NOTE: For 50 HZ applications, reduce CFM's by 17%

FIGURE 4-6

Blower coil performance ratings. (*Courtesy of Bard.*)

Drainage

Drainage arrangements must be made for the outside unit. This is typically done by installing crushed stone and sand in the area surrounding the location of the outside unit. Chapter 10 will give you all the specifications needed for drainage and other aspects of actually setting the heat-pump pad into place.

The Indoor Unit

The indoor unit for an air-source heat pump can be installed in numerous locations. Basements are an ideal place for installing inside units, but closets, attics, laundry rooms, and crawl spaces can all be used to house the interior unit. Check the manufacturer's suggestions before making your installation, but any place that allows room to work on the unit should be suitable for the installation (Fig. 4-6).

Close Proximity

The indoor unit should be installed in close proximity to the outdoor unit, the closer, the better. For example, if you were to set the outside unit so that it was near the kitchen, you should seek a location close to the kitchen for the inside unit.

Condensation Piping

Heat pumps remove moisture from homes, and this moisture is disposed of through condensation piping. The condensate line will typically drain into a portion of the plumbing system, such as a floor drain. A small trap (a U-bend in the piping) is often required on the condensate piping. When you are designing the placement of the inside unit, make sure the equipment will be high enough above the floor to allow for the installation of a trap.

Central Location

A central location in the home is the best place for the inside unit. This will allow for better air distribution. In other words, if you have the choice of putting the inside unit at one end of the home or near the middle of the home, choose the location in the middle.

Routing the Refrigeration Tubing

When routing the refrigeration tubing that connects the inside unit to the outside unit, you should avoid all sharp turns. When the tubing is run through a wall, it should be protected from chaffing with the use of a sleeve, such as a piece of plastic plumbing pipe. Plan on a sleeve that will be large enough to accommodate the tubing, the insulation used to wrap the tubing, and electrical wires that will be run with the tubing.

When planning the installation of the tubing, you should avoid situations that will require the tubing to come into contact with floor joists and wall studs. Plan to use hangers that will keep the tubing from making direct contact with structural members. This is done to reduce noise in the home when the units are running.

Oil

Oil must be available to the compressor of a heat pump. When inside and outside units are installed at different heights above or below ground level, special arrangements may be needed to assure the proper conveyance of oil. Read the manufacturer's recommendations to see if and where oil traps may be needed.

Backup Heat

Backup heat is needed with almost all air-source heat pumps. This is not much of a factor during the design stage, but make sure that the heat pump you buy has adequate arrangements for backup heat.

Duct Work

Duct work is what carries the air from a heat pump to the register outlets. It is the duct work that often causes the most trouble in designing a heat-pump system for houses that have been built without heat pumps installed. When the new construction of a house is being planned, duct work is not such a bother.

Heat pumps often require larger duct work than forced hot-air furnaces. Many people who are replacing hot-air furnaces assume, wrongly, that their existing duct work can be used effectively with the new heat pump. While some of the duct work often can be used, much of it will normally be too small.

When duct work is nonexistent in an existing home, installing it can be more than just a little troublesome. A typical layout will have a trunk line (the main duct system) running along the length of the home. The size of the duct work will get smaller as it gets longer. Trunk lines often run under the middle of homes. Most older trunk lines are formed from sheet metal. Supply ducts are cut into the trunk line and extended to the boots of floor registers. The supply ducts are frequently made of round, flexible duct material, and they are easy to work with. There is also a need for return-air duct work.

During the design phase of your work, the two most important jobs you have pertaining to duct work are sizing it correctly and finding suitable places to install it. If you are building a new home, routing the duct work will not be much of a problem. By reviewing your blueprints, you can find all sorts of places that will allow the installation of the duct work. Depending upon how complete your set of blueprints is, you may even have a page dedicated to showing the size and locations of duct work.

Computing the size of duct work is not very complicated if you have the proper charts and tables to work with. Most manufacturers of heat pumps will be happy to supply you with charts and tables for recommended sizes. The tables and charts provide assistance in sizing both trunk lines and branch ducts. There are provisions for round ducts and rectangular ducts. By looking at how many cubic feet of air movement is required, you can go across the chart and find the proper duct sizes.

Most suppliers of heat pumps will be happy to size heating systems for you. If you provide the suppliers with a set of blueprints, they can size the system and provide you with a material list of the items needed to get the job done. You can obtain sizing charts and tables from various manufacturers of heat pumps or from the American Society of Heating, Refrigeration, and Air Conditioning Engineers (ASHRAE). ASHRAE can be contacted by calling (404) 636-4404 or by writing to: ASHRAE, 1791 Tullie Circle NE, Atlanta, GA 30329.

Electrical Service

The use of a heat pump will require a 200-amp electrical service. Many older homes have 60- or 100-amp electrical services, and this is an expensive problem when converting to a heat pump. Having a licensed electrician upgrade an electrical service will cost quite a bit, so don't overlook this part of financial budgeting when designing your system. An electrician will be needed to install disconnect boxes for the heat-pump units. Due to the dangers involved in working with electricity, most homeowners should hire licensed electricians to perform all phases of electrical work.

The Full Installation

The full installation of a heat pump is a big job. This type of work typically requires the acquisition of a permit from a code enforcement office. In most jurisdictions, only licensed installers and homeowners who will perform their own installations are allowed to obtain a permit. Builders and general contractors usually are not allowed to perform heating installations, unless they are licensed heating professionals as well as builders and contractors. Consult your local code office if you have any questions pertaining to the limits set forth on new installations and replacements.

Installing Ground-Water-Source and Closed-Loop Heat Pumps

Water-source heat pumps are more expensive to install than air-source heat pumps, but they make heat pumps effective in extremely cold climates. States with extreme winter climates are not suitable choices for the installation of air-source systems. Such systems will perform, but the cost of operating them during cruel, cold winters is not effective. This is not the case with water-source systems. Since ground water remains at a fairly stable temperature throughout the year, heat pumps that use ground water for operation are stable and cost-effective in terms of operations cost. The downside is the additional cost incurred to install a water-source system. Over time, the added expense of acquisition may be replenished in the form of savings from operating expenses.

Much of the work involved with a water-source installation is similar to, if not the same as, what is used with an air-source system. There are, however, some significant differences involved with the installation of water-based heat pumps. Since many of the basic installation matters have already been covered, we will concentrate on the primary differences encountered when installing water-source systems (Figs. 5-1 to 5-3).

Open-loop design. (*Courtesy of Edwards.*)

Horizontal-loop design. (*Courtesy of Edwards.*)

Vertical-loop design. (*Courtesy of Edwards.*)

Water-Source Heat Pumps

Water-source heat pumps are not nearly as common as air-source heat pumps. One reason they are not so common is the expense of installing them. Another reason is that many consumers have never heard of them. Water-source heat pumps are probably the most cost-effective type of heat pump you can buy.

Most water-source heat pumps depend on well water for their operation. It is, however, possible to use surface water, such as a pond or river, to provide the heat source for water-source heat pumps. One big advantage to well water is its fairly constant temperature. In many parts of the country, well water temperatures stay between 48° and 50°F throughout the year. The exact temperature fluctuates some, and different states report different temperatures, but the water can be counted on to stay moderate all year (Fig. 5-4).

The fact that well water stays so warm in winter is a big advantage when using a heat pump. For example, an air-source heat pump might have to pull its heat from outside air with a temperature well below freezing. A well-water heat pump, under the same conditions, will be drawing its heat from the much warmer well water. This results in less work for the heat pump and more cash savings for the user.

While there is little question that a water-source heat pump is more cost-effective to operate than an air-source heat pump, there are downsides to water-source systems. The biggest drawback is the cost of having the system installed.

When you set out to design an effective water-source system, you will run into many of the same considerations that must be dealt with when planning an air-source system. In addition to all the overlapping factors, there are many unique aspects of water-source systems to be evaluated. Let's take the time now to examine the key elements of designing a working water-source heat-pump system that uses well water as a heat source (Fig. 5-5).

The Well

The first major design step for a heat pump that receives its heat from well water is the well itself. Do you have an existing well that can be used, or will you need to have one drilled? The expense involved with drilling wells is one serious drawback to well-water heat-pump systems (Fig. 5-6).

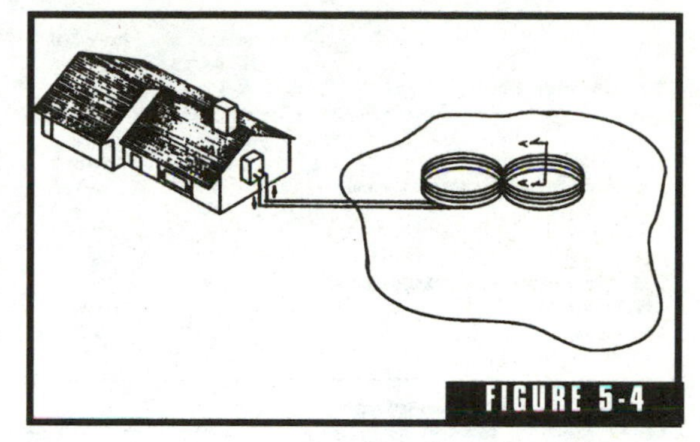

FIGURE 5-4

Vertical-loop pond system. (*Courtesy of Bard.*)

WATER COIL SELECTION GUIDE

Potential Problem	Use Copper Coil	Use Cupro-Nickel Coil
Scaling—		
Calcium and magnesium salts (hardness)	Less than 350 ppm (25 grain/gallon)	More than 350 ppm (up to sea water)
Iron oxide	Low	High
Corrosion*—		
pH	7 to 9	5-7 and 8-10
Hydrogen sulfide	Less than 10 ppm	10-50 ppm
Carbon dioxide	Less than 50 ppm	50 to 75 ppm
Dissolved oxygen	Only with pressurized water tank	All systems
Chloride	Less than 300 ppm	300 to 600 ppm
Total dissolved solids	Less than 1,000 ppm	1,000 to 1,500 ppm
Biological Growth—		
Iron bacteria	Low	High
Suspended Solids—	Low	High

* Important: If the concentration of these corrosives exceeds the maximum tabulated in the cupro-nickel column, then the potential for serious corrosion problems exists. Water treatment may be required.

FIGURE 5-5

Water coil selection guide. (*Courtesy of Bard.*)

Domestic Water Uses	Peak Demand Allowance for Pump	Individual Fixture Flow Rate
	gpm Column 1	gpm Column 2
Household Uses		
Bathtub or tub-and-shower combination	2.00	8.0
Shower only	1.00	4.0
Lavatory	.50	2.0
Toilet—flush tank	.75	3.0
Sink, kitchen—including garbage disposal	1.00	4.0
Dishwasher	.50	2.0
Laundry sink	1.50	6.0
Clothes washer	2.00	8.0
Irrigation, Cleaning and Miscellaneous		
Lawn irrigation (per sprinkler)	2.50	5.0
Garden irrigation (per sprinkler)	2.50	5.0
Automobile washing	2.50	5.0
Tractor and equipment washing	2.50	5.0
Flushing driveways and walkways	5.00	10.0
Cleaning milking equipment & milk storage tank	4.00	8.0
Hose cleaning barn floors, ramps, etc.	5.00	10.0
Swimming pool (initial filling)	2.50	5.0

FIGURE 5-6

Sizing guide for domestic water usage. (*Courtesy of Bard.*)

Not all wells are suitable for use as a heat source for a heat pump. Wells that are shallow, have a low reservoir of water, or a low flow rate will not produce satisfactory results when used as a heat source. Therefore, drilled wells are usually the only type of well that will provide dependable service (Fig. 5-7).

The recovery rate of a drilled well must be known in order to properly plan the installation of a well-water heat pump. A professional well driller can test the flow rate of a well to determine how many gallons per minute (gpm) the well is capable of producing. Since the results of this test are crucial to the effective operation of the new heat pump, it is best to engage a trained professional to perform the test (Figs. 5-8 to 5-15).

Dual Duty

If you already have a drilled well that provides potable water to your home, the well may be capable of performing dual duty. By this, I mean, the well might be able to go on providing water for domestic uses as well as providing a heat source for your new heat pump (Fig. 5-16).

When a well is called upon to perform double duty, it should have a flow rate capable of providing adequate water for both types of demand simultaneously. Not all wells have a strong enough gpm to do this, and there are ways to work around such a problem, but the ideal situation will not call for a compromise.

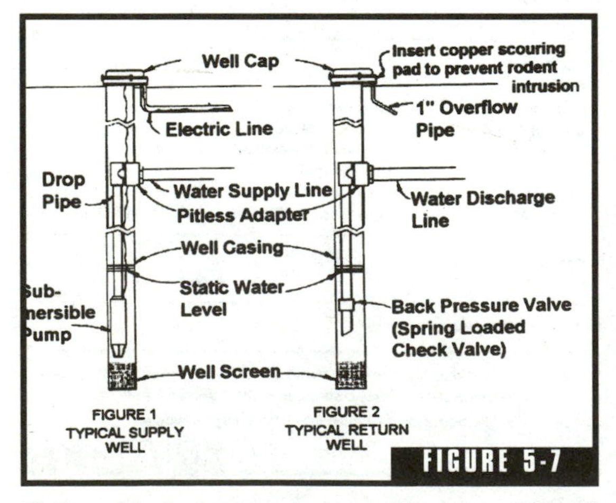

Well supply and return system. (*Courtesy of Bard.*)

Reverse-Acting Switch

If you must compromise, a reverse-acting pressure switch is the most cost-effective

WATER SYSTEM WORKSHEET

(Method applicable to submersible pumps. Consult well driller for sizing of other types of pumps)

A. WELL PUMP SIZING

Branch "A" — Well Pump .... Piping from pump in well to pressure tank.

Branch "B" — Domestic Water Supply .... Piping from tank to the fixtures throughout house.

Branch "C" — Heat Pump Water Supply .... Piping from tank through heat pump coil to drain.

1.	Determine household water needs for Table 2, Column 1 Enter here	9	gpm Branch "B"
2.	Enter gpm flow rate for heat pump to be installed from specifications (water coil rated flow).	6	gpm Branch "C"
3.	Add lines 1 and 2 for total water flow rate required.	15	gpm Branch "A"

NOTE: If piping layout has more branches, determine the flow rate for these from Table 2, Column 1, and include in total.

B. DETERMINING WATER PRESSURE REQUIREMENTS PIPE SIZING FOR EACH PIPE BRANCH — Household plumbing, Branch "B", may be assumed to have a total pressure requirement of 30 psig (69 ft. hd.)

		BRANCH "A"	BRANCH "C"	
4.	Tentatively select a pipe size and enter here. Table 4 or 5	1-1/4	1	inch pipe
5.	Consult Table 7 and enter equivalent feet of pipe for one elbow of the size selected in step 4 above using gpm of the branch. Enter here.	7	6	equiv. feet
6.	From the piping layout, determine the number of fittings needed for the branch. Enter here.	4	12	fittings
7.	Multiply line 5 by line 6. Enter total here.	28	72	equiv. feet
8.	From piping layout, determine total lineal feet of pipe in the branch. Enter here.	118	140	lineal feet
9.	Add lines 7 and 8. Enter here.	146	212	total feet
10.	Consult Table 4 and 5 for pipe size and total gpm needed for each branch determined from Section A above and enter friction loss here.	3.5	2.5	ft. hd./100 ft.
11.	Multiply line 10 by line 9, divide by 100 and enter here as total piping friction loss.	5.1	5.3	ft. hd.
12.	Consult heat pump specifications for the unit to be installed and enter unit pressure drop here. Branch C only. (Water coil pressure drop).		5.8	ft. hd.
13.	Pressure drop for constant flow valve and solenoid valve. Tables 8 and 9.		38.2	ft. hd.
14.	Calculate total pressure drop, Branch "C", by adding lines 11 and 12. Enter here. Branch "A" pressure drop is the same as line 11. Branch "A" should be entered here.	5.1	49.3	ft. hd.
15.	Multiply line 14 by .433 to convert to psig. Enter here.	2.2	21.3	psig

16. From the piping layout, determine parallel flow among the branches. Beginning at the well pump, add the friction loss in psig for the well pump branch (Branch "A") to the branch having the higher pressure drop (Branch "B" or Branch "C"). Note: if more than three branches are required by the piping layout, select that branch which has the highest pressure drop and add this pressure drop to Branch "A". Enter on line 16 the number obtained as total piping pressure loss due to pipe friction ______ 32.2 ______ psig

17. Add 20 psig to line 16 to obtain the pressure switch cut out point. Enter this value here. ______ 52.2 ______ psig — tank cutout setting

18. Multiply line 17 by 2.31 to convert to ft. hd. Enter here. ______ 120.6 ______ ft. hd.

19. Pump requirements will be: ______ 15 ______ gpm at ______ 120.6 ______ ft. hd. (line 18)

plus ______ 60 ______ feet lift (Vertical distance to water in well)

Total Pressure ______ 180.6 ______ ft. hd.

C. PRESSURE TANK SIZING (Applicable to bladder or diaphragm type tanks only — recommended type.)

20.	Enter desired minimum off time of the well pump in minutes and fractions of minutes. Never less than two minutes.	2	minutes
21.	Enter pressure switch cut-in point psig. At least as great a pressure as required for cut-in pressure on line 16.	32.2	psig
22.	Enter pressure cut-out psig from line 17. Usually 20 psig higher than value in line 21.	52.2	psig
23.	Multiply line 3 by line 20 to determine minimum drawdown (acceptable volume).	30	gallons
24.	Refer to Table 3 or pressure tank specifications for a specific model of tank(s) to select nominal capacity of tank needed, using information from lines 21, 22 and 23 above.	120	gallons

FIGURE 5-8

Water system worksheet sample. (*Courtesy of Bard.*)

WATER SYSTEM WORKSHEET

(Method applicable to submersible pumps. Consult well driller for sizing of other types of pumps)

A. **WELL PUMP SIZING**

Branch "A" — Well Pump .... Piping from pump in well to pressure tank.

Branch "B" — Domestic Water Supply .... Piping from tank to the fixtures throughout house.

Branch "C" — Heat Pump Water Supply .... Piping from tank through heat pump coil to drain.

1. Determine household water needs for Table 2, Column 1 Enter here ____________ gpm Branch "B"
2. Enter gpm flow rate for heat pump to be installed from specifications (water coil rated flow). ____________ gpm Branch "C"
3. Add lines 1 and 2 for total water flow rate required. ____________ gpm Branch "A"

NOTE: If piping layout has more branches, determine the flow rate for these from Table 2, Column 1, and include in total.

B. **DETERMINING WATER PRESSURE REQUIREMENTS PIPE SIZING FOR EACH PIPE BRANCH — Household plumbing, Branch "B", may be assumed to have a total pressure requirement of 30 psig (69 ft. hd.)**

	BRANCH "A"	BRANCH "C"	
4. Tentatively select a pipe size and enter here. Table 4 or 5	____________	____________	inch pipe
5. Consult Table 7 and enter equivalent feet of pipe for one elbow of the size selected in step 4 above using gpm of the branch. Enter here.	____________	____________	equiv. feet
6. From the piping layout, determine the number of fittings needed for the branch. Enter here.	____________	____________	fittings
7. Multiply line 5 by line 6. Enter total here.	____________	____________	equiv. feet
8. From piping layout, determine total lineal feet of pipe in the branch. Enter here.	____________	____________	lineal feet
9. Add lines 7 and 8. Enter here.	____________	____________	total feet
10. Consult Table 4 and 5 for pipe size and total gpm needed for each branch determined from Section A above and enter friction loss here.	____________	____________	ft. hd./100 ft.
11. Multiply line 10 by line 9, divide by 100 and enter here as total piping friction loss.	____________	____________	ft. hd.
12. Consult heat pump specifications for the unit to be installed and enter unit pressure drop here. Branch C only. (Water coil pressure drop).		____________	ft. hd.
13. Pressure drop for constant flow valve and solenoid valve. Tables 8 and 9. (or water regulating valves)		____________	ft. hd.
14. Calculate total pressure drop, Branch "C", by adding lines 11 and 12. Enter here. Branch "A" pressure drop is the same as line 11. Branch "A" should be entered here.	____________	____________	ft. hd.
15. Multiply line 14 by .433 to convert to psig. Enter here.	____________	____________	psig

16. From the piping layout, determine parallel flow among the branches. Beginning at the well pump, add the friction loss in psig for the well pump branch (Branch "A") to the branch having the higher pressure drop (Branch "B" or Branch "C"). Note: if more than three branches are required by the piping layout, select that branch which has the highest pressure drop and add this pressure drop to Branch "A". Enter on line 16 the number obtained as total piping pressure loss due to pipe friction ____________ psig
17. Add 20 psig to line 16 to obtain the pressure switch cut out point. Enter this value here. ____________ psig — tank cutout setting
18. Multiply line 17 by 2.31 to convert to ft. hd. Enter here. ____________ ft. hd.
19. Pump requirements will be: ____________ gpm at ____________ ft. hd. (line 18)

    plus ____________ feet lift (Vertical distance to water in well)

    Total Pressure ____________ ft. hd.

C. **PRESSURE TANK SIZING (Applicable to bladder or diaphragm type tanks only — recommended type.)**

20. Enter desired minimum off time of the well pump in minutes and fractions of minutes. ____________ minutes

    Never less than two minutes.
21. Enter pressure switch cut-in point psig. At least as great a pressure as required for cut-in pressure on line 16. ____________ psig
22. Enter pressure cut-out psig from line 17. Usually 20 psig higher valve in line 21. ____________ psig
23. Multiply line 3 by line 20 to determine minimum drawdown (acceptable volume). ____________ gallons
24. Refer to Table 3 or pressure tank specifications for a specific model of tank(s) to select nominal capacity of tank needed, using information from lines 21, 22 and 23 above. ____________ gallons

FIGURE 5-9

Water system worksheet. (*Courtesy of Bard.*)

PRESSURE DROP CALCULATIONS TO SELECT CIRCULATION PUMP

Transfer fluid requirements for closed-loop, earth-coupled heat pump systems varies with fluid temperature and heat pump size. To determine the circulation pump size requirement, the system flow rate requirements (GPM for heat pump used) and total system pressure drop in feet of head loss. From these two pieces of information a circulation pump can be selected from the pump manufacturer's performance curves.

The fluid (water) flow rate and water coil pressure drop are found in the manufacturer's heat pump specifications or TABLE 2 for Bard water source heat pumps. The head loss for different pipe materials and sizes per 100 feet are found in TABLE 5 of this section and a quick pump selection table for flow rates that match Bard water source heat pumps are in TABLE 4 of this section. See blank form F1125 in the back of this manual.

Following are two examples of how to determine the head loss of earth loops. First example will be a series horizontal system and the second example will be a parallel vertical system.

EXAMPLE 1:

A. Series horizontal system.
B. Bard WPV36C water source heat pump to be used.
C. Heat pump water flow requirements is 7 GPM with a 7.86 foot head loss See TABLE 2.
D. Earth loop 1200 feet 1-1/2 inches SDR 17 polybutylene pipe.
E. 20 feet 1 inch copper pipe connecting earth loop to water source heat pump.
F. The circulation pumping system lay out to be similar to FIGURE 16.
G. Noburst antifreeze used.

CALCULATING PUMP WORKSHEET

1. Find the Bard heat pump model used in TABLE 2. MODEL WPV36C
2. Enter water coil head loss (TABLE 2): 7.86 ft hd
3. Continue across TABLE 2 to find GPM flow required for this heat pump. 7 GPM
4. Count each elbow, tee, reducer, air scoop, flowmeter, etc., as 3 feet of pipe equivalent. Add the equivalent feet of pipe to the actual feet of pipe used. The total length is used to determine the piping heat loss below.

Pipe Type and Size	No. Elbows, Tees Devices, Etc.*		Equiv. Ft. of Pipe	Actual Pipe Used	Total Pipe Length
		x 3			
1" copper	20	x 3	60	20	80
1-1/2" PB SDR17	None	x 3	None	1200	1200
		x 3			
		x 3			

*If the pipe is bent at a 2 foot radius or larger, DO NOT figure the curve as an elbow.

5. PIPING HEAD LOSS for different types of pipe at GPM flow rate of water source heat pump. NOTE: For parallel earth loop system, figure for only one loop.

Pipe Type and Size	Total Pipe Length		Piping** Head Loss (Table 3)		
	(	÷ 100)		=	ft.hd.
1" copper	(80	÷ 100)	4	=	3.2 ft.hd.
1-1/2" PB SDR17	(1200	÷ 100)	.4	=	4.8 ft.hd.
	(	÷ 100)		=	ft.hd.
	(	÷ 100)		=	ft.hd.

**For a parallel earth loop, divide the heat pump GPM (line 3) by number of loops to determine flow rate through each individual loop to select piping head loss. ft.hd.

SUBTOTAL 15.86 ft.hd.

6. Multiply SUBTOTAL by multiplier (see TABLE 5) to obtain TOTAL HEAD LOSS FOR SYSTEM using antifreeze solution. TOTAL HEAD LOSS 21.09 ft.hd.

7. PUMP SELECTION: Use Table 4 and flow rate, (line 3). Select the pump output which is larger or equal to the TOTAL HEAD LOSS FOR SYSTEM (line 5 or 6). GPM-1 (Circulating Pump Model) 1 (No. Pumps)

If the TOTAL HEAD LOSS calculated in line 6 is greater than the pump outputs listed in Table 4, go to the pump manufacturer's performance curves and find the required GPM flow for the heat pump. Pump performances are listed for each pump model at different flow rates.

Series pump performance is simply a TOTAL OF THE INDIVIDUAL PUMP PERFORMANCE. If one pump can overcome 10 feet of heat loss, two can overcome 20 feet of head loss, three can overcome 30 feet of head loss, etc.

***REMEMBER: UNDER NO CIRCUMSTANCES MIX DIFFERENT PUMP SIZES WHEN USING PUMPS IN SERIES ***

FIGURE 5-10

Pressure drop calculations. (*Courtesy of Bard.*)

TABLE 2
WATER COIL PRESSURE DROP (Water Only)

Model	WPV24 WPV24C	WQS30A WPV30B WPV30C	WQS36A WPV36B WPV36C	WQS42A WPV53B WPV42C	WPV62B WPV60C
GPM	Ft. Hd.	Ft. Hd.	Ft. Hd.	Ft. Hd.	Ft. Hd.
4	6.93	5.78	–	–	–
5	8.08	7.39	5.08	–	–
6	9.50	12.24	6.36	2.31	–
7	10.85	14.78	7.86	3.44	–
8	–	22.18	9.59	4.67	7.16
9	–	–	11.56	6.01	8.92
10	–	–	13.75	7.44	10.75
11	–	–	–	9.01	12.71
12	–	–	–	10.63	14.79
13	–	–	–	–	17.22
14	–	–	–	–	19.88
15	–	–	–	–	22.89

TABLE 4
SELECTION OF CIRCULATION PUMP OR PUMPS

	Grundfos* Pump Models	No. of Pumps	Water Flow Rate Required in G.P.M.						
			4	6	8	10	12	14	16
Pump Output (Feet of Head) @ GPM @ Top of Column	UP26-96F	1	26.5	25	22.5	20	18	16	14
	UP26-96F	2	53	50	45	40	36	32	28
	UP26-99F	1	30.5	29	28	27	25	23.5	22
	UP26-99F	2	61	58	56	54	50	47	44
	Loop Pump Models								
	GPM-1**	1		29	28	27	25	23	22
	GPM-2**	2		58	56	54	50	47	44

* Other models of circulation pumps may be used. Consult the manufacturer's specifications.
** See manual 2100-212 GPM series loop pump modules for complete details.

TABLE 3
MINIMUM FLOW IN PIPE FOR TURBULANCE (GPM*)

Nominal Pipe	Size (Pipe ID)	Water at 40° F	GS4 25% Volume 25° F	Noburst 30% Volume 25°F	Methanol 20% Volume 25°F
PE (SDR-11)					
3/4"	(0.86)	1.1	1.9	3.1	2.4
1"	(1.077)	1.3	2.3	3.9	3.1
1 1/4"	(1.385)	1.7	3.0	5.0	3.9
1 1/2"	(1.554)	1.9	3.5	5.7	4.4
2"	(1.943)	2.4	4.3	7.0	5.5
PE (SCH)					
3/4"	(0.824)	1.0	1.9	3.0	2.3
1"	(1.049)	1.3	2.4	3.8	3.0
1 1/4"	(1.380)	1.7	3.1	5.0	3.9
1 1/2"	(1.610)	2.0	3.6	5.8	4.6
2"	(2.067)	2.5	4.6	7.6	5.9
PB (SDR-17,IPS)					
1 1/2"	(1.676)	2.1	3.7	6.0	4.8
2"	(2.095)	2.6	4.6	7.6	5.9
PB (SDR-13.5,CTS)					
1"	(0.957)	1.2	2.1	3.5	2.7
1 1/4"	(1.171)	1.4	2.6	4.3	4.9
1 1/2"	(1.385)	1.7	3.1	5.0	3.9
2"	(1.811)	2.2	4.0	6.6	5.1

* For each separate loop.
NOTE: When selecting pipe size for parallel flow, it is necessary to maintain turbulent flow in the earth coil for heat transfer. The table above lists the minimum flows for turbulence.

FIGURE 5-11A

Sizing data. (*Courtesy of Bard.*)

TABLE 5
PIPING FEET OF HEAD LOSS AT DIFFERENT FLOW RATES PER 100 FEET

Pipe Size and Material	GPM Flow Rate										
	DI	1	2	3	4	5	6	8	10	12	14
Connection Hose 1"	1.05	*	*	*	1.33	1.95	2.68	4.43	6.53	8.99	11.77
PVC 3/4" - 200 PSI		*	*	*	3.7	5.7	*	*	*	*	*
PVC 1" - 200 PSI		*	*	*	1.0	1.9	2.7	4.2	6.3	8.9	11.8
Copper 3/4"		*	*	*	4.3	6.3	*	*	*	*	*
Copper 1"		*	*	*	1.5	1.9	2.7	4.5	6.9	9.6	12.8
PE3408 (Polyethylene)	DI										
1. SDR-11 3/4	0.860	0.31	1.03	2.07	3.41	5.03	*	*	*	*	*
2. SDR-11 1	1.077	0.11	0.36	0.71	1.18	1.73	2.38	3.92	*	*	*
3. SDR-11 1 1/4	1.358	*	0.12	0.24	0.39	0.58	0.79	1.31	1.93	2.65	3.47
4. SDR-11 1 1/2	1.554	*	*	0.13	0.21	0.31	0.42	0.69	1.02	1.40	1.83
5. SDR-11 2	1.943	*	*	*	0.07	0.11	0.15	0.24	0.35	0.48	0.63
6. SCH 40 3/4	0.824	0.38	1.26	2.54	4.18	6.16	8.46	*	*	*	*
7. SCH 40 1	1.049	0.12	0.40	0.81	1.33	1.96	2.69	4.45	*	*	*
8. SCH 40 1 1/4	1.380	*	0.11	0.22	0.36	0.54	0.74	1.21	1.79	2.46	3.21
9. SCH 40 1 1/2	1.610	*	*	0.11	0.18	0.26	0.35	0.58	0.86	1.18	1.55
10. SCH 40 2	2.067	*	*	*	*	0.08	0.11	0.18	0.26	0.36	0.47
PB2110 (Polybutylene)	DI										
11. SDR-17, IPS 1 1/2	1.676	*	*	0.09	0.15	0.21	0.29	0.48	0.71	0.98	1.28
12. SDR-17, IPS 2	2.095	*	*	*	0.05	0.07	0.10	0.17	0.25	0.34	0.44
13. SDR-13.5, Cts 1	0.957	0.19	0.62	1.25	2.06	3.03	4.16	*	*	*	*
14. SDR-13.5, Cts 1 1/4	1.171	*	0.24	0.48	0.79	1.17	1.60	2.64	*	*	*
15. SDR-13.5, Cts 1 1/2	1.385	*	0.11	0.22	0.36	0.53	0.72	1.19	1.76	2.41	3.20
16. SDR-13.5, Cts 2	1.811	*	*	0.06	0.10	0.15	0.20	0.33	0.49	0.68	0.88

NOTE: 1. These head losses are for water at 40°F temperature.
2. Count each elbow, tee, reducer, air scoop, flow meter, etc., as 3 feet of equivalent pipe length and add to actual measured pipe length for total length.
3. To adjust the total earth loop piping head loss for other antifreezes and water solutions at 25°F, multiply pressure loss on line 6 for water by: Noburst - 1.33, GS4 - 1.18, Methanol - 1.25

FIGURE 5-11B

Sizing data. (*Courtesy of Bard.*)

Friction loss tables and for common pipe diameters and materials. Figures given are friction loss in feet of head per one hundred feet of pipe. Doubling the diameter of a pipe increases its capacity four times, not two times.

TABLE 4

	1/2" ID = .622"				3/4" ID = .824"				1" ID = 1.049"			
	Steel		Plastic		Steel		Plastic		Steel		Plastic	
GPM	Ft.	Lbs.	Ft.	Lbs.	Ft.	Lbs.	Ft.	Lbs.	Ft.	Lbs.	Ft.	Lbs.
2	4.8	2.1	4.1	1.8								
3	10.0	4.3	8.7	3.8	2.5	1.1	2.2	1.0				
4	17.1	7.4	14.8	6.4	4.2	1.8	3.7	1.6				
5	25.8	11.2	22.2	9.6	6.3	2.7	5.7	2.5	1.9	.8	1.8	.8
6	36.5	15.8	31.2	13.5	8.9	3.9	8.0	3.5	2.7	1.2	2.5	1.1
7	48.7	21.1	41.5	18.0	11.8	5.1	10.6	4.6	3.6	1.6	3.3	1.4
8	62.7	27.2	53.0	23.0	15.0	6.5	13.5	5.9	4.5	2.0	4.2	1.8
9	78.3	34.0	66.0	28.6	18.8	8.2	16.8	7.3	5.7	2.5	5.2	2.3
10	95.9	41.6	80.5	34.9	23.0	10.0	20.4	8.9	6.9	3.0	6.3	2.7
12					32.6	14.1	28.6	12.4	9.6	4.2	8.9	3.9
14					43.5	18.9	38.0	16.5	12.8	5.6	11.8	5.1
16					56.3	24.4	48.6	21.1	16.5	7.2	15.1	6.6
18					70.3	30.5	60.5	26.3	20.6	8.9	18.7	8.1
20					86.1	37.4	73.5	31.9	25.1	10.9	22.8	9.9
22					104.0	45.1			30.2	13.1	27.1	11.8
24									35.6	15.5	31.1	13.5
25									38.7	16.8	34.6	15.0
30									54.6	23.7	48.1	20.9
35									73.3	31.8	64.3	27.9
40									95.0	41.2	82.0	35.6

Areas above the heavy lines are recommended for normal operation.

FIGURE 5-12A

Sizing data. (*Courtesy of Bard.*)

TABLE 5

	1 1/4" ID = 1.380"				1 1/2" ID = 1.610"				2" ID = 2.067			
	Steel		Plastic		Steel		Plastic		Steel		Plastic	
GPM	Ft.	Lbs.	Ft.	Lbs.	Ft.	Lbs.	Ft.	Lbs.	Ft.	Lbs.	Ft.	Lbs.
10	1.8	.8	1.7	.7								
12	2.5	1.1	2.3	1.0	1.2	.5	1.1	.5				
14	3.3	1.4	3.1	1.3	1.5	.7	1.4	.6				
16	4.2	1.8	4.0	1.7	2.0	.9	1.9	.8				
18	5.2	2.3	4.9	2.1	2.4	1.1	2.3	1.0				
20	6.3	2.7	6.0	2.6	2.9	1.3	2.8	1.2				
25	9.6	4.2	9.1	3.9	4.5	2.0	4.3	1.9	1.3	.6	1.3	.6
30	13.6	5.9	12.7	5.5	6.3	2.7	6.0	2.6	1.8	.8	1.8	.8
35	18.2	7.9	16.9	7.3	8.4	3.6	8.0	3.5	2.4	1.0	2.4	1.0
40	23.5	10.2	21.6	9.4	10.8	4.7	10.2	4.4	3.1	1.3	3.0	1.3
45	29.4	12.8	28.0	12.2	13.5	5.9	12.5	5.4	3.9	1.7	3.8	1.6
50	36.0	15.6	32.6	14.1	16.4	7.1	15.4	6.7	4.7	2.0	4.6	2.0
60	51.0	22.1	45.6	19.8	23.2	10.1	21.6	9.4	6.6	2.9	6.4	2.8
70	68.8	29.9	61.5	26.7	31.3	13.6	28.7	12.5	8.9	3.9	8.5	3.7
80	89.2	38.7	77.9	33.8	40.5	17.6	36.8	16.0	11.4	5.0	10.9	4.7
90	112.0	48.6	96.6	41.9	51.0	22.1	45.7	19.8	14.2	6.2	13.6	5.9
100	138.0	59.9			62.2	27.0	56.6	24.6	17.4	7.6	16.5	7.2
120					88.3	38.3			24.7	10.7	23.1	10.0
140					119.0	51.6			33.2	14.4	30.6	13.2
160					156.0	67.7			43.0	18.7	39.3	17.1
180									54.1	23.5	48.9	21.2
200									66.3	28.8	59.4	25.8
220									80.0	34.7		
240									95.0	41.2		
260									111.0	48.2		

Areas above the heavy lines are recommended for normal operation.

FIGURE 5-12B

Sizing data. (*Courtesy of Bard.*)

TABLE 7—FRICTION LOSSES THROUGH FITTINGS IN TERMS OF EQUIVALENT LENGTHS OF PIPE

Type Fitting and Application	Pipe and Fitting Material ①	Equivalent Length of Pipe Nominal Size Fitting and Pipe						
		1/2	3/4	1	1-1/4	1-1/2	2	2-1/2
Insert Coupling	Plastic	3	3	3	3	3	3	3
Threaded Adapter Plastic or Copper to Thread	Copper	1	1	1	1	1	1	1
	Plastic	3	3	3	3	3	3	3
90° Standard Elbow	Steel	2	3	3	4	4	5	6
	Copper	2	3	3	4	4	5	6
	Plastic	4	5	6	7	8	9	10
Standard Tee Flow Thru Run	Steel	1	2	2	3	3	4	5
	Copper	1	2	2	3	3	4	5
	Plastic	4	4	4	5	6	7	8
Stadard Tee Flow Thru Side	Steel	4	5	6	8	9	11	14
	Copper	4	5	6	8	9	11	14
	Plastic	7	8	9	12	13	17	20
Gate Valve (or Ball)	②	2	3	4	5	6	7	8
Swing Check Valve	②	4	5	7	9	11	13	16

Friction loss tables for fittings (Table 7). Figures given are friction losses in terms of equivalent length (in feet) of straight pipe.
① Loss figures are based on equivalent lengths of indicated pipe material and
② Loss figures are for screwed valves and are based on equivalent lengths of steel pipe.

FIGURE 5-13A

Sizing data. (*Courtesy of Bard.*)

way to control the water flow between domestic use and heat-pump use. The switch is installed in the water pipe between the heat pump and the pressure tank in the well system (Figs. 5-17 and 5-18).

The switch can be programmed to cut off the heat pump when available water pressure drops to a preselected level. For example, if two people in your home were taking a shower simultaneously, the switch might cut the heat pump off to allow enough water pressure to run both shower heads.

If the well system is equipped with an adequate pump and a pressure tank, as most well systems are, a reverse-acting pressure switch is a good compromise, when needed.

An Alternative

An alternative to a reverse-acting pressure switch is the installation of a stronger pump and a larger pressure tank. The combination of a pump with a higher gpm rating and a larger reserve of water in the pressure tank can overcome the need for a reverse-acting pressure switch.

If you opt for a larger pump, be sure that the pump will not pump water faster than the well can replenish it. For example, assume your well has a gpm rating of 4 and a pump with a gpm rating of 3 is installed. Installing a pump with a gpm rating of 5 would have the ability to drain the well. Since the well only produces 4 gallons of water per minute, pumping water at a rate of 5 gallons per minute will not work for very long (Fig. 5-19).

FRICTION LOSS THROUGH CONSTANT FLOW VALVES

	PSI	FT. H_2O
All GPM Ratings	15	34.7

Minimum Water Regulating Valve Pressure Drop

	1/2"		3/4"		1"	
GPM	PSI	FT. HD	PSI	FT. HD.	PSI	FT. HD.
4.0	4.0	9.2	1.0	2.3	—	—
5.0	6.0	13.9	1.3	3.0	—	—
10.6	9.5	22.0	3.0	6.9	1.5	3.5
12.7	15.0	34.6	4.2	9.7	2.0	4.6

FRICTION LOSS THROUGH SOLENOID VALVE (FT. H_2O)

GPM	3/4"	1"
5	5.8	3.5
10	6.3	3.6
15	8.3	4.6
20	12.0	7.5
25	16.7	12.1
30	23.1	18.5

FIGURE 5-13B

Sizing data. (*Courtesy of Bard.*)

The advantage of a larger pump and pressure tank is that the heat pump will not have to shut down during periods of peak use. The disadvantage is the cost of the pump and the larger pressure tank. Peak demands on water for domestic use is typically minimal. Taking a shower or filling a washing machine are two times when the heat pump might have to cut off, but these activities don't take long or occur frequently throughout a 24-hour period. Therefore, a reverse-acting pressure switch will often provide adequate service at a reduced cost.

Mineral Content

If the well has a high mineral content, the effects of the well water can be detrimental to the heat-pump system. High concentrations of acid, iron, calcium, magnesium, and similar materials can reduce the life expectancy of a heat-pump system.

Many plumbing contractors and most water-conditioning companies will test well water free of charge. They will come to your home and take a sample of the water. Some companies will do an in-home test of the water while you watch. Others will send the water off to a lab to have it analyzed.

If there are high concentrations of corrosive materials in the water, water-conditioning equipment can be installed to neutralize the water. The cost of this equipment and its installation is substantial, so have your well water tested before you make a firm commitment to a water-source heat pump that will use the well water has a heat source.

Modern heat pumps are being made with materials that resist the effects of most water containing impurities. Unless the presence of impurities in a well is unusually high, there will probably be no need for water-conditioning equipment to be installed. If a well does show potential problems when tested, consult with various manufacturers to see if any water treatment is needed. By providing manufacturers with test results, they will be able to give definitive answers on what, if any, type of conditioning equipment is needed.

Where Does the Water Go?

Where does the water go when the heat pump has extracted its heat? It returns to ground, normally. It is common for a second well, a return well, to be drilled into the same vein of water that provides water to

CALCULATING PUMP WORKSHEET

1. Find the Bard heat pump model used in TABLE 2. MODEL ______________
2. Enter water coil head loss (TABLE 2): ______ ft. hd.
3. Continue across TABLE 2 to find GPM flow required for this heat pump. ________ GPM
4. Count each elbow, tee, reducer, air scoop, flowmeter, etc., as 3 feet of pipe equivalent. Add the equivalent feet of pipe to the actual feet of pipe used. The total length is used to determine the piping heat loss below.

Pipe Type and Size	No. Elbows, Tees Devices, Etc.		Equiv. Ft. of Pipe	Actual Pipe Used	Total Pipe Length
______	______	x 3	______	______	______
______	______	x 3	______	______	______
______	______	x 3	______	______	______
______	______	x 3	______	______	______
______	______	x 3	______	______	______
______	______	x 3	______	______	______

*If the pipe is bent at a 2 foot radius or larger, DO NOT figure the curve as an elbow.

5. PIPING HEAD LOSS for different types of pipe at GPM flow rate of water source heat pump. NOTE: For parallel earth loop system, figure for only one loop.

Pipe Type and Size	Total Pipe Length	Piping** Head Loss (Table 3)	
______	(______ ÷ 100)	______	______ ft.hd.
______	(______ ÷ 100)	______	______ ft. hd.
______	(______ ÷ 100)	______	______ ft. hd.
______	(______ ÷ 100)	______	______ ft. hd.
______	(______ ÷ 100)	______	______ ft. hd.

**For a parallel earth loop, divide the heat pump GPM (line 3) by number of loops to determine flow rate through each individual loop to select piping head loss. SUBTOTAL ______ ft. hd.

x ______

6. Multiply SUBTOTAL by multiplier (see TABLE 5) to obtain TOTAL HEAD LOSS FOR SYSTEM using antifreeze solution. TOTAL HEAD LOSS ______ ft. hd.

7. PUMP SELECTION: Use Table 4 and flow rate, (line 3). Select the pump output which is larger or equal to the TOTAL HEAD LOSS FOR SYSTEM (line 5 or 6).

If the TOTAL HEAD LOSS calculated in line 6 is greater than the pump outputs listed in Table 4, go to the pump manufacturer's performance curves and find the required GPM flow for the heat pump. Pump performances are listed for each pump model at different flow rates.

Series pump performance is simply a TOTAL OF THE INDIVIDUAL PUMP PERFORMANCE. If one pump can overcome 10 feet of heat loss, two can overcome 20 feet of head loss, three can overcome 30 feet of head loss, etc.

***REMEMBER: UNDER NO CIRCUMSTANCES MIX DIFFERENT PUMP SIZES WHEN USING PUMPS IN SERIES ***

F1125-595

FIGURE 5-14

Worksheet for calculating pump size. (*Courtesy of Bard.*)

TABLE 2
WATER COIL PRESSURE DROP (Water Only)

Model	WPV24 WPV24C	WQS30A WPV30B WPV30C	WQS36A WPV36B WPV36C	WQS42A WPV53B WPV42C	WPV62B WPV60C
GPM	Ft. Hd.	Ft. Hd.	Ft. Hd.	Ft. Hd.	Ft. Hd.
4	6.93	5.78	–	–	–
5	8.08	7.39	5.08	–	–
6	9.50	12.24	6.36	2.31	–
7	10.85	14.78	7.86	3.44	–
8	–	22.18	9.59	4.67	7.16
9	–	–	11.56	6.01	8.92
10	–	–	13.75	7.44	10.75
11	–	–	–	9.01	12.71
12	–	–	–	10.63	14.79
13	–	–	–	–	17.22
14	–	–	–	–	19.88
15	–	–	–	–	22.89

TABLE 4
SELECTION OF CIRCULATION PUMP OR PUMPS

Pump Output (Feet of Head) @ GPM @ Top of Column	Grundfos* Pump Models	No. of Pumps	Water Flow Rate Required in G.P.M.						
			4	6	8	10	12	14	16
	UP26-96F	1	26.5	25	22.5	20	18	16	14
	UP26-96F	2	53	50	45	40	36	32	28
	UP26-99F	1	30.5	29	28	27	25	23.5	22
	UP26-99F	2	61	58	56	54	50	47	44
	Loop Pump Models								
	GPM-1**	1		29	28	27	25	23	22
	GPM-2**	2		58	56	54	50	47	44

* Other models of circulation pumps may be used. Consult the manufacturer's specifications.

** See manual 2100-212 GPM series loop pump modules for complete details.

TABLE 3
MINIMUM FLOW IN PIPE FOR TURBULANCE (GPM*)

Nominal Pipe Size	(Pipe ID)	Water at 40° F	GS4 25% Volume 25° F	Noburst 30% Volume 25°F	Methanol 20% Volume 25°F
PE (SDR-11)					
3/4"	(0.86)	1.1	1.9	3.1	2.4
1"	(1.077)	1.3	2.3	3.9	3.1
1 1/4"	(1.385)	1.7	3.0	5.0	3.9
1 1/2"	(1.554)	1.9	3.5	5.7	4.4
2"	(1.943)	2.4	4.3	7.0	5.5
PE (SCH)					
3/4"	(0.824)	1.0	1.9	3.0	2.3
1"	(1.049)	1.3	2.4	3.8	3.0
1 1/4"	(1.380)	1.7	3.1	5.0	3.9
1 1/2"	(1.610)	2.0	3.6	5.8	4.6
2"	(2.067)	2.5	4.6	7.6	5.9
PB (SDR-17,IPS)					
1 1/2"	(1.676)	2.1	3.7	6.0	4.8
2"	(2.095)	2.6	4.6	7.6	5.9
PB (SDR-13.5,CTS)					
1"	(0.957)	1.2	2.1	3.5	2.7
1 1/4"	(1.171)	1.4	2.6	4.3	4.9
1 1/2"	(1.385)	1.7	3.1	5.0	3.9
2"	(1.811)	2.2	4.0	6.6	5.1

* For each separate loop.

NOTE: When selecting pipe size for parallel flow, it is necessary to maintain turbulent flow in the earth coil for heat transfer. The table above lists the minimum flows for turbulence.

FIGURE 5-15A

Sizing data. (*Courtesy of Bard.*)

TABLE 5
PIPING FEET OF HEAD LOSS AT DIFFERENT FLOW RATES PER 100 FEET

Pipe Size and Material	GPM Flow Rate										
	DI	1	2	3	4	5	6	8	10	12	14
Connection Hose 1"	1.05	*	*	*	1.33	1.95	2.68	4.43	6.53	8.99	11.77
PVC 3/4" - 200 PSI		*	*	*	3.7	5.7	*	*	*	*	*
PVC 1" - 200 PSI		*	*	*	1.0	1.9	2.7	4.2	6.3	8.9	11.8
Copper 3/4"		*	*	*	4.3	6.3	*	*	*	*	*
Copper 1"		*	*	*	1.5	1.9	2.7	4.5	6.9	9.6	12.8
PE3408 (Polyethylene)	DI										
1. SDR-11 3/4	0.860	0.31	1.03	2.07	3.41	5.03	*	*	*	*	*
2. SDR-11 1	1.077	0.11	0.36	0.71	1.18	1.73	2.38	3.92	*	*	*
3. SDR-11 1 1/4	1.358	*	0.12	0.24	0.39	0.58	0.79	1.31	1.93	2.65	3.47
4. SDR-11 1 1/2	1.554	*	*	0.13	0.21	0.31	0.42	0.69	1.02	1.40	1.83
5. SDR-11 2	1.943	*	*	*	0.07	0.11	0.15	0.24	0.35	0.48	0.63
6. SCH 40 3/4	0.824	0.38	1.26	2.54	4.18	6.16	8.46	*	*	*	*
7. SCH 40 1	1.049	0.12	0.40	0.81	1.33	1.96	2.69	4.45	*	*	*
8. SCH 40 1 1/4	1.380	*	0.11	0.22	0.36	0.54	0.74	1.21	1.79	2.46	3.21
9. SCH 40 1 1/2	1.610	*	*	0.11	0.18	0.26	0.35	0.58	0.86	1.18	1.55
10. SCH 40 2	2.067	*	*	*	*	0.08	0.11	0.18	0.26	0.36	0.47
PB2110 (Polybutylene)	DI										
11. SDR-17, IPS 1 1/2	1.676	*	*	0.09	0.15	0.21	0.29	0.48	0.71	0.98	1.28
12. SDR-17, IPS 2	2.095	*	*	*	0.05	0.07	0.10	0.17	0.25	0.34	0.44
13. SDR-13.5, Cts 1	0.957	0.19	0.62	1.25	2.06	3.03	4.16	*	*	*	*
14. SDR-13.5, Cts 1 1/4	1.171	*	0.24	0.48	0.79	1.17	1.60	2.64	*	*	*
15. SDR-13.5, Cts 1 1/2	1.385	*	0.11	0.22	0.36	0.53	0.72	1.19	1.76	2.41	3.20
16. SDR-13.5, Cts 2	1.811	*	*	0.06	0.10	0.15	0.20	0.33	0.49	0.68	0.88

NOTE: 1. These head losses are for water at 40°F temperature.
2. Count each elbow, tee, reducer, air scoop, flow meter, etc., as 3 feet of equivalent pipe length and add to actual measured pipe length for total length.
3. To adjust the total earth loop piping head loss for other antifreezes and water solutions at 25°F, multiply pressure loss on line 6 for water by: Noburst - 1.33, GS4 - 1.18, Methanol - 1.25

FIGURE 5-15B

Sizing data. (*Courtesy of Bard.*)

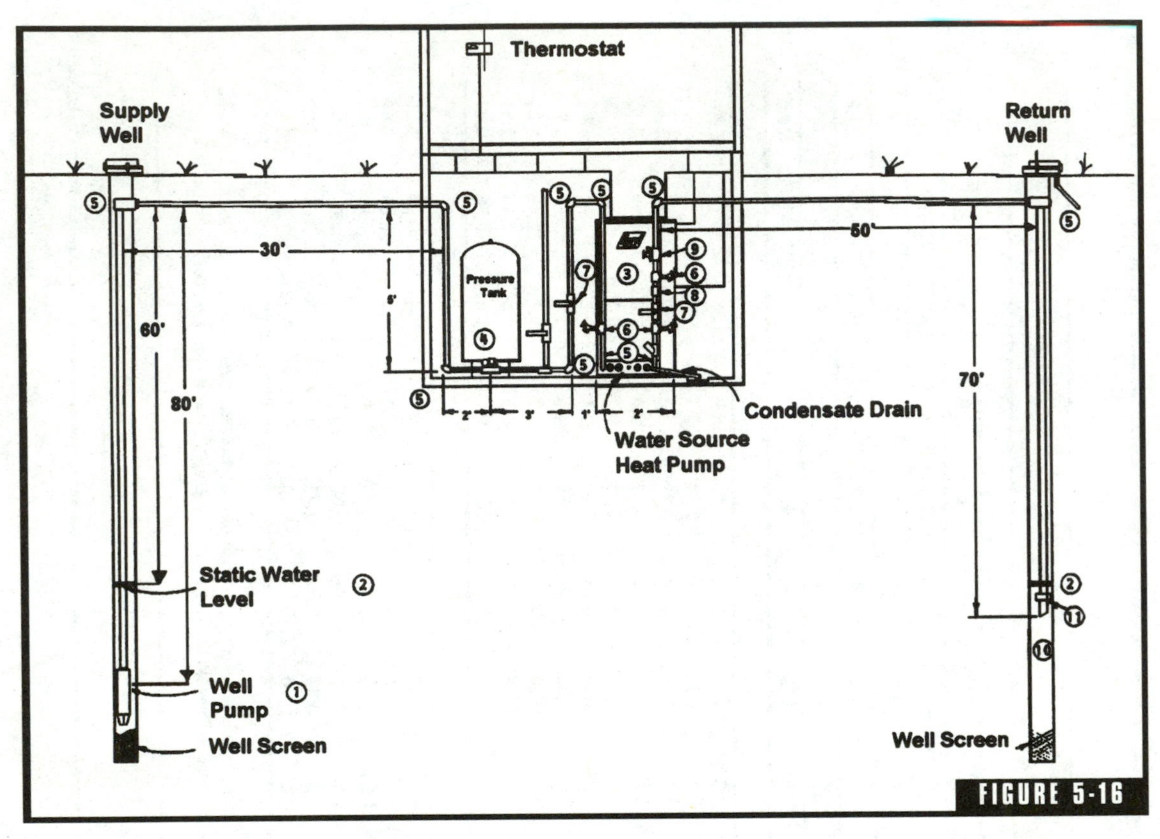

FIGURE 5-16

Typical return well system. (*Courtesy of Bard.*)

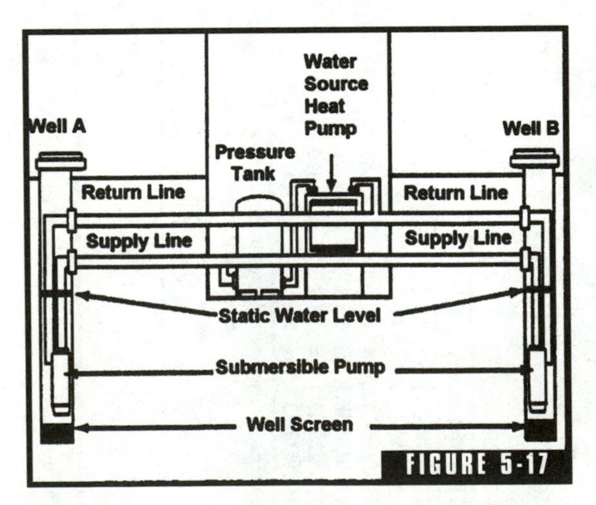

FIGURE 5-17

Two-well reversible system. (*Courtesy of Bard.*)

PRESSURE REQUIRED ON BACK PRESSURE VALVE (Spring-Loaded Check Valve) TO MAINTAIN 10 PSI ON DISCHARGE LINE

Depth to Static Water Level (Feet)	Spring Release Pressure (PSI)
0	10
10	14
20	18
30	23
40	27
50	31
60	36
70	40
80	45
90	49
100	53

FIGURE 5-18

Pressure requirements. (*Courtesy of Bard.*)

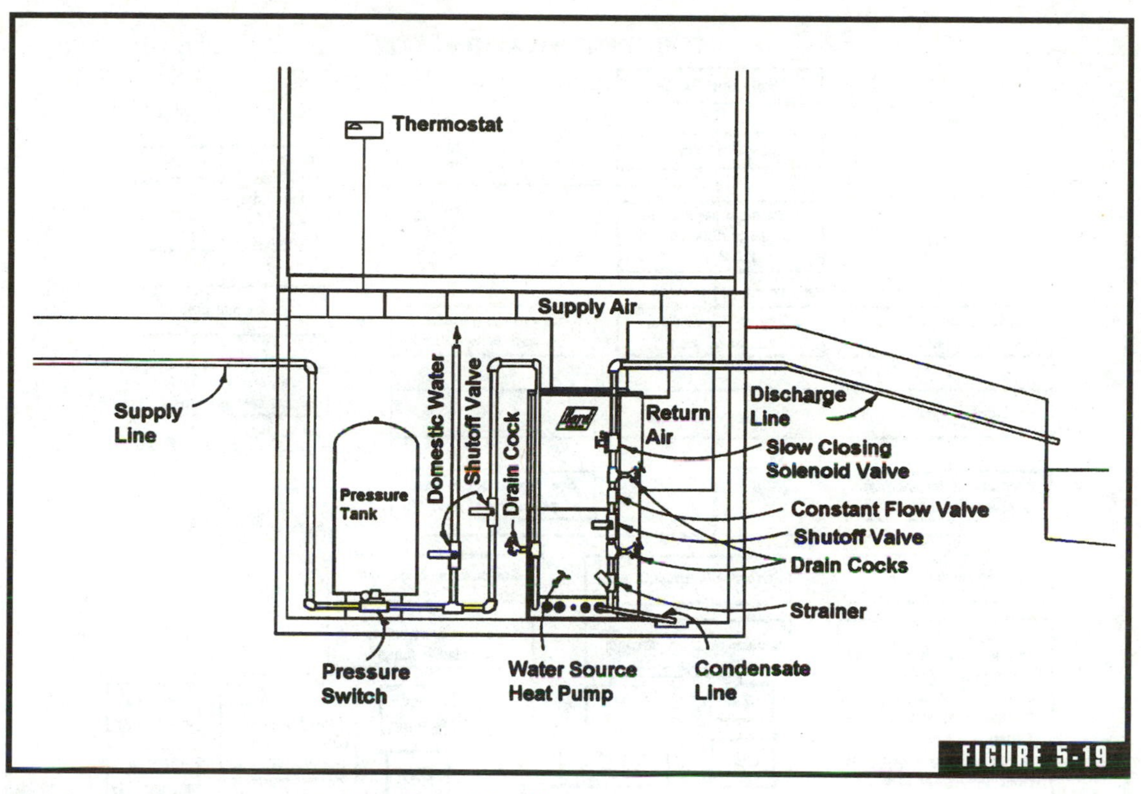

Standard piping arrangement. (*Courtesy of Bard.*)

the supply well. However, this practice may not be acceptable in all states, since some states are more protective in their code restrictions than others. Before you commit to using a well-water heat pump, check with local code enforcement office for details on any restrictions that may apply to its use or the use of return wells.

When a return well is drilled, it must be installed far enough away from the supply well so as not to create a thermal imbalance. The advantage to using well water as a heat source is its fairly constant temperature. If return water is dumped back into the aquifer that supplies the supply well, it is possible that the water being returned will lower the temperature of the well water in the supply well. This would defeat the primary advantage of a well-water heat pump. It is also necessary to evaluate soil conditions when using buried systems (Figs. 5-20 to 5-22).

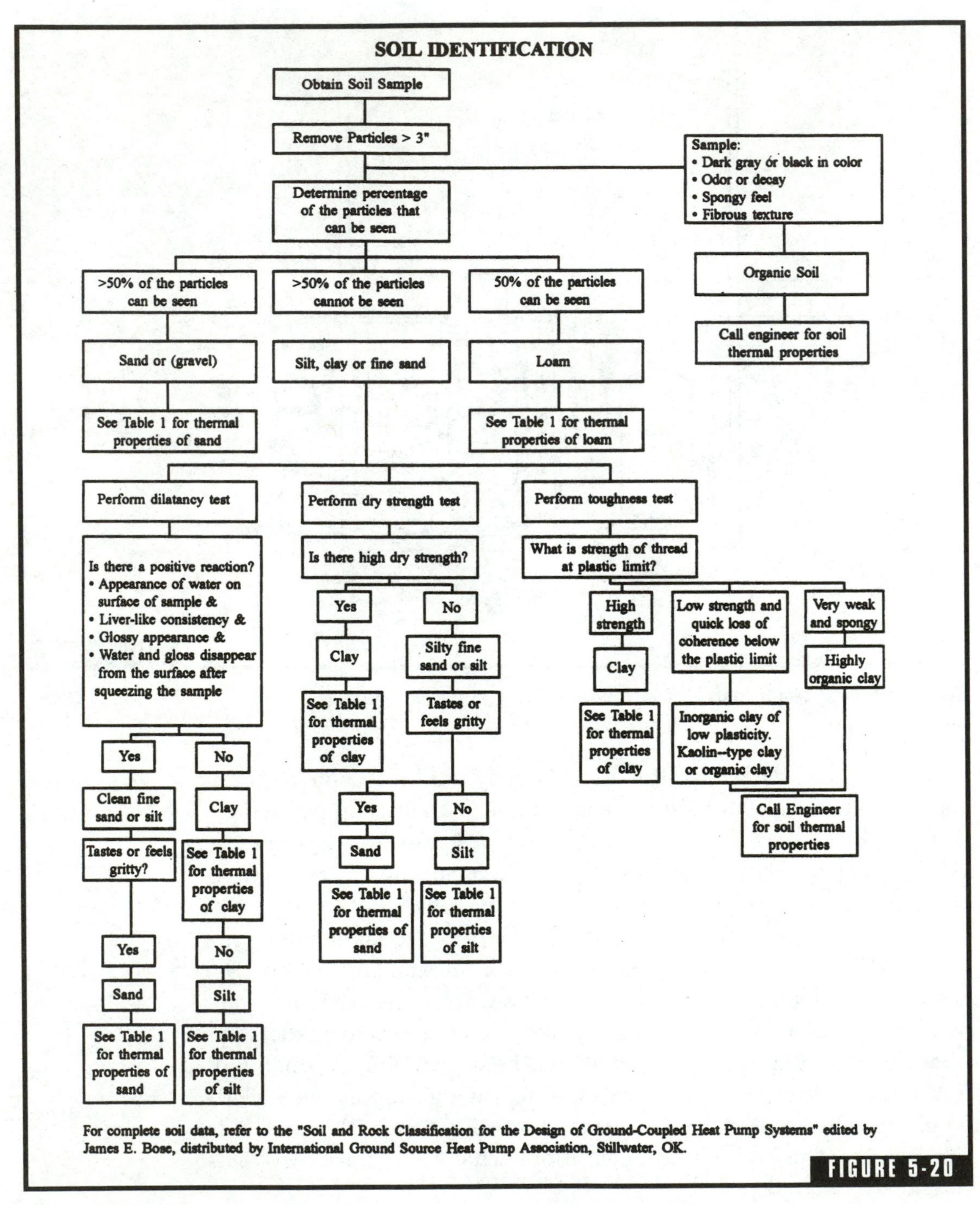

FIGURE 5-20

Soil identification. (*Courtesy of Bard.*)

SOIL THERMAL PROPERTIES

Thermal Texture Class	Thermal WM•K	Conductivity Btu/ft hr °F	Thermal CM2/sec	Diffusivity ft2/day
Sand or (gravel)	0.77	0.44	0.0045	0.42
Silt	1.67	0.96	--	--
Clay	1.11	0.64	0.0054	0.50
Loam	0.91	0.52	0.0049	0.46
Saturated Sand	2.50	1.44	0.0093	0.86
Saturated Silt or Clay	1.67	0.96	0.0066	0.61

FIGURE 5-21

Thermal properties of soil. (*Courtesy of Bard.*)

FIELD INDENTIFICATION OF INORGANIC SOIL TEXTURAL TYPES

	GENERAL CHARACTERISTICS				
Textural type	Main fractions*	Appearance, dry conditions	Rubbed between fingers, dry conditions	Squeezed in palm of hand, moist conditions at field capacity	Rolling into ribbon, moist conditions
Sandy soil	At least 85% sand particles	Crumbly with no clods or lumps; individual soil grains visible to the naked eye	Gritty: soil grains readily felt	Can form cast; crumbles with least amount of handling	Cannot form ribbon
Sandy loam soil	At least 50% sand particles; not more than 20% clay	Mainly crumbly and loose; grains readily seen and felt	Gritty: soil grains readily felt	Can form cast that will bear careful handling	Cannot form ribbon
Loam soil	At least 80% sand silt in about equal proportion; not more than 20% clay	Mainly crumbly; some clods or lumps	Fairly smooth but some gritty feeling; lumps easily broken	Can form cast that can be handled freely	Cannot form ribbon
Silt loam soil	At least 50% silt; not more than 20% clay	Quite cloddy but some crumbly materials	Lumps easily broken and easily pulverized; thereafter floury texture and soft feel	Can form coast that can be handled freely; wet soil runs together and puddles	Cannot form perfect ribbon; has broken surface, cracks appear
Clay loam soil	20 to 30% clay	Fine-textured soil; quite cloddy but some crumbly material	Lumps hard; not easily broken	Can form cast that can be handled freely; soil plastic	Can form perfect ribbon but breaks easily
Clay soil	30 to 100% clay	Fine-textured soil; breaks into very hard clods	Lumps very hard; difficult, if not impossible, to break	Can form cast that can be handled freely; soil plastic	Can form ribbon that will support its own weight
Silt soil	At least 80% silt				

*Based on the textural classification of the U.S. Bureau of Public Roads.
Refer to text for: shaking (dilatancy) test, shine test, dry-stength test.
Source: Soil Thermal Characteristics in Relation to Underground Power Cables. 14

FIGURE 5-22

Soil texture types. (*Courtesy of Bard.*)

Closed-Loop Heat Pumps

If you want a water-source heat pump but don't want to (or can't) drill a well, you might consider using a closed-loop heat pump (Fig. 5-23). Closed-loop systems consist of a series of pipes that are installed below ground. The ground and the water in the pipes work to provide a heat pump with its needed heat source. Some people refer to closed-loop systems as water-source heat sources, and others call them earth-source heat sources.

Closed-loop systems are relatively new, and their design can become quite complicated (Fig. 5-24). Since there is a lot of exchange between temperatures in the ground around the loop, one must be careful not to allow the changes in temperature to become too drastic. This is usually accomplished by installing an expansive piping system.

Horizontal Loops

Horizontal loops of piping are often used with closed-loop systems. The layout of the piping is done at depths below the local frost line (Figs. 5-25 and 5-26). It is often recommended that the loop be installed at a depth of 6 feet. The overall area of ground needed to install a horizontal loop varies. Many conditions contribute to the size and design of the loop. Soil conditions, temperatures, and other factors must be considered when engineering the loop.

Vertical Loops

Vertical loops can be used in place of horizontal loops when a limited amount of surface area is available (Figs. 5-27 and 5-28). With a vertical loop, the piping is buried in the ground vertically. Due to the depths reached with vertical loops, the likelihood of warmer soils being present is good, and this is better for the efficiency of the heat pump.

Both horizontal and vertical closed-loop systems require a special pumping kit. The kit is normally offered by manufacturers of heat pumps as an optional accessory item.

Antifreeze is used in both types of closed-loop systems, but less of it is needed with vertical loops. Since vertical loops typically extend into warmer soil, less antifreeze is needed to pull the heat source (Figs. 5-29 to 5-32).

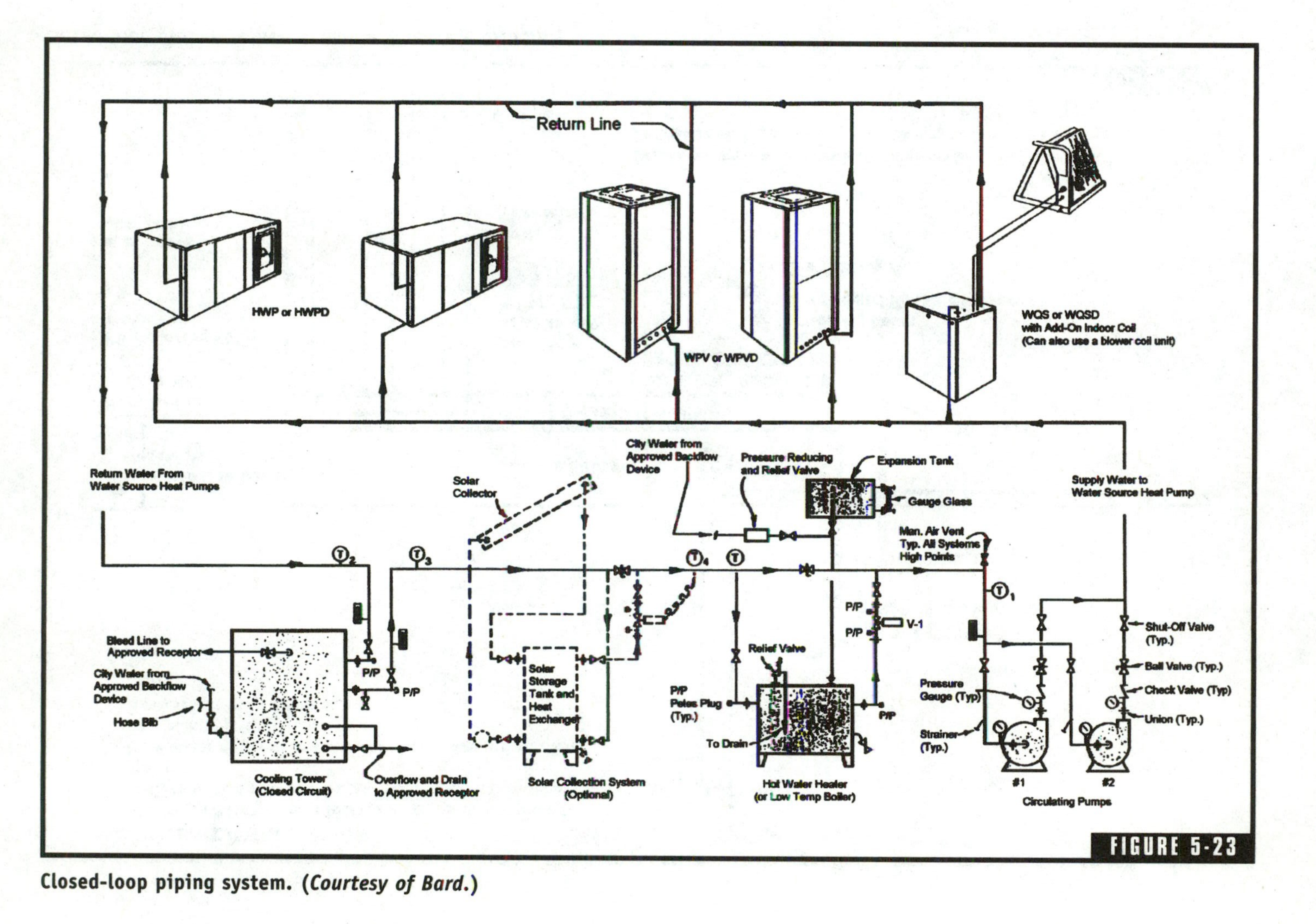

Closed-loop piping system. (*Courtesy of Bard.*)

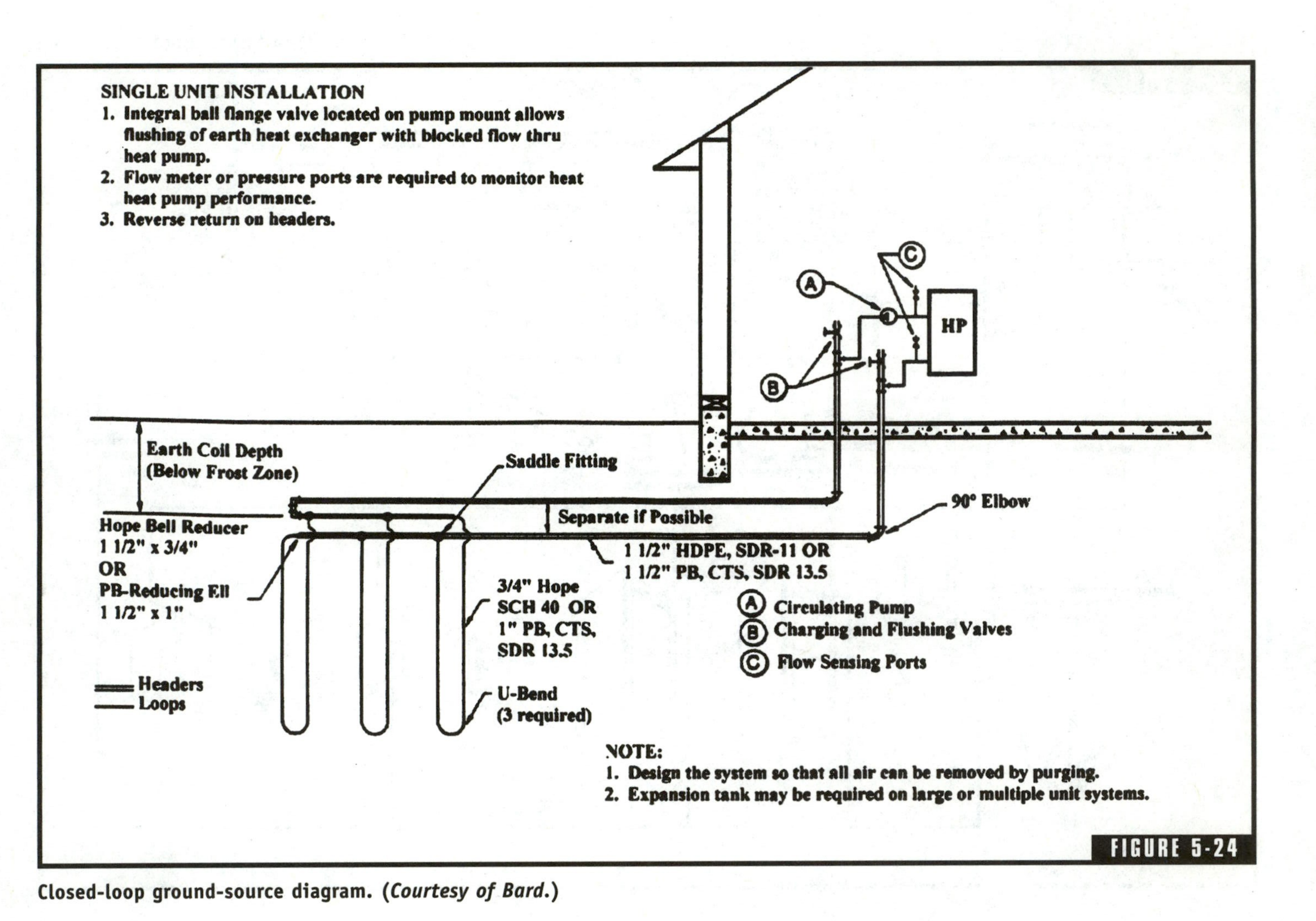

Closed-loop ground-source diagram. (*Courtesy of Bard.*)

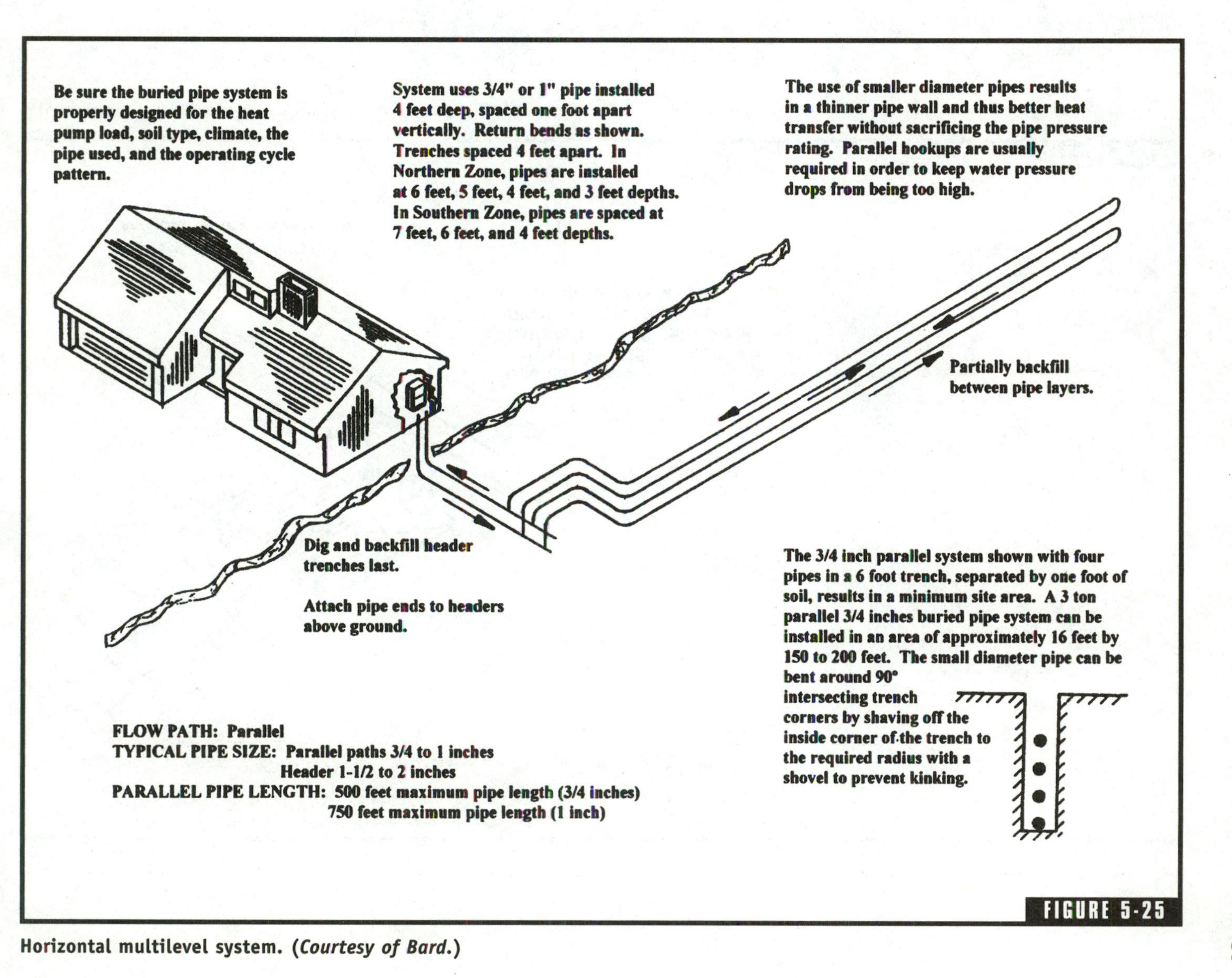

Horizontal multilevel system. (*Courtesy of Bard.*)

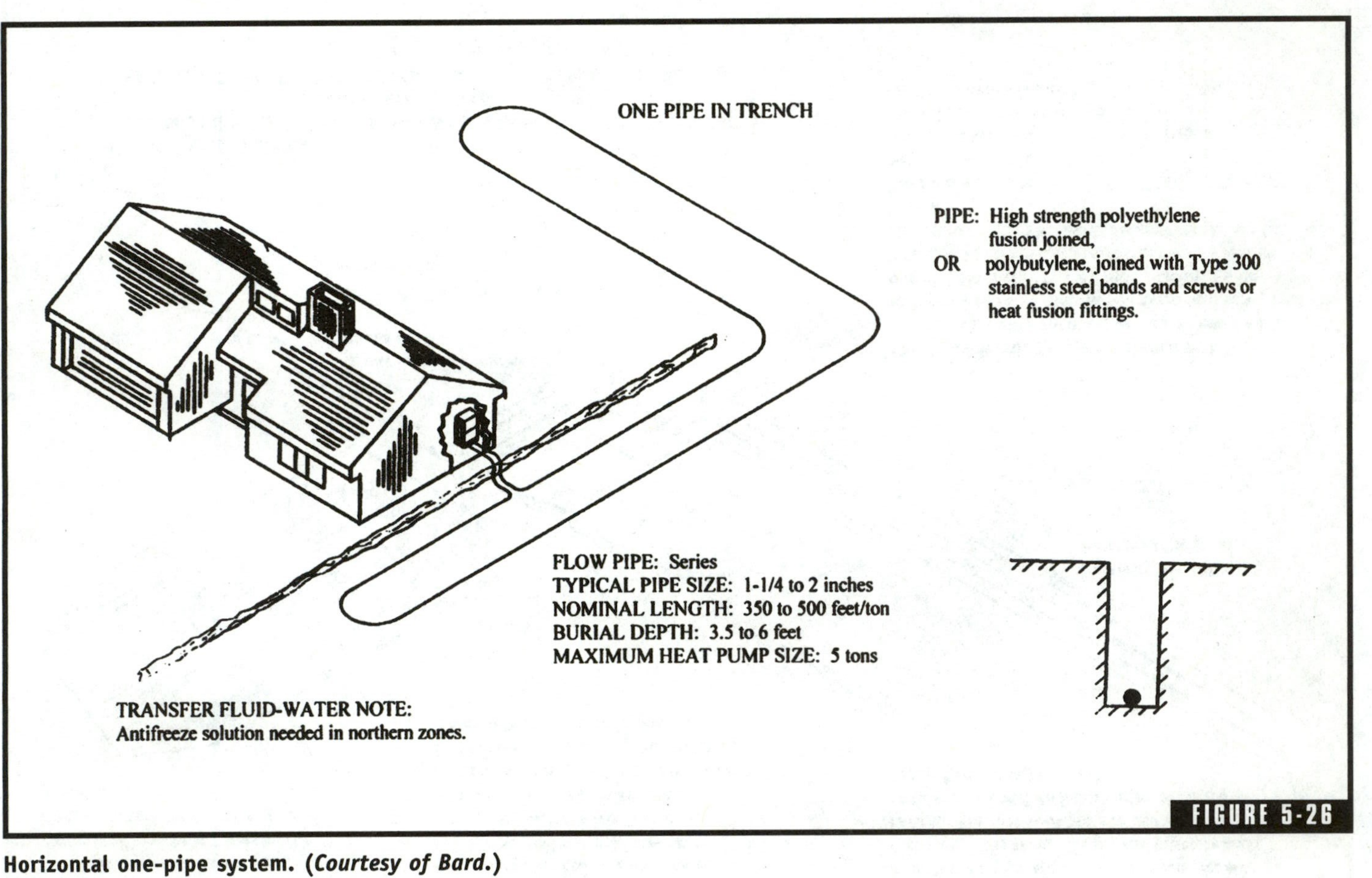

FIGURE 5-26

Horizontal one-pipe system. (*Courtesy of Bard.*)

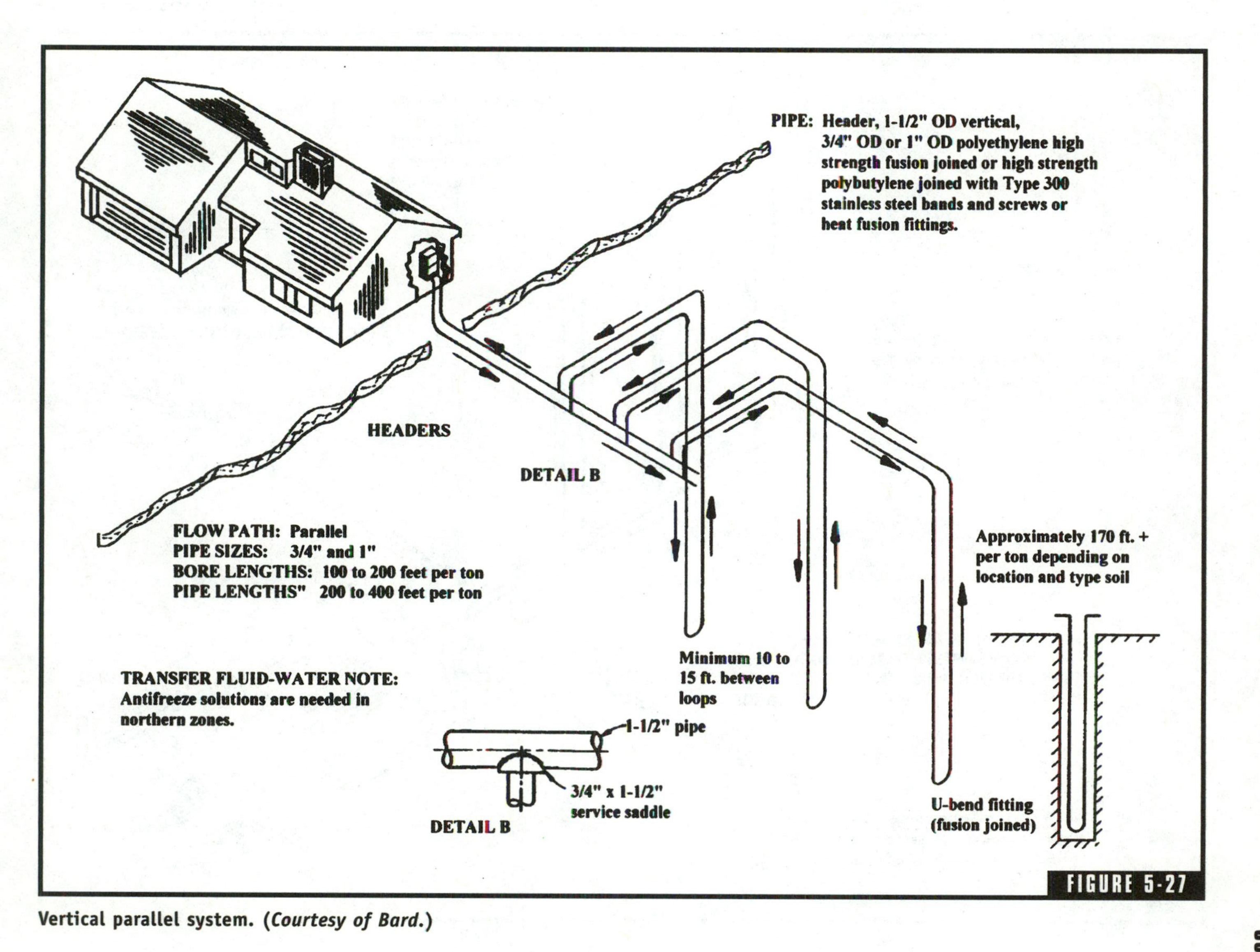

FIGURE 5-27

Vertical parallel system. (*Courtesy of Bard.*)

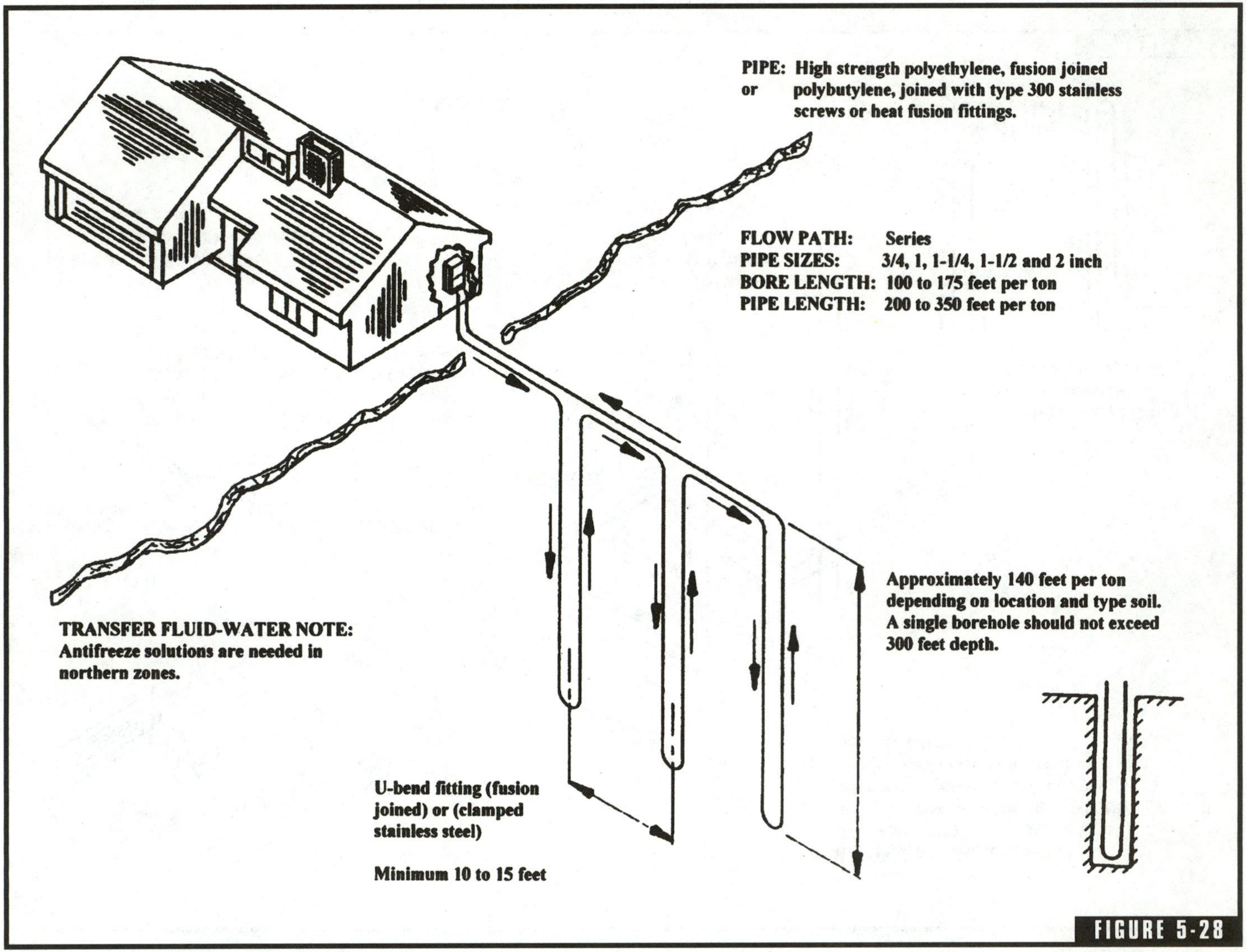

PIPE: High strength polyethylene, fusion joined
or polybutylene, joined with type 300 stainless screws or heat fusion fittings.

FLOW PATH: Series
PIPE SIZES: 3/4, 1, 1-1/4, 1-1/2 and 2 inch
BORE LENGTH: 100 to 175 feet per ton
PIPE LENGTH: 200 to 350 feet per ton

Approximately 140 feet per ton depending on location and type soil. A single borehole should not exceed 300 feet depth.

TRANSFER FLUID-WATER NOTE:
Antifreeze solutions are needed in northern zones.

U-bend fitting (fusion joined) or (clamped stainless steel)

Minimum 10 to 15 feet

FIGURE 5-28

Vertical series system. (*Courtesy of Bard.*)

Locating the Closed Loop

Where is the best place to install a closed loop? Some places are better than others for the installation of closed loops, especially if a horizontal design is used. Let's take a moment to look at some creative places to put your piping.

A Greenhouse

If you are planning to build a greenhouse at anytime in the future, you might consider installing a horizontal-loop system where the greenhouse can be built over it. The greenhouse will collect a lot of heat that can be transferred to the loop, through the earth. In the hot months of summer, the greenhouse will have to be shaded and vented to allow the heat pump to function at its best.

If you decide on this approach, be advised that care must be taken not to damage the underground loop when constructing the greenhouse. Additionally, if problems ever arise with the underground piping, getting to the loop components for repair work will be difficult. A portable greenhouse, one with a metal frame and a plastic cover, is an ideal solution to this problem.

A Vegetable Garden

Burying the closed loop beneath a large vegetable garden is a great idea. In summer, the foliage from the garden plants will help shade the ground over the loop, keeping it cooler. In winter, the plants will be gone and the bare ground will help the loop to absorb heat from the sun's rays.

RULE OF THUMB

When using an antifreeze solution, the minimum water flow rate for the selected heat pump will need to be increased 40% to have the same heat transfer.

Example: A 4 GPM flow would need to be increased to 5.6 GPM for approximately the same heat transfer properties.

FIGURE 5-29

Water flow requirements when antifreeze is used in a system. (*Courtesy of Bard.*)

RULE OF THUMB

To adjust the total earth loop piping head loss for other antifreezes and water solutions at 25°F, multiply pressure loss on line 6 for water by:

Fluid	Multiplier
30% Noburst (Propylene glycol)	1.33
20% Methanol Alcohol	1.25
27% Potassium Acetate (GS4)	1.18

FIGURE 5-30

Fluid muliplier. (*Courtesy of Bard.*)

Assumed fluid properties for Table 3 at 25°F flow temperature*.

Solution	% Volume	Viscosity (Centipoise)	Density (Lb./Ft.3)
Water	100	1.55	62.4
Noburst	30	4.7	64.2
GS4	25	3.1	68.3
Methanol	20	3.5	60.8

*Water at 40°F

FIGURE 5-31

Fluid properties. (*Courtesy of Bard.*)

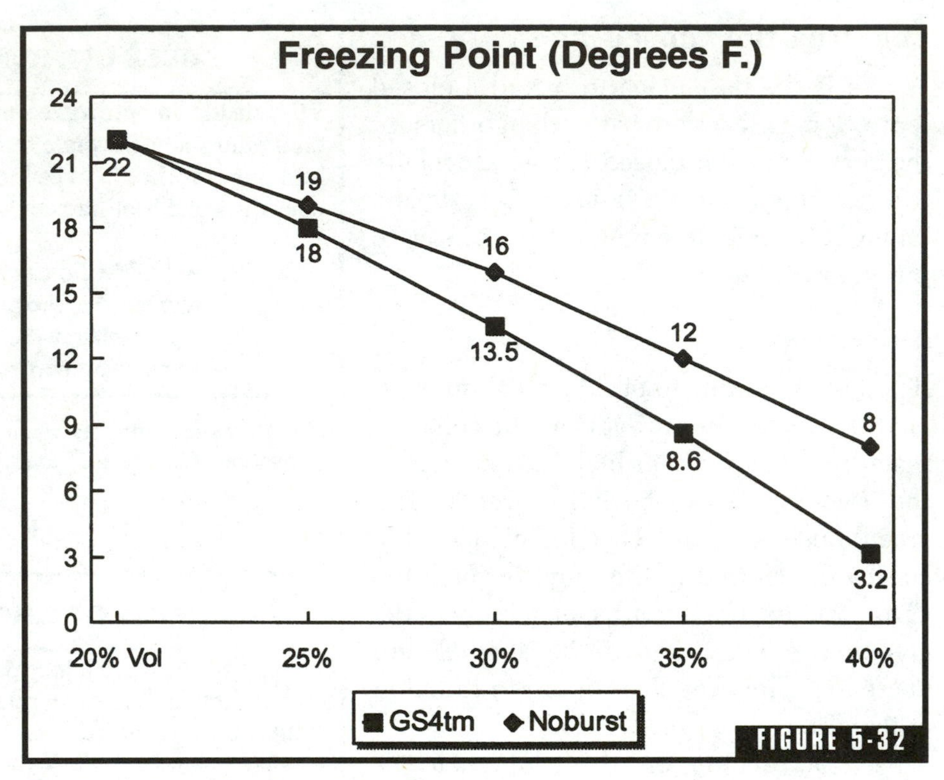

Freezing points. (*Courtesy of Bard.*)

Near a Septic Field

Putting a horizontal loop near a septic field may not seem like a good idea at first, but it can actually be a very good location for the coil. Waste leaving a septic tank and entering a drainfield contains heat. As the waste-water perks into the earth, the heat is given off. If you have your coils in the proximity of the distribution lines of a septic field, there is a good possibility for extra heat gain (Fig. 5-33).

Locations to Avoid

Just as there are some locations that provide good installation traits for a closed-loop system, there are some locations to avoid. Since the ground coil is made of pipes, there is a possibility for breakage and leaks. This possibility is escalated in some installation sites.

When you are planning the location of a horizontal coil, you should avoid areas that may be encroached upon by the roots of trees. Just as tree roots will invade sewer pipes, clogging and breaking them, they can also invade the piping in an underground loop.

Don't install the underground loop in an area that will be subject to traffic from heavy vehicles. The movement of vehicles over the piping could result in stress on the pipe and ultimately broken pipes.

If an underground loop ever develops a leak, finding and fixing the leak can be very difficult, even under the best of conditions. If the loop happens to be installed under a permanent structure, such as a house, the job can become impractical, if not impossible (Fig. 5-34).

RULE OF THUMB
Minimum Diameters for Boreholes

Nominal Pipe Size	Single U-Bend	Double U-Bend
3/4"	3-1/4"	4-1/2"
1"	3-1/2"	5-1/2"
1-1/4"	4"	5-3/4"
1-1/2"	4-3/4"	6"
2"	6"	7"

FIGURE 5-33

Minimum diameters for boreholes. (*Courtesy of Bard.*)

Would anyone really build a house over their underground coil? I've read that the practice is used to take advantage of the natural heat loss present under a home, but I've never seen it done. And, I wouldn't recommend doing it.

Locations for vertical loops are less troublesome. Since the pipe or pipes are installed vertically, much like a drilled well, there is less risk of damage to the piping. The vertical installation is less likely to be

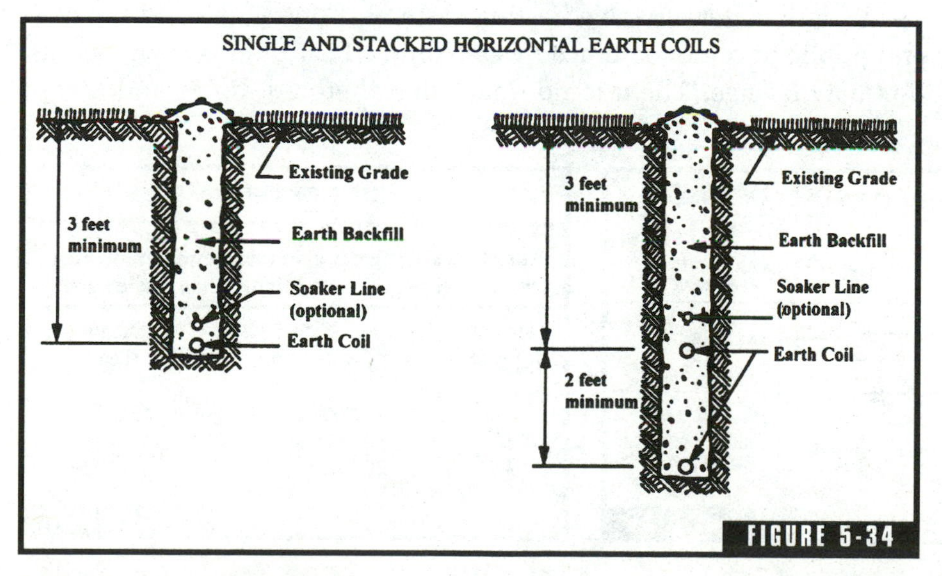

Earth coils. (*Courtesy of Bard.*)

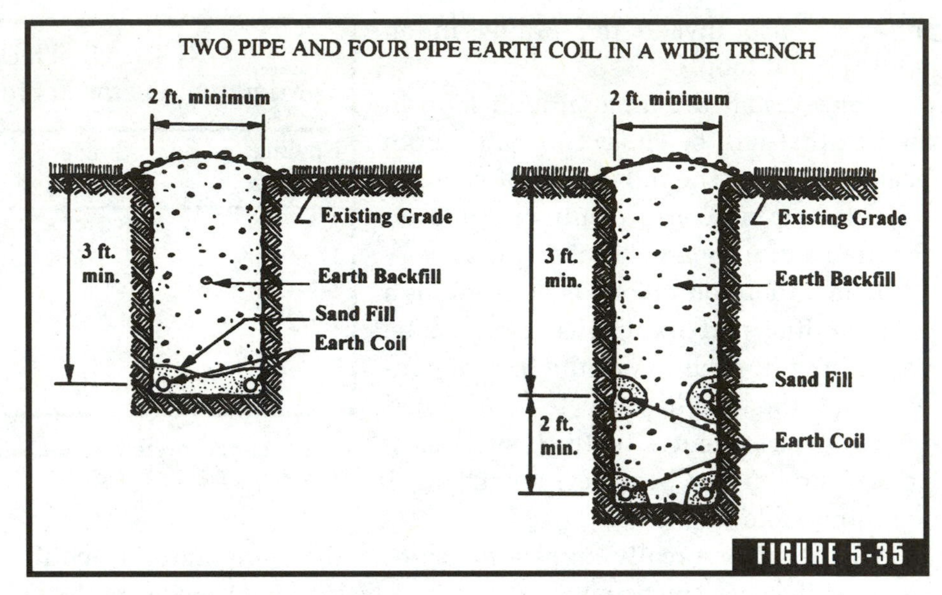

Earth coils. (*Courtesy of Bard.*)

damaged by ground movement. The trenching for a system is very important (Figs. 5-35 to 5-41).

If you work in regions where winter temperatures fall and stay below 27°F for extended periods of time, water-source heat pumps are viable options for effective heating systems. Contractors and the general public have not embraced water-source heat pumps as of yet, but this may change. There is no doubt that there are times and places

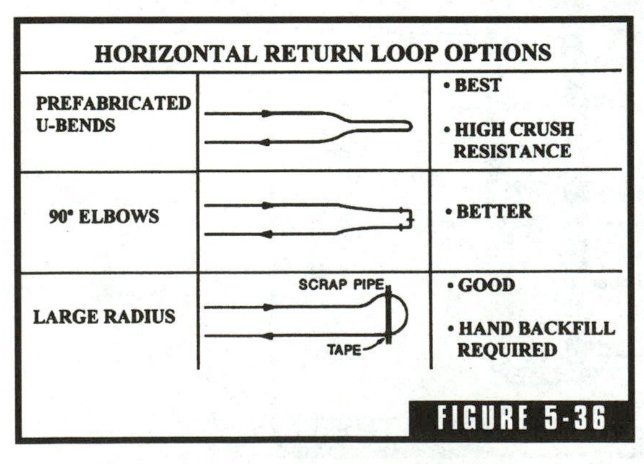

Return-loop options. (*Courtesy of Bard.*)

RULE OF THUMB

Trench length is reduced in the following proportion when multiple pipes are installed in a single trench

Number of Pipes	Trench/Pipe (Feet)	Depths (Feet)
1	500/500	5
2	300/600	4, 6
4	200/800	3, 4, 5, 6

FIGURE 5-37

Trench sizing. (*Courtesy of Bard.*)

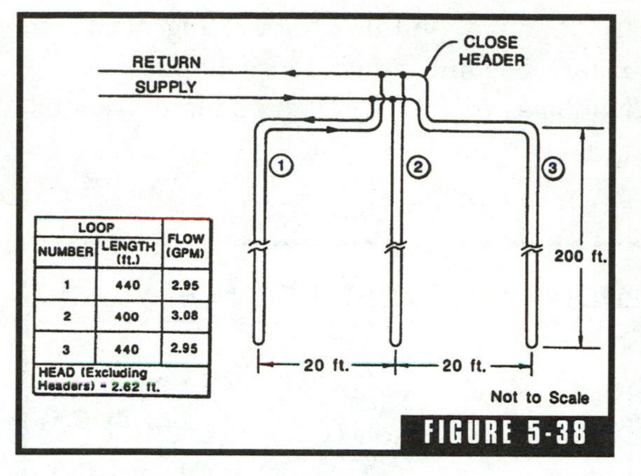

Recommended supply and return layout. (*Courtesy of Bard.*)

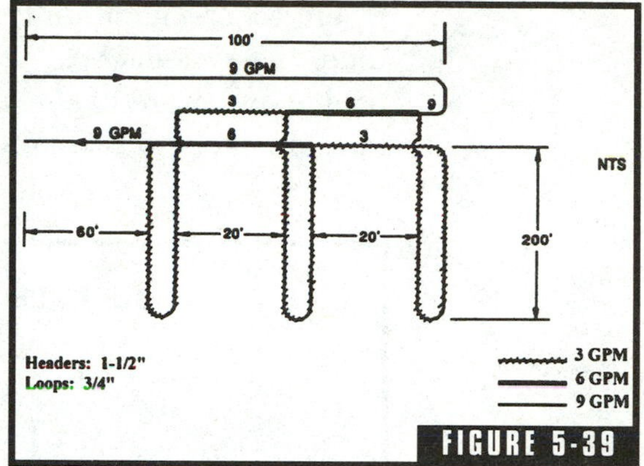

A supply and return layout that should not be used. (*Courtesy of Bard.*)

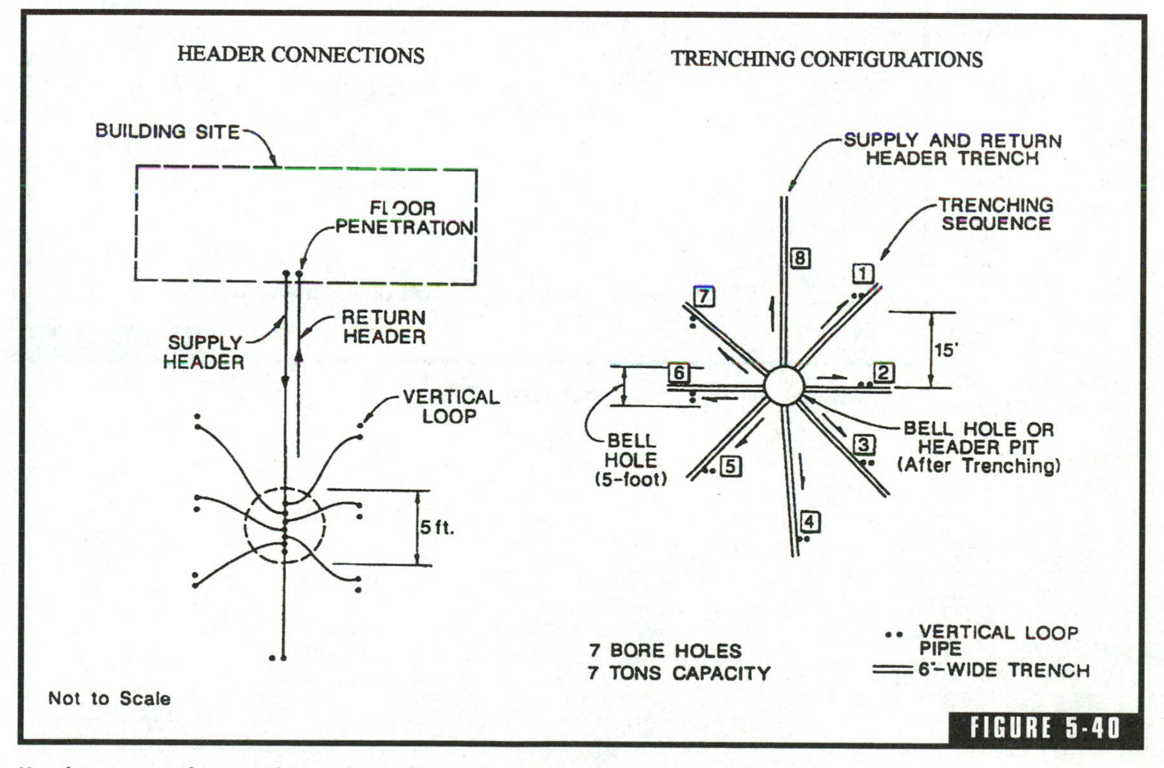

Header connections and trench configurations. (*Courtesy of Bard.*)

where water-source units are suitable solutions for heating needs. In time, the systems will probably become much more popular. Once people are aware of the advantages offered by water-source systems, they are likely to catch on quickly.

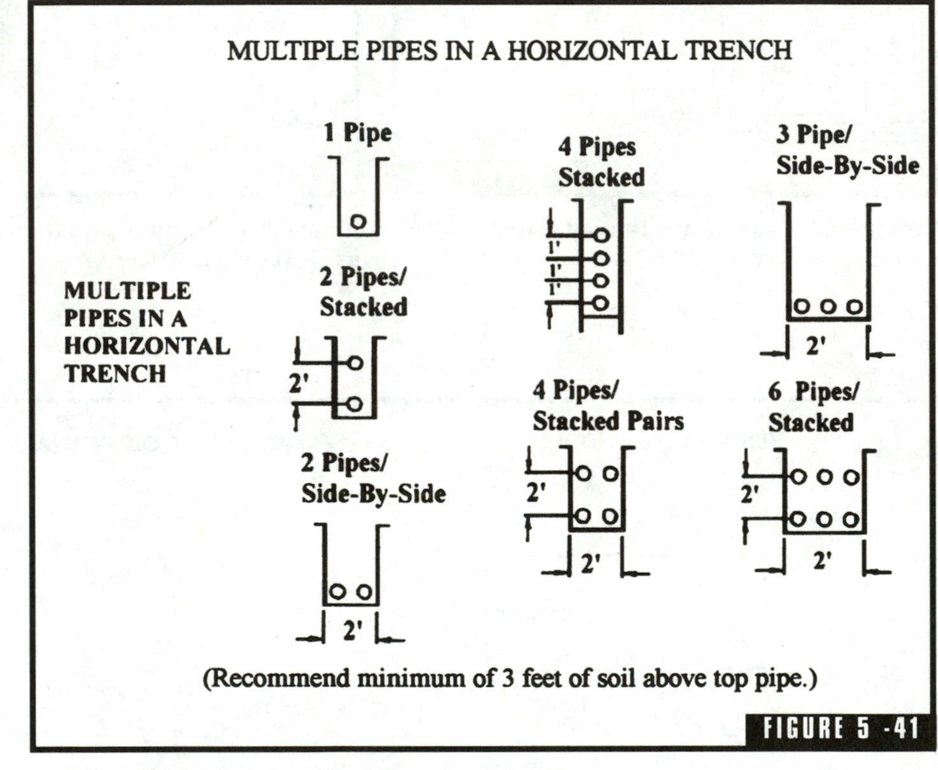

FIGURE 5-41

Pipe installation in a trench. (*Courtesy of Bard.*)

CHAPTER

6

Troubleshooting Heat Pumps

This chapter is all about troubleshooting heat pumps and their components. If you've ever had to endure a sweltering summer day or a cold winter night when your heat pump wouldn't work, you'll appreciate the value of this chapter. For those of you who have been lucky enough not to experience a heating or cooling failure, there is no time like the present to prepare for the unexpected. If you are a builder who relies on independent contractors, this chapter will at the very least help you to understand what your service people are talking about, and the information here may allow you to detect problems on your own. Even if you are an experienced service technician, you may find a few tips here that will make your job easier.

Many people don't associate troubleshooting with the installation of a new heat pump. In their minds, troubleshooting and repair work is something that is done when heat pumps age and become less dependable. The truth is, a lot of troubleshooting and repair work is done in conjunction with new installations.

Any professional installer will be able to tell you many stories about new parts that were defective and new equipment that simply didn't do what it was supposed to do. Anyone who has worked in the field for very long will have had their share of problems with new installations. Your need for easy-to-understand, effective troubleshooting techniques is of paramount importance (Figs. 6-1 and 6-2).

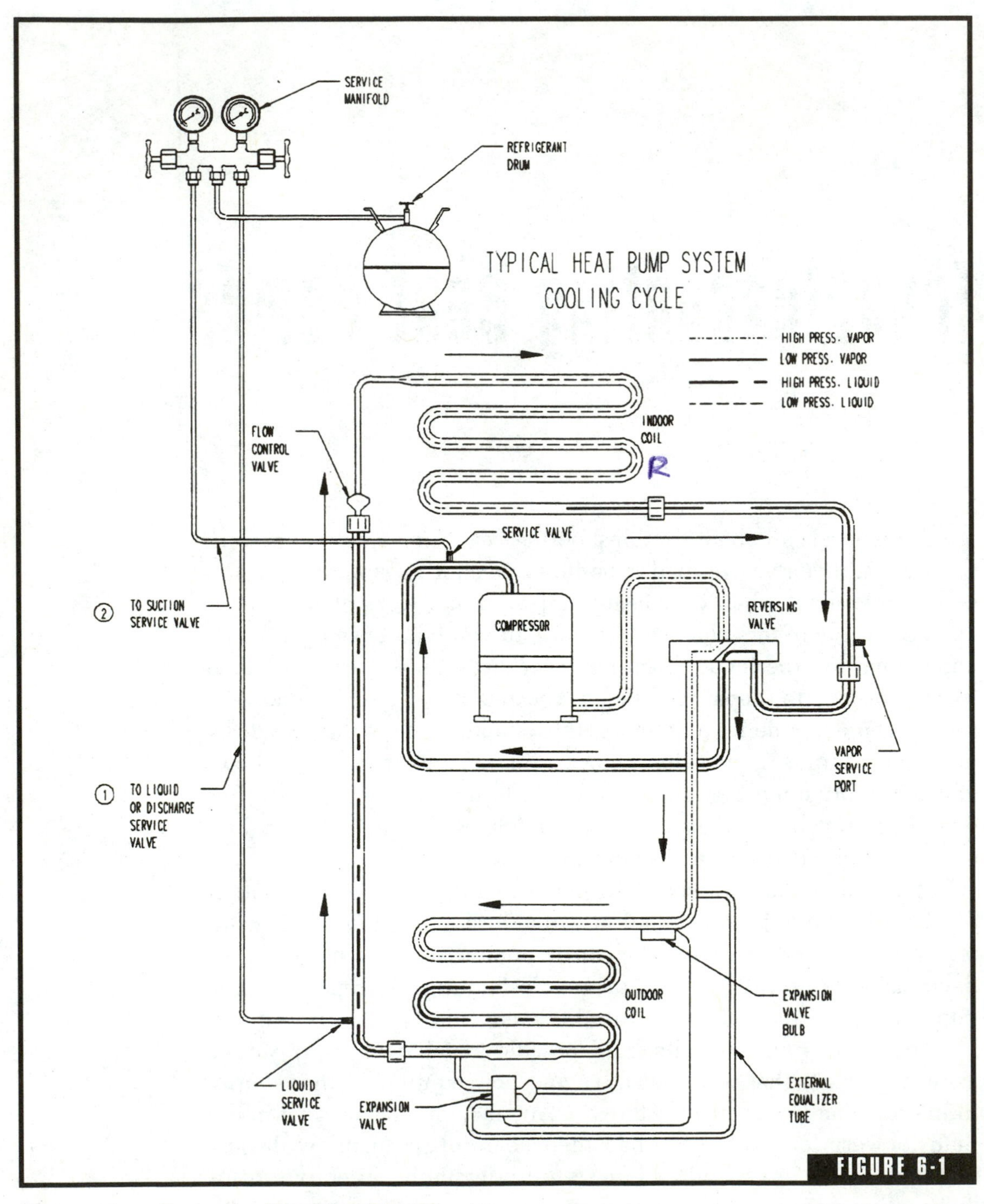

Heat-pump cooling cycle. (*Courtesy of Bard.*)

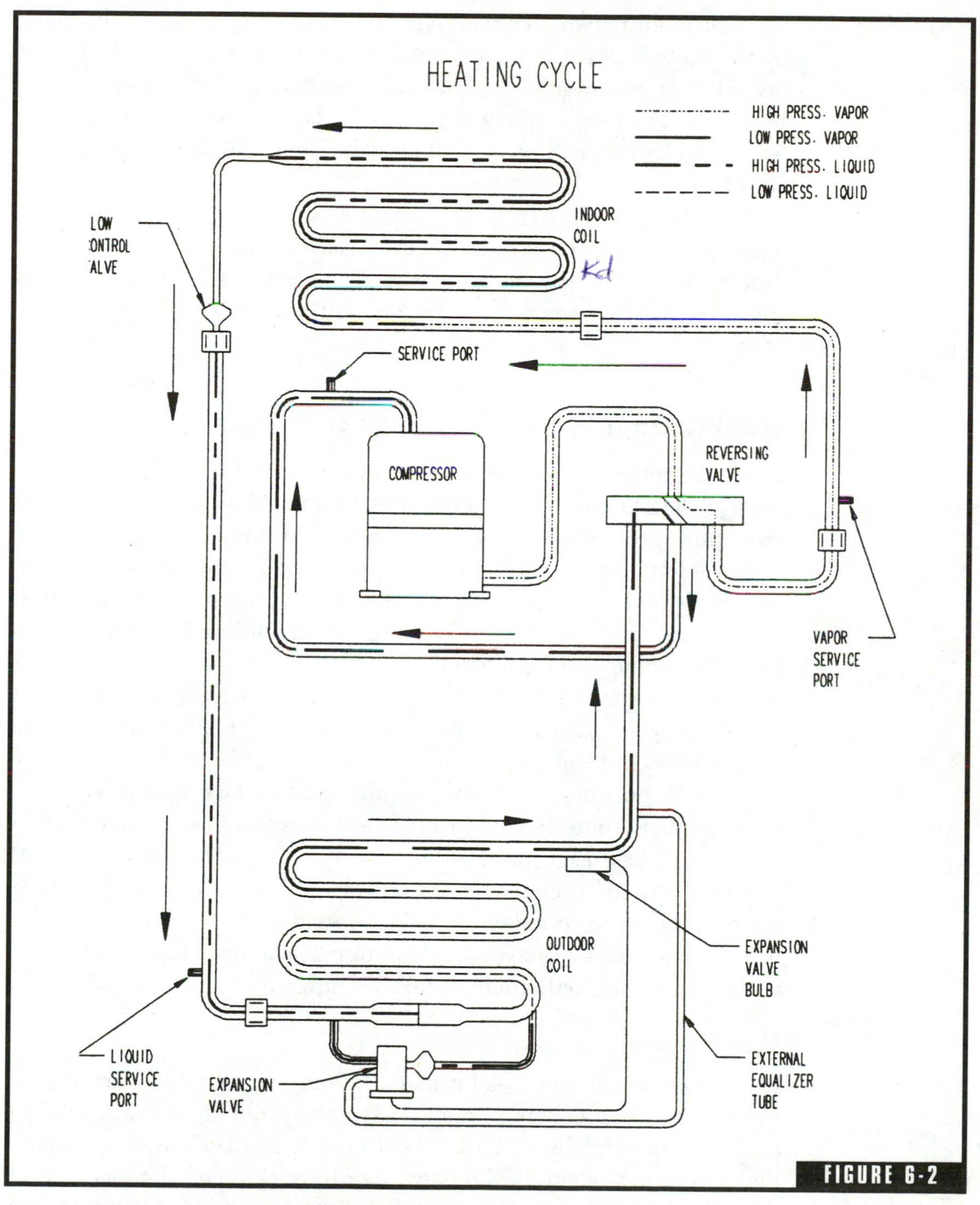

FIGURE 6-2

Heat-pump heating cycle. (*Courtesy of Bard.*)

You should always consult and adhere to recommendations made by the manufacturer of the heat pump you are working with. Most, if not all, heat pumps are supplied with instructions for troubleshooting problems that may occur with the unit. Those suggestions, coupled with this chapter, will more than prepare you for finding the causes of your problems.

Troubleshooting techniques for air-source heat pumps are somewhat different from the procedures used with water-source heat pumps. We are going to cover both types of heat pumps, but since air-source units are the most popular, we will start with them.

Troubleshooting Air-Source Heat Pumps

Troubleshooting air-source heat pumps is not difficult if you have enough data to consult. In many ways, troubleshooting is a trial-and-error procedure. Professionals learn how to minimize wasted time by avoiding unneeded troubleshooting steps. While this is fine for professionals, it is often best for inexperienced people to take the troubleshooting process slowly. This may mean following step-by-step instructions that fail to be fruitful, but in the end, the right moves will be made. If you, as an inexperienced installer, try to skip through certain steps, you could wind up losing more time than if you had gone by the book, so to speak.

One of the worst problems you are likely to face with a new heat pump is that it simply won't run. The frustration level reached when a new unit has been installed and won't run can be extreme. What would you do first if a new heat pump wouldn't cut on? Well, kicking it won't solve your problem, so let's see what will (Fig. 6-3). If you will be working with a water-source heat pump, you can skip ahead to the section that deals with them. It follows immediately after this section.

Won't Run

If you have a unit that won't run at all, there is probably a very simple solution to the problem. In many cases, the problem is a lack of electricity. A fuse might be blown or a circuit breaker may be tripped. Perhaps someone cut off the power while working on the installation and failed to cut it back on. Maybe someone moved the disconnect

TROUBLESHOOTING PROCEDURES

Symptom	Possible Causes	What to Check	How to Check or Repair
Compressor contactor does not energize (cooling or heating)	Control circuit wiring	Check for R connection at outdoor unit, and 24V between R-C.	Run R connection to outdoor unit to power heat pump control
	Compressor lock out	1. Check for 24V between L1-C on heat pump control 2. Check across high pressure switch	1. If no voltage between L1-C turn, thermostat off and on again to reset high pressure switch. 2. If high pressure switch is open and will not reset, replace high pressure switch.
	Compressor short cycle protection	Check for 24V between CC-C and Y-C on heat pump control	1. If no voltage between CC-C jumper speed up terminal and within 10 seconds power should appear between CC-C. Remove speed up jumper after 10 seconds.
	Heat pump control defective	Check all other possible causes. Manual 2100-065.	Replace heat pump control.
	Contactor defective	Check for open or shorted coil winding	Replace contactor.
Fan outdoor motor does not run (cooling or heating except during defrost)	Motor defective	Check for open or shorted motor winding	Replace motor.
	Motor capacitor defective	Check capacitor rating. Check for open or shorted capacitor.	Replace capacitor.
	Heat pump control defective	Check across fan relay on heat pump control. (Com-NC)	Replace pump control.
Reversing valve does not energize (heating only).	Reversing valve solenoid coil defective	Check for open or shorted coil.	Replace solenoid coil.
	Heat pump control defective	Check for 24V between RV-C and B-C.	1. Check control circuit wiring. 2. Replace heat pump control.
Unit will not go into defrost (heating only)	Heat pump control defective	Disconnect defrost thermostat and jumper across sensor terminals. This should cause the unit to go through a defrost cycle within one minute.	1. If unit goes though defrost cycle, check defrost thermostat. 2. If unit does not go through defrost cycle, replace heat pump control.
Unit will not come out of defrost (heating only)	Heat pump control defective	Jumper across speed up terminals. This should cause the unit to come out of defrost within one minute.	1. If unit comes out of defrost cycle, check defrost thermostat. 2. If unit does not come out of defrost cycle, replace heat pump control.

FIGURE 6-3

Troubleshooting guide for heat pumps. (*Courtesy of Bard.*)

lever when you weren't looking. At any rate, check all fuses, circuit breakers, and disconnect switches to make sure they are as they should be.

If everything checks out to be normal, check the voltage in the condensing unit. Wait! Don't attempt this step unless you are familiar with electrical meters and working with electricity. These troubleshooting steps will, in many cases, involve working with electricity. The power from electrical wiring can be deadly. If you are not skilled in working with electricity, defer to a licensed professional for all work involving electrical testing. If you find that the proper voltage is present for the condensing unit, you are working your way through the many possibilities for the cause of your problem.

Before you go too far with technical tests, try a very simple test. Go to the indoor thermostat and check all of the settings. Is the selector lever set on the proper cycle? In other words, if the selector is set for cooling on a cold winter day, the heat pump is not going to run. Has the thermostat been set at a temperature setting that would not call for the heat pump to cut on? For example, if the selector is set in the heating mode and the thermostat temperature setting is below the present temperature in the home, the heat pump is not going to work. These are extremely simple problems to check for, but they are just as capable of keeping the heat pump from running as some serious types of defects.

Assuming that all of the settings for the thermostat are set properly, you should check the control to see if it is out of calibration or defective. Any new thermostat should have been packed with its own set of instructions and troubleshooting tips. Due to the variances between thermostats, it is important that you follow the instructions given for your particular model.

If you have a high-pressure control, and you probably don't, it could be stuck in the open position. Refer to the manufacturer's paperwork and check the control.

A bad transformer can keep your heat pump from starting. If you understand electrical wiring, you should check to see that the transformer is wired correctly. Consult the manufacturer's recommendations for troubleshooting the transformer and follow them. It may be necessary to replace the transformer.

It is possible that the cause of your problem, if you haven't found it by now, is the contacts in the compressor. External overloads can be overcome by replacing the contacts. Internal overloads require that the compressor be replaced.

If you are dealing with new equipment, the cause of your problem is most likely a simple one. Take your time in checking all the possible causes, as described above, and you should have very little trouble finding the root cause of your problem.

Low Liquid Pressure

A reading of low liquid pressure and a unit that will not cool properly is a sign of trouble. The cause of the trouble can be one of many. There may not be an adequate amount of refrigerant in the system. If the system is low on refrigerant, it would be wise to check the entire system for leaks.

The problem could be caused by something as simple as a dirty filter or coil. Inspect the coil and filter to see that they are clean and allowing a clear air flow.

Remember that we have talked about check valves being installed backward? Well, that could be the cause for a unit that is not cooling well and that is giving a reading of low liquid pressure. Inspect the check valve to see that it is installed in the right direction. If it's installed properly, you may ultimately have to check to see that it is working correctly. A bad expansion valve can also cause a unit to give the same trouble symptoms caused by a bad check valve.

There are two other possible causes that are capable of causing a heat pump to cool poorly and produce a reading of low liquid pressure. A restriction in the liquid line could be at fault. If that's not the case, you should divert your attention to the compressor valves. Rely on the instructions provided by the manufacturer of your equipment in working with the compressor.

We have just discussed a combination of low liquid pressure and a lack of cooling ability. Since heat pumps heat homes as well as cool them, we must consider the problem of having a low liquid pressure and insufficient heat. When this is the case, you have to change your troubleshooting procedure.

If the system does not have a suitable charge of refrigerant in it, you could experience a low liquid pressure and a lack of satisfactory heat. Check the refrigerant and add some if necessary. When refrigerant is low, there is a chance some part of the system has a leak in it. You should check the system out thoroughly to discover any hidden leaks.

Restrictions in the refrigeration lines can also contribute to a low liquid pressure and a poor heating experience. Look over your refrigerant lines and see if any visible restrictions are present.

The valves in the system could also be at fault. Inspect the check valve and expansion valve. Make sure the check valve is installed with the arrow on the outside of the body pointing in the direction of the intended flow.

If you are particularly unlucky, you may find that the compressor valves are bad. This will probably call for a replacement compressor, but as always, refer to the suggestions provided by the manufacturer of the equipment.

High Liquid Pressure

A unit that is producing a reading of high liquid pressure and failing to cool air satisfactorily might be suffering from a dirty outside coil. Since this is an easy cause to detect or rule out, it is the most sensible step to take first. Inspect the outside coil and flush it out if necessary.

If the coil is not responsible for your problem, there are some other aspects of the outside unit you should check. Observe the fan in the outdoor unit to see that it is turning in the proper direction. Also, check to see that there are no loose blades on the fan. Obviously, you must be careful not to get fingers caught in a moving fan, so take necessary precautions when working around the blade.

Once you have ruled out the coil and the fan, it is time to check out the motor. Is it running at the proper revolutions per minute (rpm)? Is the run capacitor defective? After you have investigated the motor, move on to the refrigeration lines.

If the outside unit is not at fault, check the charge of refrigerant in the system; it may be overcharged. There is also a chance that noncondensables have found their way into the system and will have to be purged out.

Now that we have covered the troubleshooting for the cooling side of the unit, let's talk about the same problem on the heating side. When the unit has a high liquid pressure and is not heating properly, check the indoor coil and filter. If these items are dirty, they may be causing your problem.

If the filter and coil check out okay, turn your attention to the refrigerant charge. If the system is overcharged, you may have found the cause of your problem.

Noncondensables in the system can also cause a heat pump to heat poorly and to run with a high liquid pressure. If the system has been tainted by noncondensables, you will have to purge them from the piping.

If the blower motor on the heat pump is not sized properly or is running backward, you could consider it the cause of your problem. Check the motor to ensure it is of the proper size, is running at the right speed, and is turning in the correct direction.

Heat registers that are closed could be your only defect. Improper air flow can cause a high liquid pressure and a lack of heat. Check the registers to be sure that they are open and unobstructed. At the same time, check the return grill to make sure it is not blocked.

If you have connected your new heat pump to existing duct work that served some other type of equipment previously, there may be a problem with the size of the duct work. This, of course, could be a major problem. If you have ruled out all of the other possibilities, you may have to delve into the duct work and check its sizing.

SATURATED SUCTION TEMPERATURE (R-22)	
Saturated Suction Suction Pressure PSIG	Temperature (°F)
50	26
53	28
55	30
58	32
61	34
63	36
65	38
67	39
70	41
73	43
76	45
79	47
82	49
86	51

FIGURE 6-4

Saturated suction temperature. (*Courtesy of Bard.*)

Low Suction Pressure

What should you look for when your unit has low suction pressure and is not cooling properly? One of the first places to look is at the filter. If the filter is dirty and blocking air flow, your heat pump could experience low suction pressure and a lack of cooling ability. Any restriction in air flow could cause this type of problem (Figs. 6-4 to 6-7).

Outdoor Ambient Temperature (°F Dry Bulb)	Return Air Temperature °F—Wet Bulb			
	59	63	67	71
105	1	1	5	
95	1	3	(8)	20
90	1	7	14	26
85	3	9	19	33
80	8	14	25	39
75	10	20	30	42

SYSTEM SUPERHEAT

FIGURE 6-5

System superheat. (*Courtesy of Bard.*)

If refrigerant is not being returned to the compressor as it should be, the heat pump will exhibit low suction pressure and poor cooling performance. Test the refrigerant charge and consider looking for leaks if the charge is low.

A defective expansion valve or a fan in the outside unit that is running backward could be the cause of the problem. While it is unlikely, you may find that the motor for the fan is not sized properly. A defective run capacitor could also be the culprit. Any of these causes could be the root of your problem.

If you are suffering from low suction pressure and a lack of heat, check the outside coil to see if it is dirty or obstructed. While you are outside, check the fan and the motor on the outside unit. Make sure

TUBING CHART

TUBING CHART

Basic Condensing Unit Model	Refrigerant Line Length (Ft.) 0 - 20		21 - 60		61 - 100	
	Liquid	Suction	Liquid	Suction	Liquid	Suction
24UHP	3/8"	5/8"	3/8"	3/4"	3/8"	3/4"
30UHP	3/8"	5/8"	3/8"	3/4"	3/8"	3/4"
36UHP	3/8"	5/8"	3/8"	3/4"	1/2"	7/8"
42UHP	3/8"	7/8"	3/8"	7/8"	1/2"	1-1/8"
48UHP	3/8"	7/8"	3/8"	7/8"	1/2"	1-1/8"
60UHP	3/8"	7/8	3/8"	7/8"	1/2"	1-1/8"

The basis for selection is to maintain adequate velocity which assures adequate oil return to the compressor, an acceptable pressure drop to assure compressor capacity, and minimum tubing costs.

These recommendations are based on the use of standard refrigeration tubing.

Line sizes listed are outside tube dimensions.

These suggestions do not include consideration for additional pressure drop due to elbows, valves, or reduced joint sizes.

FIGURE 6-6

Tubing chart. (*Courtesy of Bard.*)

that the fan is not running backward and the motor is running at the proper speed. Check the run capacitor to see that it is doing its job.

Other causes of low suction pressure and a lack of heat could be defective expansion valves, restricted refrigerant lines, or improper refrigerant charges.

High Suction Pressure

Suppose the heat pump is not cooling efficiently and is showing signs of high suction pressure. What should you do? Testing the refrigerant charge may prove that the system is overcharged. You may find that the check valve or the reversing valve is defective or installed improperly. There is also a possibility that noncondensables have invaded the system. As something of a worst-case scenario, the compressor valves may be bad.

When you are experiencing high suction pressure and a lack of heat, you may find that the refrigerant lines are overcharged. There is also a possibility that noncondensables are infecting the system. A bad check valve or reversing valve could also be at the root of the problem. And it is possible that the compressor valves are bad.

Operating Pressures Are Normal

When operating pressures are normal and a heat pump will not heat or cool properly, the list of possible causes is not so long. Since we have been dealing with cooling problems first, let's continue that approach.

Assume that you have a heat pump that is running at normal operating pressures, but it is not producing enough cool air. What are you going to look for? Air leaks are one cause you could look for, but you should begin by looking at the electrical wiring, assuming that you are experienced enough to work with electrical wiring and components safely. Inspect all wiring, circuits, and components to see that there are no defects present. Make sure the wires are not loose and that they are connected to their terminals properly. If the wiring checks out, look for air leaks (Figs. 6-8 to 6-14).

Air leaks aren't likely, but it is possible a section of duct work has come loose. If the installer failed to secure supplies to the trunk line or supply boots, the supply duct may have dropped out of place. This could cause cool air to be blown into some area other than the intended cooling zone.

Don't overlook the possibility that supply registers are closed. If you have young children, the fascination of closing off all the floor registers may be more than their curious fingers can resist.

If you have proved that the wiring is up to snuff and that air leakage is not a problem, there is only one other conclusion to draw, and it is not a pretty one. In the event that the simple checks have turned up empty, you must consider the fact that the heat pump may not have been sized properly. This is certainly not a thought you would want to conjure, but it is a possibility.

Total System Charge for Split Systems

Outdoor Section	Indoor Section	Total R-22 Charge (Oz.)
24UHPSC	BC24C	121 oz.
	SA36AS-A	123 oz.
30UHPSC	BC36C	136 oz.
	A36AS-A	123 oz.
	A37AS-A	144 oz.
36UHPSC	BC36C	177 oz.
	A37AS-A	186 oz.
42UHPSC	BC48C	180 oz.
	A61AS-A	186 oz.
48UHPSB	BC48C	202oz.
	A61AS-A	206 oz.
48UHPSC	BC48C	206 oz.
	A61AS-A	220 oz.
60UHPSC	BC60C	244 oz.
	A61AS-A	218 oz.

The above includes 25' of 3/8" diameter liquid line. For other than 25' and other tube sizes, adjust the total charge according to the following schedule.

Liquid Line Diameter	Oz. R-22 Per Ft.①
3/8"	.6
1/2"	1.2

Installer Note: Stamp or mark the final system charge determined above on the outdoor unit serial plate.

① These values should only be applied during initial system charging. System operating charge should be adjusted in cooling mode for optimum performance outlined in the installation instructions for that model outdoor section.

FIGURE 6-7

Total system charge for split system. (*Courtesy of Bard.*)

When an undersized unit is suspected, you should go back over the system and size it again. This is the only way to be certain that the equipment is not of the wrong tonnage. If you sized the system yourself, it may very well pay to call in someone else to check over your sizing calculations and installation.

If your system is running at routine pressure and is not providing adequate heat, check the outside thermostat. The cause of your problem may be as simple as a thermostat setting that does not meet your needs.

Assuming that the thermostat is not causing your lack of heat, check the duct work to see that there is proper air flow and no leaks. Again, check the registers to be sure they are open. Inspect any dampers that have been installed in the duct work to be sure that they

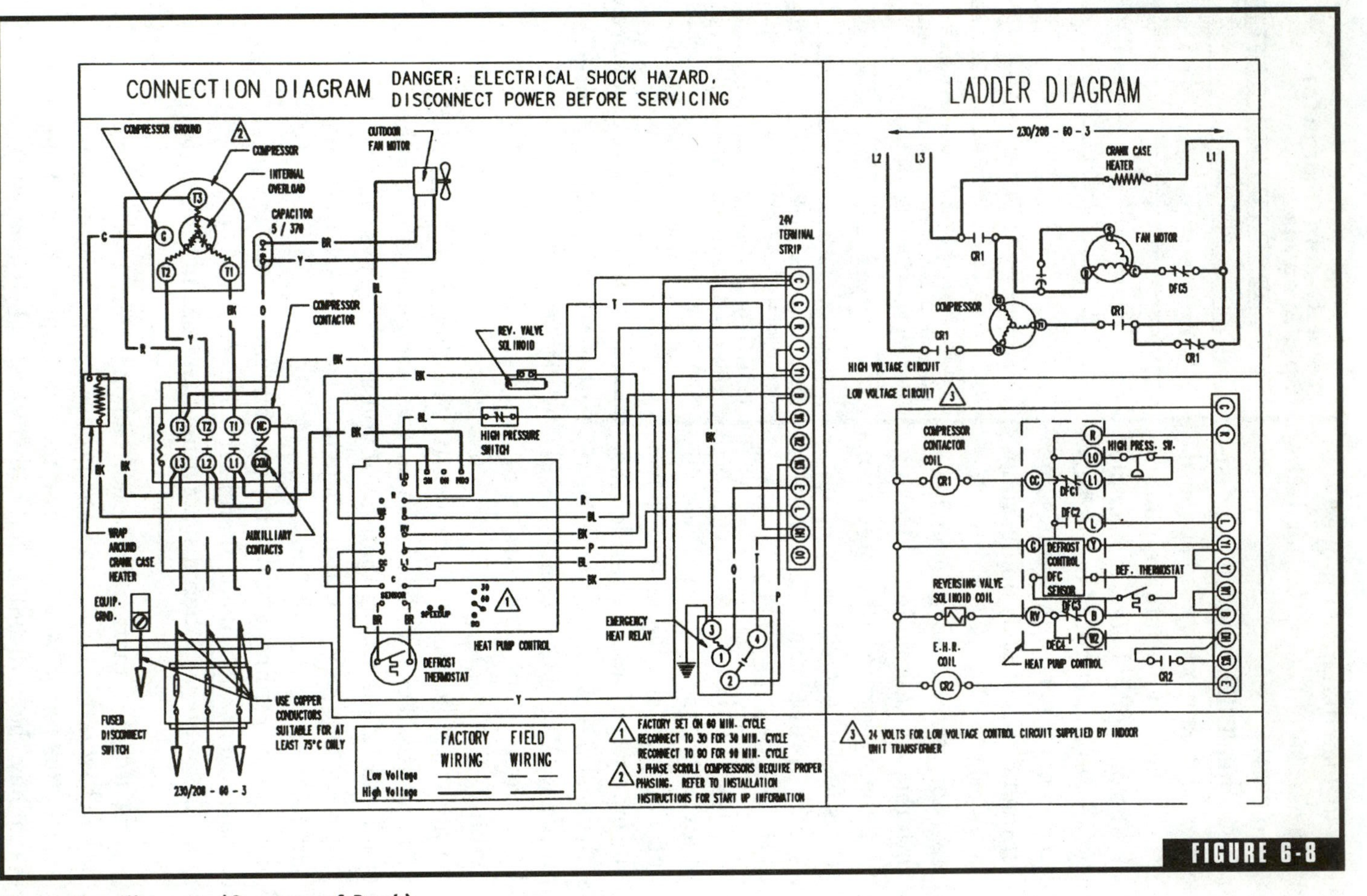

Connection diagram. (*Courtesy of Bard.*)

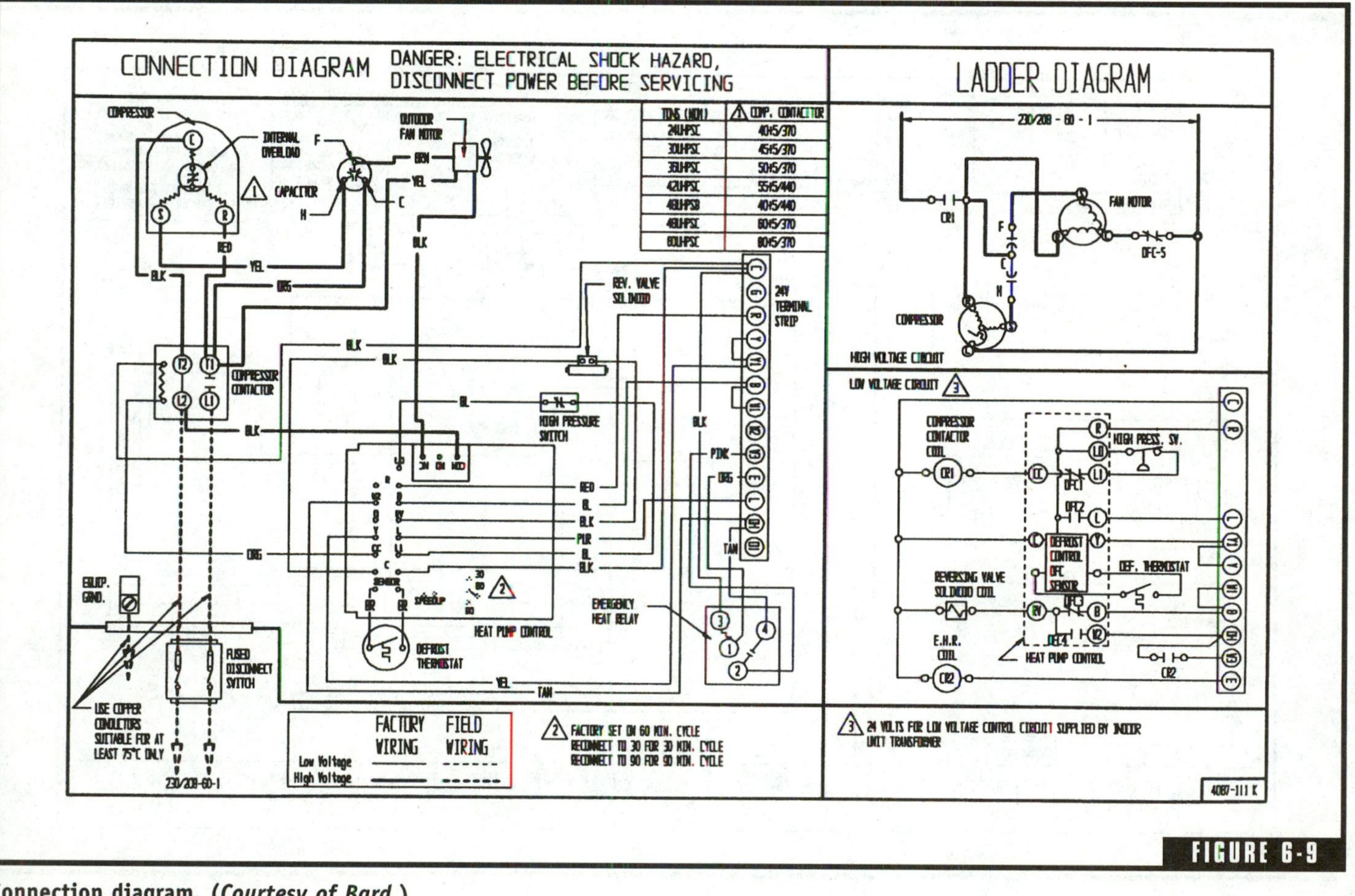

FIGURE 6-9

Connection diagram. (*Courtesy of Bard.*)

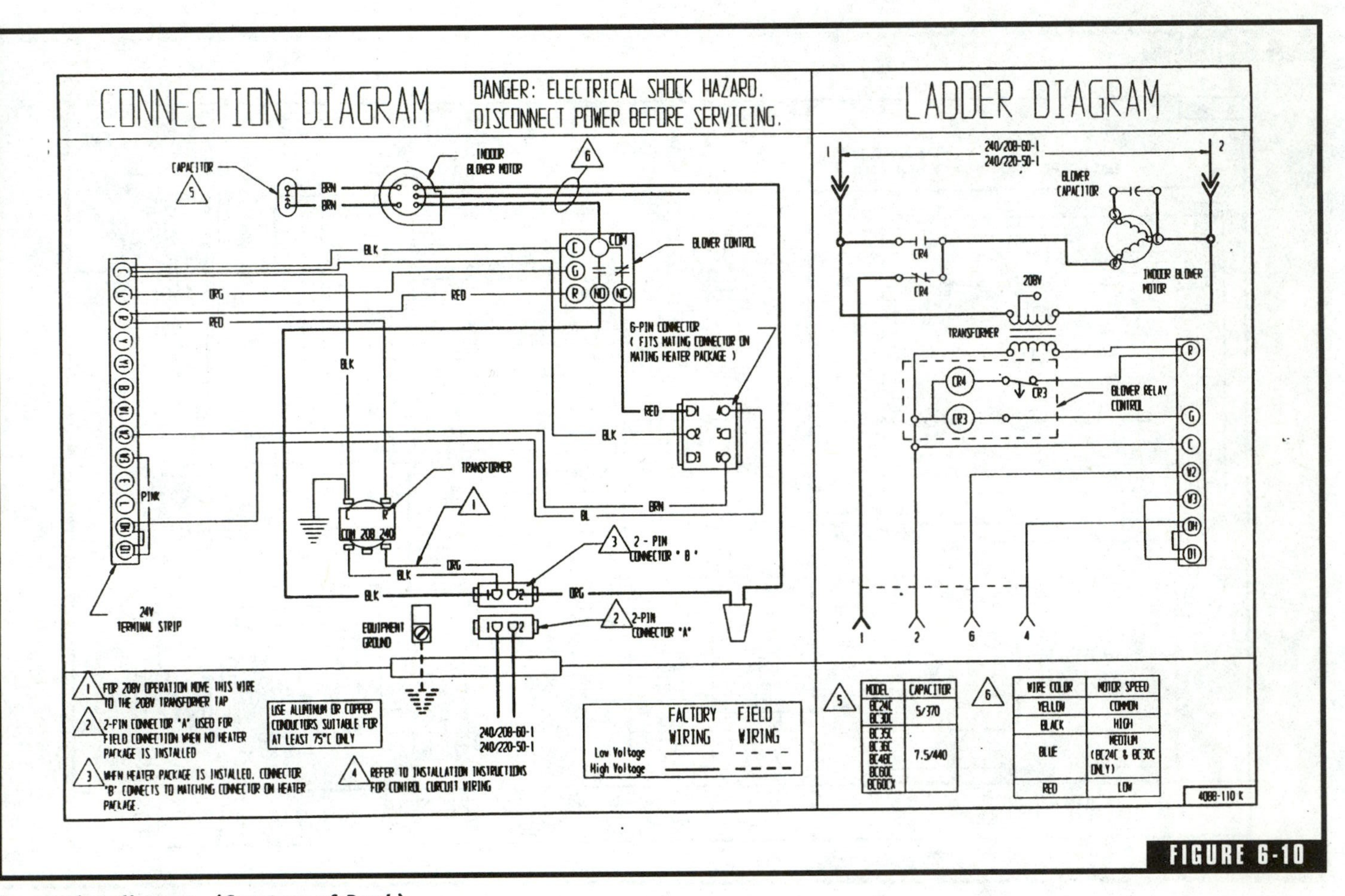

Connection diagram. (*Courtesy of Bard.*)

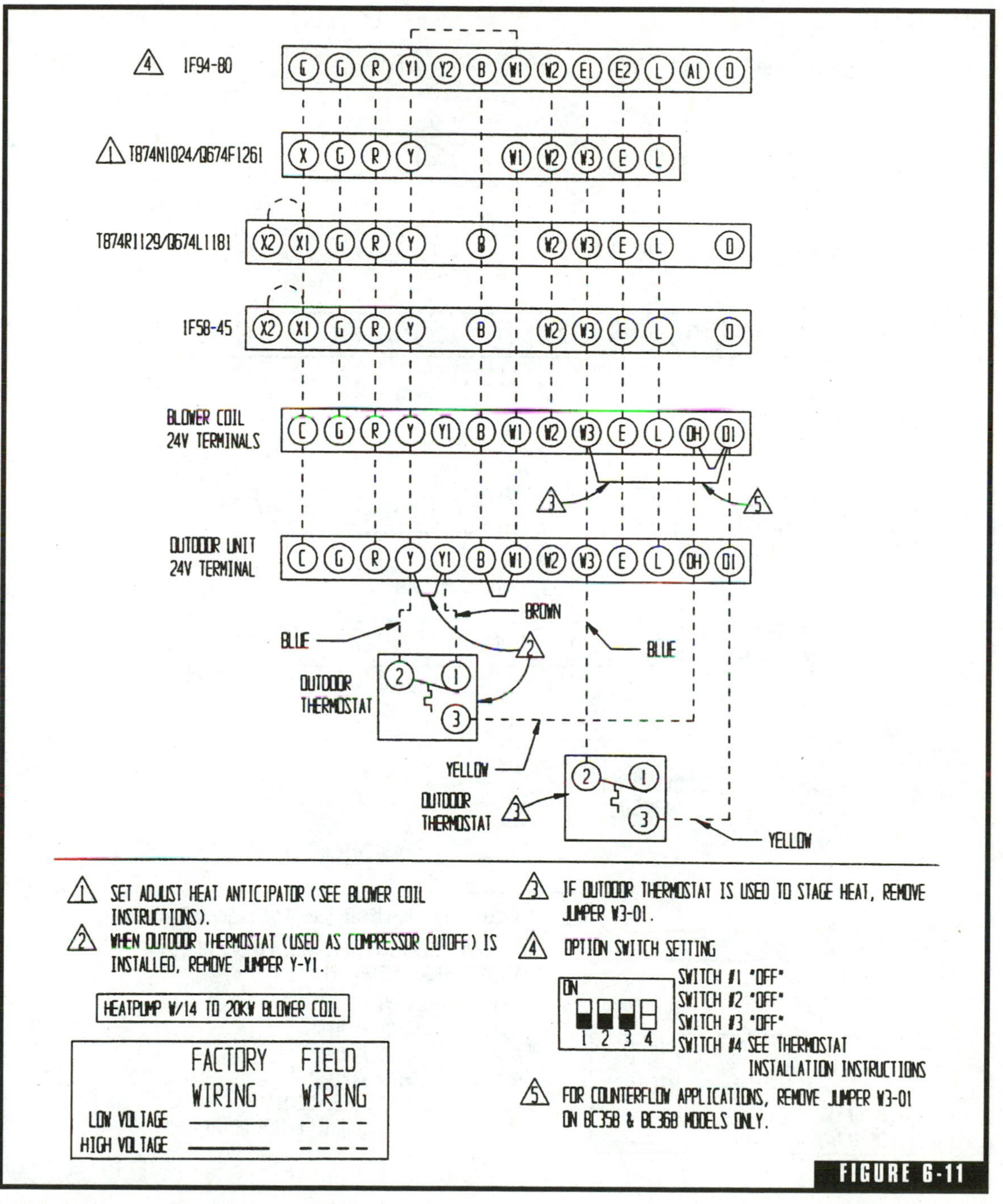

Wiring diagram. (*Courtesy of Bard.*)

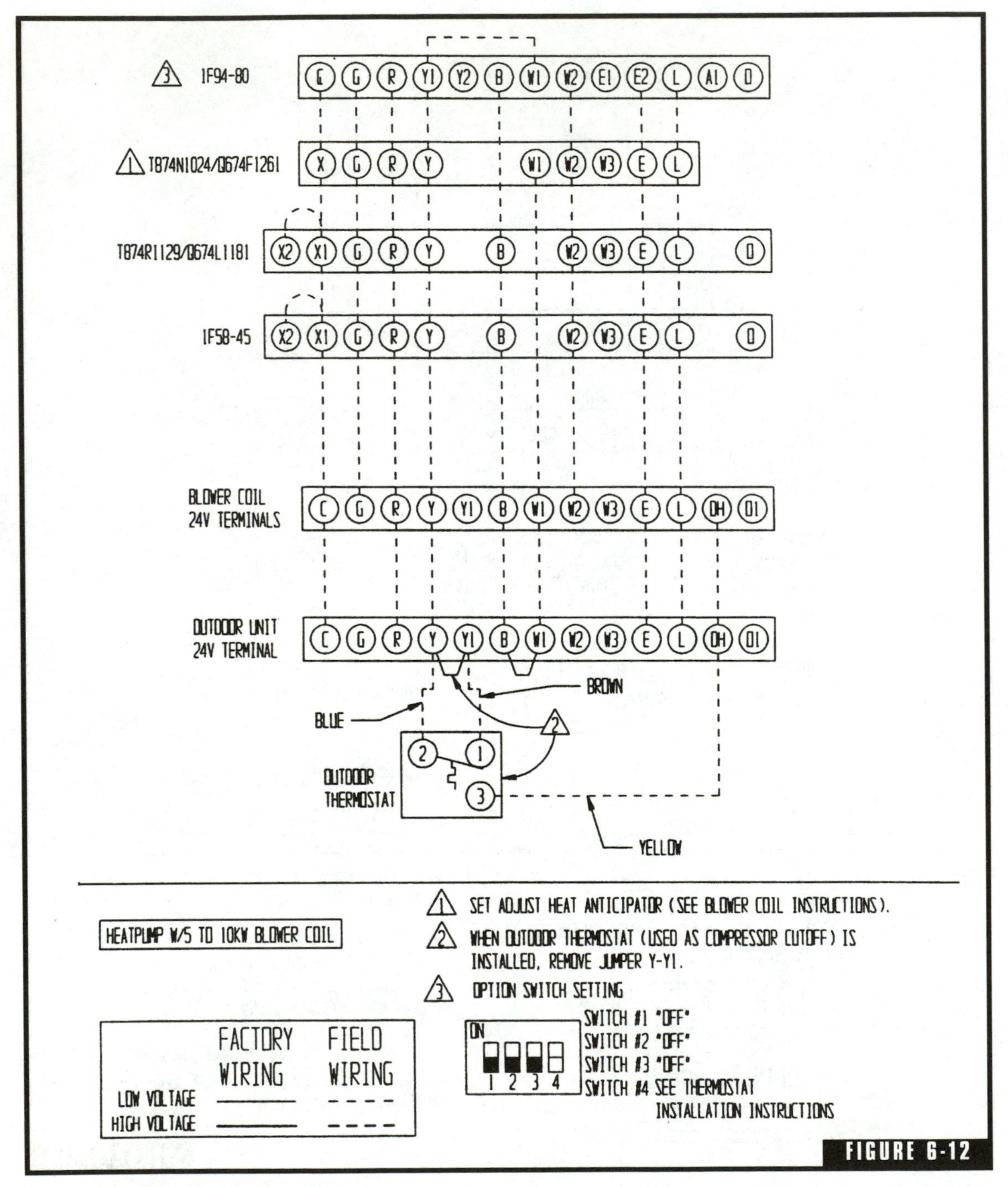

Wiring diagram. (*Courtesy of Bard.*)

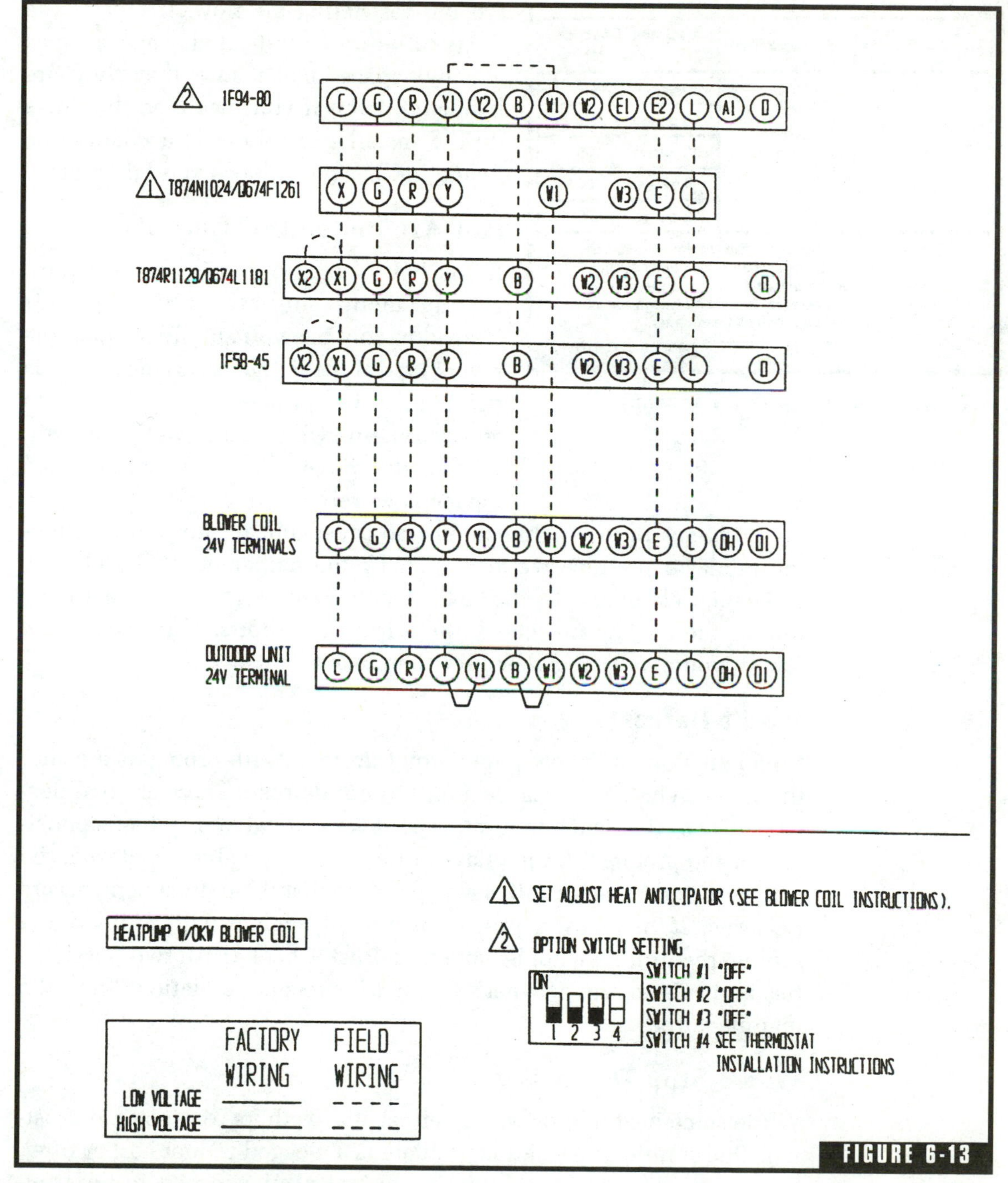

Wiring diagram. (*Courtesy of Bard.*)

Transformer VA	FLA @ 240V	Maximum Distance In Feet (1)
55	2.3	20 gauge – 45
		18 gauge – 60
		16 guage – 100
		14 gauge – 160
		12 gauge – 250

(1) For split systems, this is the maximum distance between the indoor section and outdoor section, and between the indoor section and thermostat each could be up to 90 feet for 18 gauge and 65 feet for 20 gauge on 40VA transformer.

FIGURE 6-14

Low voltage wiring. (*Courtesy of Bard.*)

are not restricting air flow. If any of the ducts run through unheated space, such as a crawl space, make sure that they are insulated. Even if you insulated the ducts during installation, there is a chance the insulation has come loose and fallen off.

Hot Air instead of Cool Air

If you have a heat pump in the air-conditioning mode and are getting hot air instead of cool air, you should suspect the reversing valve. If the reversing valve is not functioning properly, it is possible to get warm air when you want cool air. Inspect the valve relay to make sure the device is working as it should.

Problems in the wiring could also cause a heat pump in the cooling mode to produce warm air. Check the amperage of the electric resistance elements. If you get an amperage reading, move to the indoor thermostat and check the wiring for shorts. If no shorts are found, you may have a defective thermostat.

Won't Defrost

When an air-source heat pump won't defrost itself, problems are sure to follow. What would cause a unit to not defrost? There are two possible causes that come to mind. The reversing valve may be responsible for the problem. If the relays of the reversing valve are closed, the problem could manifest. It is also possible that the defrost controls are defective. If the pressure pipe is obstructed, such as with a kink in the tubing, the unit may not be able to defrost. Check these two possibilities and refer to your owner's manual for recommendations from the manufacturer.

Won't Stop Defrosting

While some heat pumps won't defrost at all, others won't stop defrosting. One simple cause of this problem is a clogged indoor coil or filter. Since this is easy to check for, it is a logical place to begin the troubleshooting process. If the coil or filter is dirty, it will have to be cleaned or replaced.

The temperature bulb that controls the defrost cycle may be loose or uninsulated. This is certainly worth a look. Defective defrost controls may be causing the never-ending defrost cycle, and they may have to be repaired or replaced. The expansion valve could also cause a problem with the defrosting cycle, so it should be checked out. If none of these causes seem to be likely, check the refrigerant charge to be sure it is at an acceptable level.

Partial Defrosting

There may be times when an air-source heat pump will only obtain partial defrosting during the defrost cycle. When this is the case, there are two relatively simple causes that may be responsible and one not-so-simple cause to look out for. Let's hear the bad news first. When a unit is not defrosting completely, the compressor valves may be defective. However, before you jump to the conclusion that the compressor is shot, check the refrigerant charge to see that it is where it should be (Figs. 6-15 and 6-16).

If there are no abnormalities in the refrigerant charge, check the defrost controls and circuits. This will require testing for pressure and temperature. Refer to recommendations of the manufacturer of your particular piece of equipment for guidance.

Compressor Cuts Off Early

If you have a heat pump where the compressor cuts off early during a defrost cycle, you should check the indoor coil and filter. If these items are dirty or restricted, clean them, make sure good air flow is possible, and test the equipment to see if your simple repair has made a difference.

Assuming that the problem is not associated with a dirty coil or filter, check the expansion valve to see if it may be defective. You will be checking for pressure and temperature, as per the manufacturer's recommendations.

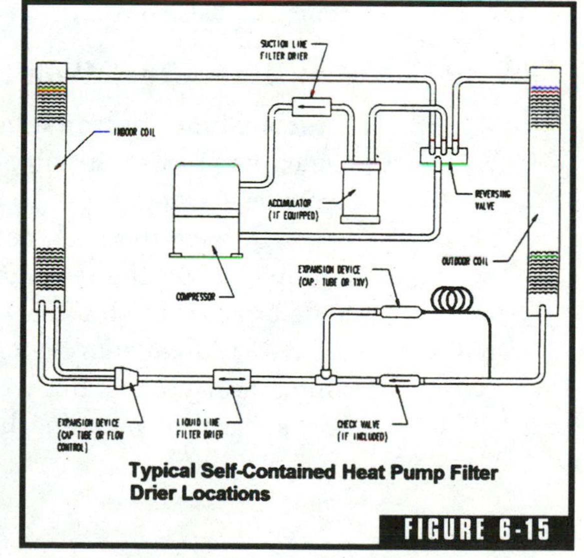

Heat pump filter/drier locations. (*Courtesy of Bard.*)

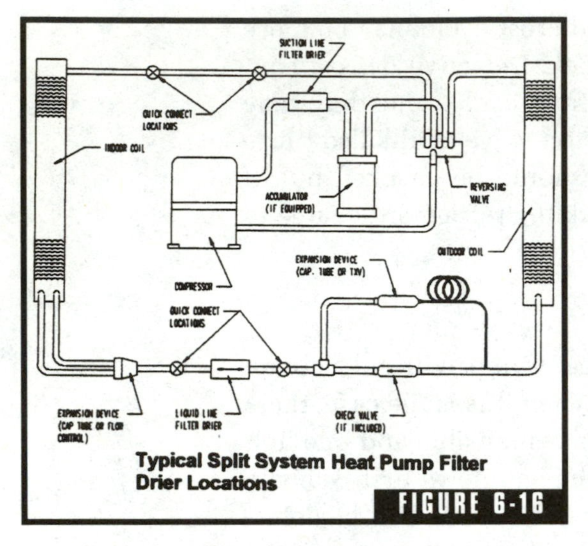

Typical Split System Heat Pump Filter Drier Locations

FIGURE 6-16

Locations for filter/driers in split systems. (*Courtesy of Bard*.)

Your problem could be caused by a bad check valve or even the reversing valve. Inspect these valves to make sure they are installed and operating properly.

Auxiliary Heat

The auxiliary heat in a heat pump may stay on after a defrost cycle is completed. If this happens, the thermostat for the heat is the most logical place to begin your troubleshooting. With the heat pump in a defrost cycle, check out the thermostat. You may have to replace it.

Another cause for the auxiliary heat not cutting off could be a defective relay. If you have the knowledge to work with relays, check out the one for the emergency heat. If it is defective, a replacement will be required.

The components and circuitry associated with the auxiliary heat may require your attention. The three things to check are the de-ice control, the control relay, and the contactors. Field repair or replacement may be required.

Auxiliary Heat Won't Come On

If the auxiliary heat in a heat pump won't come on during a defrost cycle, the problem is going to be caused by either the defrost components or their circuitry. To determine the exact cause of the problem, you will have to check out all of the components and their wiring. Follow the specific recommendations of the equipment manufacturer to determine which of the components is not working properly.

We are done with the primary troubleshooting steps involving air-source heat pumps, but we are far from done with this chapter. The next subject to be approached is that of water-source heat pumps.

Troubleshooting Water-Source Heat Pumps

Troubleshooting water-source heat pumps is similar in many ways to troubleshooting air-source heat pumps (Fig. 6-17). However, due to the different operating procedures for the two types of systems, there are obviously some variations in the troubleshooting process. The water-source heat pump involves components not found in an air-source system. There are also aspects of an air-source system that are not present in a water-source system. For this reason, we will go over the same basic problems for water-source heat pumps that we have just studied for air-source heat pumps. There will be many similarities in some circumstances, and none in others.

Won't Run

If you have a unit that won't run at all, there is probably a very simple solution to the problem. In many cases, the problem is a lack of electricity. A fuse might be blown or a circuit breaker may be tripped. Perhaps someone cut off the power while you were working on the installation and failed to cut it back on. Maybe someone moved the disconnect lever when you weren't looking. At any rate, check all fuses, circuit breakers, and disconnect switches to make sure they are as they should be.

If everything checks out to be normal, check the voltage for the unit. Wait! Don't attempt this step unless you are familiar with electrical meters and working with electricity. These troubleshooting steps will, in many cases, involve working with electricity. The power from electrical wiring can be deadly. If you are not skilled in working with electricity, defer to a licensed professional for all work involving electrical testing. If you find that the proper voltage is present for the condensing unit, you are working your way through the many possibilities for the cause of your problem.

Before you go too far with technical tests, try a very simple test. Go to the indoor thermostat and check all of the settings. Is the selector lever set on the proper cycle? In other words, if the selector is set for cooling on a cold winter day, the heat pump is not going to run. Has the thermostat been set at a temperature setting that would not call for the heat pump to cut on? For example, if the selector is set in the heating mode and the thermostat temperature setting is below the present

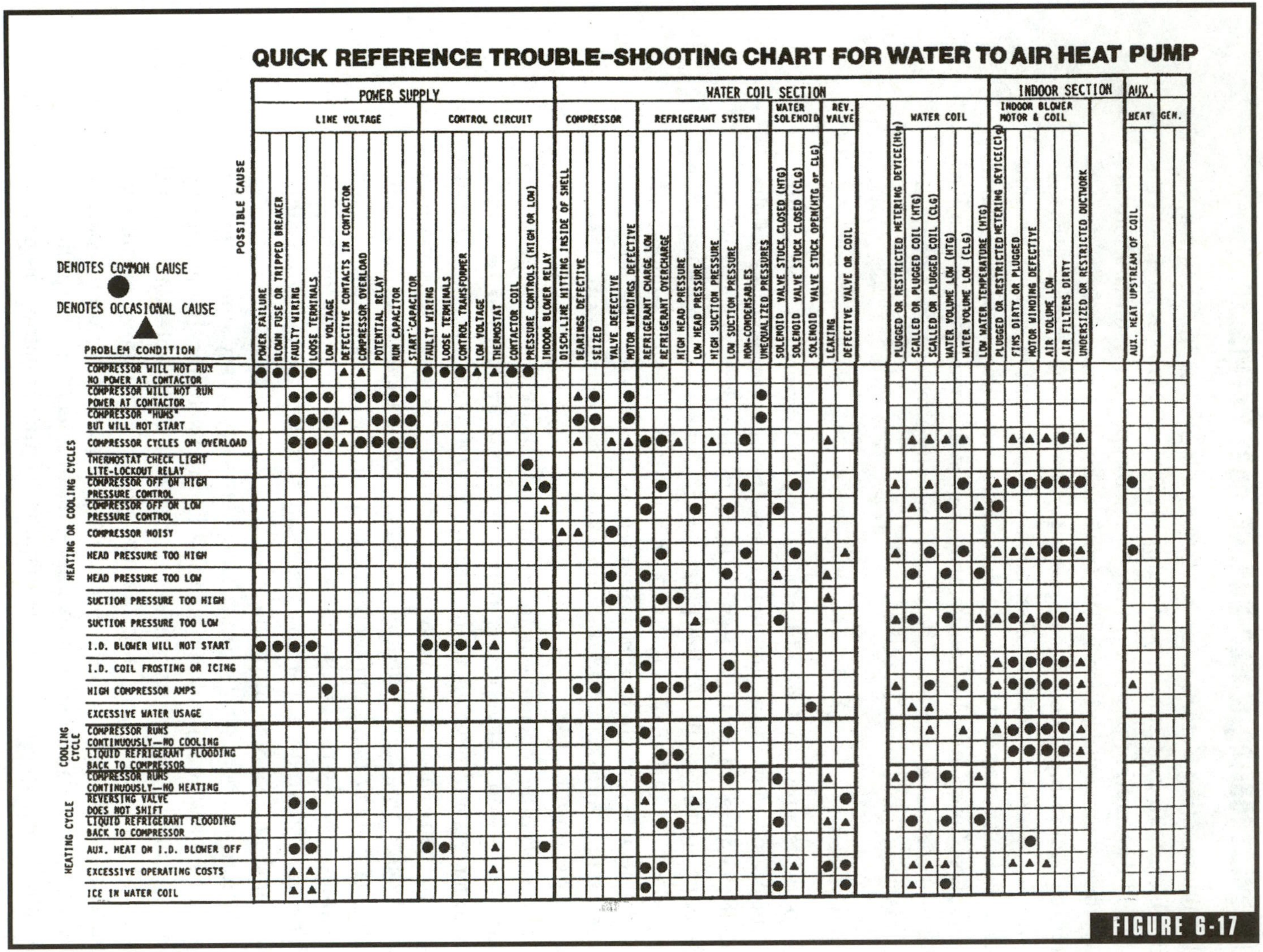

QUICK REFERENCE TROUBLE-SHOOTING CHART FOR WATER TO AIR HEAT PUMP

DENOTES COMMON CAUSE ●

DENOTES OCCASIONAL CAUSE ▲

POSSIBLE CAUSE — POWER SUPPLY

Cycle	Problem condition	Line voltage: Power failure	Line voltage: Blown fuse or tripped breaker	Line voltage: Faulty wiring	Line voltage: Loose terminals	Line voltage: Low voltage	Line voltage: Defective contacts in contactor	Line voltage: Compressor overload	Line voltage: Potential relay	Line voltage: Run capacitor	Line voltage: Start capacitor	Control circuit: Faulty wiring	Control circuit: Loose terminals	Control circuit: Control transformer	Control circuit: Low voltage	Control circuit: Thermostat	Control circuit: Contactor coil	Control circuit: Pressure controls (high or low)	Control circuit: Indoor blower relay
Heating or cooling cycles	Compressor will not run no power at contactor	●	●	●	●		▲	▲				●	●	●	▲	▲	●	●	
Heating or cooling cycles	Compressor will not run power at contactor			●	●	●		●	●	●	●								
Heating or cooling cycles	Compressor "hums" but will not start			●	●	●	▲		●	●	●								
Heating or cooling cycles	Compressor cycles on overload			●	●	●	▲	●	●	●	●								
Heating or cooling cycles	Thermostat check light lite-lockout relay																	●	
Heating or cooling cycles	Compressor off on high pressure control																	▲	●
Heating or cooling cycles	Compressor off on low pressure control																		▲
Heating or cooling cycles	Compressor noisy																		
Heating or cooling cycles	Head pressure too high																		
Heating or cooling cycles	Head pressure too low																		
Heating or cooling cycles	Suction pressure too high																		
Heating or cooling cycles	Suction pressure too low																		
Heating or cooling cycles	I.D. blower will not start	●	●	●	●							●	●	●	▲	▲			●
Heating or cooling cycles	I.D. coil frosting or icing																		
Heating or cooling cycles	High compressor amps					●				●									
Heating or cooling cycles	Excessive water usage																		
Cooling cycle	Compressor runs continuously—no cooling																		
Cooling cycle	Liquid refrigerant flooding back to compressor																		
Heating cycle	Compressor runs continuously—no heating																		
Heating cycle	Reversing valve does not shift			●	●														
Heating cycle	Liquid refrigerant flooding back to compressor																		
Heating cycle	Aux. heat on I.D. blower off			●	●							●	●			▲			●
Heating cycle	Excessive operating costs			▲	▲											▲			
Heating cycle	Ice in water coil			▲	▲														

POSSIBLE CAUSE — WATER COIL SECTION

Cycle	Problem condition	Compressor: Disch. line hitting inside of shell	Compressor: Bearings defective	Compressor: Seized	Compressor: Valve defective	Compressor: Motor windings defective	Refrigerant system: Refrigerant charge low	Refrigerant system: Refrigerant overcharge	Refrigerant system: High head pressure	Refrigerant system: Low head pressure	Refrigerant system: High suction pressure	Refrigerant system: Low suction pressure	Refrigerant system: Non-condensables	Refrigerant system: Unequalized pressures	Water solenoid: Solenoid valve stuck closed (HTG)	Water solenoid: Solenoid valve stuck closed (CLG)	Water solenoid: Solenoid valve stuck open (HTG or CLG)	Rev. valve: Leaking	Rev. valve: Defective valve or coil
Heating or cooling cycles	Compressor will not run no power at contactor																		
Heating or cooling cycles	Compressor will not run power at contactor		▲	●		●								●					
Heating or cooling cycles	Compressor "hums" but will not start		●	●		●								●					
Heating or cooling cycles	Compressor cycles on overload		▲		▲	▲	●	●	▲		▲		●					▲	
Heating or cooling cycles	Thermostat check light lite-lockout relay																		
Heating or cooling cycles	Compressor off on high pressure control							●					●			●			
Heating or cooling cycles	Compressor off on low pressure control						●			●		●			●				
Heating or cooling cycles	Compressor noisy	▲	▲		●														
Heating or cooling cycles	Head pressure too high							●					●			●			▲
Heating or cooling cycles	Head pressure too low				●		●					●			▲			▲	
Heating or cooling cycles	Suction pressure too high				●			●	●									▲	
Heating or cooling cycles	Suction pressure too low						●			▲					●				
Heating or cooling cycles	I.D. blower will not start																		
Heating or cooling cycles	I.D. coil frosting or icing						●					●							
Heating or cooling cycles	High compressor amps		●	●		▲		●	●		●		●						
Heating or cooling cycles	Excessive water usage																●		
Cooling cycle	Compressor runs continuously—no cooling				●		●					●							
Cooling cycle	Liquid refrigerant flooding back to compressor							●	●										
Heating cycle	Compressor runs continuously—no heating				●		●					●			●			▲	
Heating cycle	Reversing valve does not shift						▲			▲									●
Heating cycle	Liquid refrigerant flooding back to compressor							●	●						●			▲	▲
Heating cycle	Aux. heat on I.D. blower off																		
Heating cycle	Excessive operating costs						●	●							▲	▲		●	●
Heating cycle	Ice in water coil						●								●				●

POSSIBLE CAUSE — WATER COIL / INDOOR SECTION / AUX.

Cycle	Problem condition	Water coil: Plugged or restricted metering device (Htg)	Water coil: Scalled or plugged coil (HTG)	Water coil: Scalled or plugged coil (CLG)	Water coil: Water volume low (HTG)	Water coil: Water volume low (CLG)	Water coil: Low water temperature (HTG)	Indoor blower motor & coil: Plugged or restricted metering device (Clg)	Indoor blower motor & coil: Fins dirty or plugged	Indoor blower motor & coil: Motor winding defective	Indoor blower motor & coil: Air volume low	Indoor blower motor & coil: Air filters dirty	Indoor blower motor & coil: Undersized or restricted ductwork	Aux. heat: Aux. heat upstream of coil	Gen.
Heating or cooling cycles	Compressor will not run no power at contactor														
Heating or cooling cycles	Compressor will not run power at contactor														
Heating or cooling cycles	Compressor "hums" but will not start														
Heating or cooling cycles	Compressor cycles on overload		▲	▲	▲	▲			▲	▲	▲	●	▲		
Heating or cooling cycles	Thermostat check light lite-lockout relay														
Heating or cooling cycles	Compressor off on high pressure control	▲		▲		●		▲	●	●	●	●	●	●	
Heating or cooling cycles	Compressor off on low pressure control		▲		●		▲	●							
Heating or cooling cycles	Compressor noisy														
Heating or cooling cycles	Head pressure too high	▲		●		●		▲	▲	▲	●	●	▲	●	
Heating or cooling cycles	Head pressure too low		●		●		●								
Heating or cooling cycles	Suction pressure too high														
Heating or cooling cycles	Suction pressure too low	▲	●		●		▲	▲	●	▲	●	●	▲		
Heating or cooling cycles	I.D. blower will not start														
Heating or cooling cycles	I.D. coil frosting or icing							▲	●	●	●	●	▲		
Heating or cooling cycles	High compressor amps	▲		●		●		▲	●	●	●	●	▲	▲	
Heating or cooling cycles	Excessive water usage		▲	▲											
Cooling cycle	Compressor runs continuously—no cooling			▲		▲		▲	●	●	●	●	▲		
Cooling cycle	Liquid refrigerant flooding back to compressor								●	●	●	●	▲		
Heating cycle	Compressor runs continuously—no heating	▲	●		●		▲								
Heating cycle	Reversing valve does not shift														
Heating cycle	Liquid refrigerant flooding back to compressor		●		●		●								
Heating cycle	Aux. heat on I.D. blower off									●					
Heating cycle	Excessive operating costs		▲	▲	▲				▲	▲	▲				
Heating cycle	Ice in water coil		▲		●										

FIGURE 6-17

Troubleshooting guide. (*Courtesy of Bard.*)

temperature in the home, the heat pump is not going to work. These are extremely simple problems to check for, but they are just as capable of keeping the heat pump from running as some serious types of defects.

Assuming that all of the settings for the thermostat are set properly, you should check the control to see is it is out of calibration or defective. Any new thermostat should have been packed with its own set of instructions and troubleshooting tips. Due to the variances between thermostats, it is important that you follow the instructions given for your particular model.

A bad transformer can keep your heat pump from starting. If you understand electrical wiring, you should check to see that the transformer is wired correctly, has the proper voltage coming through it, and is not burned out. Consult the manufacturer's recommendations for troubleshooting the transformer and follow them. It may be necessary to replace the transformer.

If you are dealing with new equipment, the cause of your problem is most likely a simple one. Take your time in checking all the possible causes, as described above, and you should have very little trouble finding the root cause of your problem (Figs. 6-18 to 6-27).

All Heat

If you have a problem where a heat pump gives you all heat and no cool air, the problem is almost guaranteed to be associated with the reversing valve. The reversing valve is probably wired incorrectly. Check the wiring at both the heat pump and the thermostat. If you are convinced there is nothing wrong with the wiring, you must troubleshoot the valve itself and may have to replace it.

Not Enough Cool Air

If you are not getting enough cool air from a heat pump, there are several possible causes for the problem. Start with a simple check. Inspect the supply registers to see that they are open. Inquisitive children will sometimes close registers just to see if they can do it. Obviously, if the registers are closed, a sufficient volume of cool air cannot fill your home. You may also want to check the dampers in the duct work to make sure they are open and operating properly.

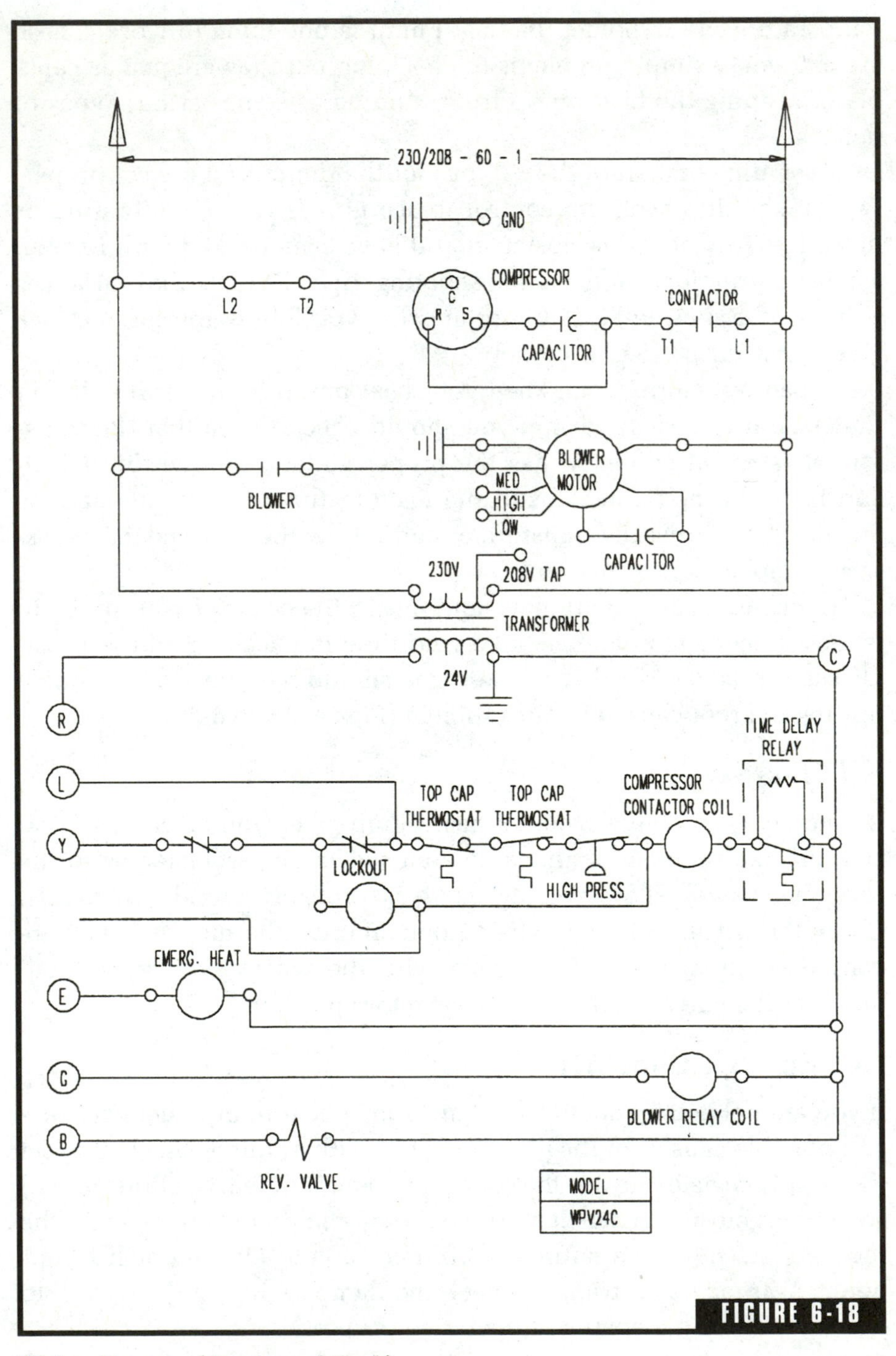

FIGURE 6-18

Wiring diagram. (*Courtesy of Bard.*)

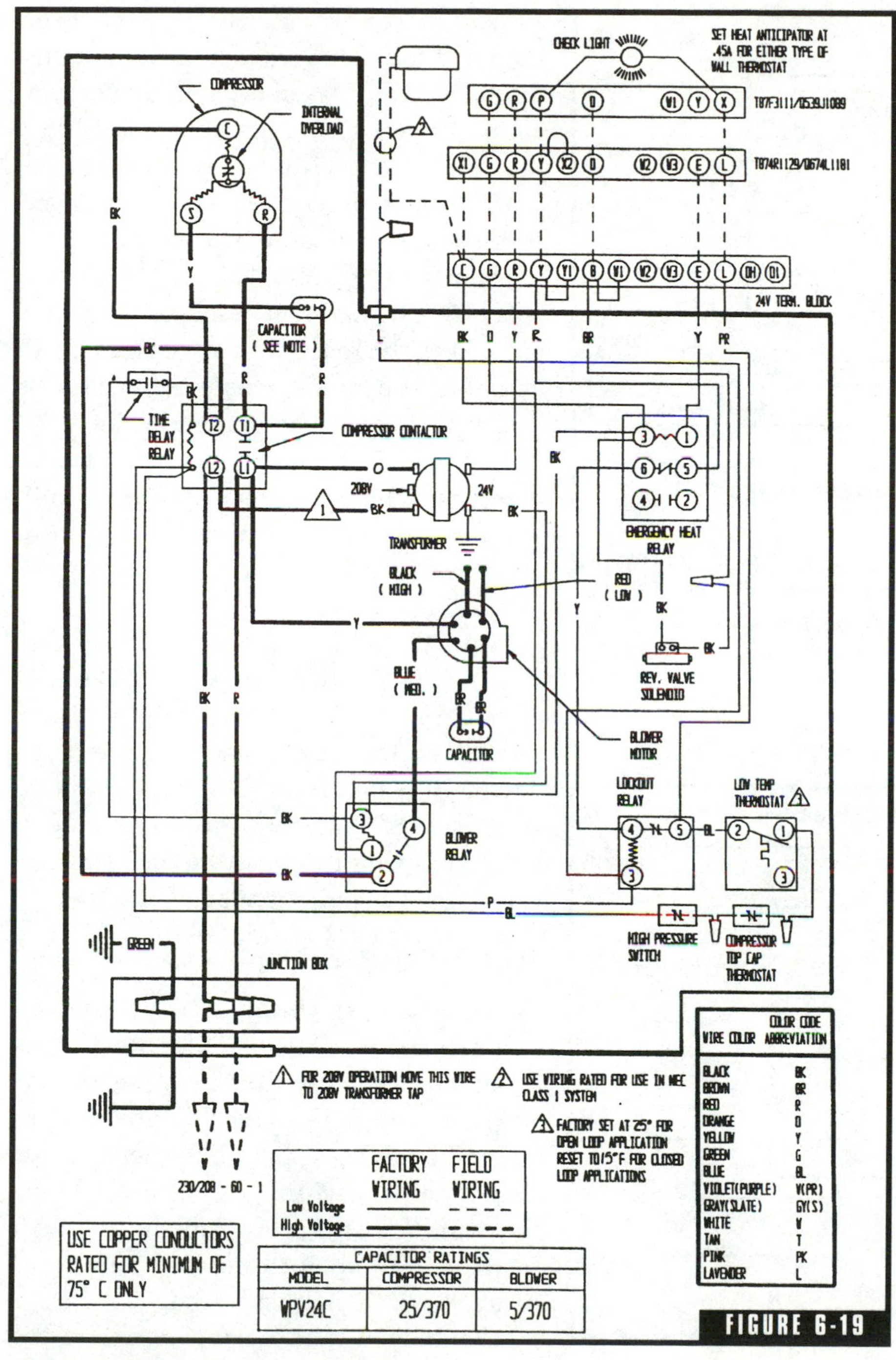

Wiring diagram. (*Courtesy of Bard.*)

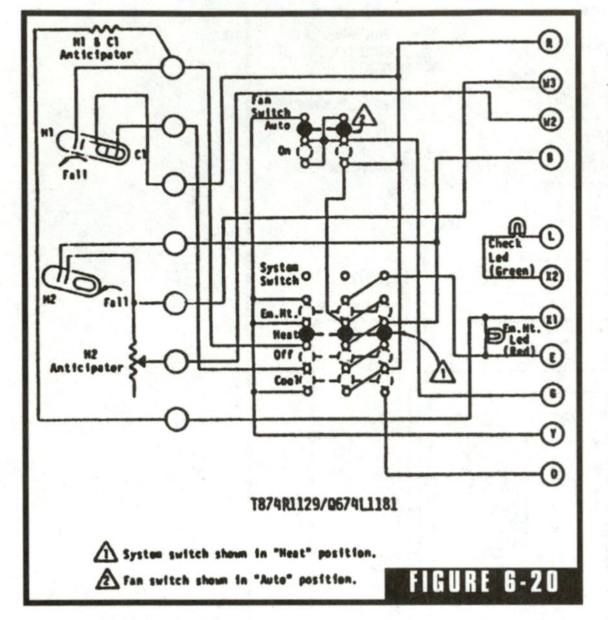

Wiring diagram. (*Courtesy of Bard.*)

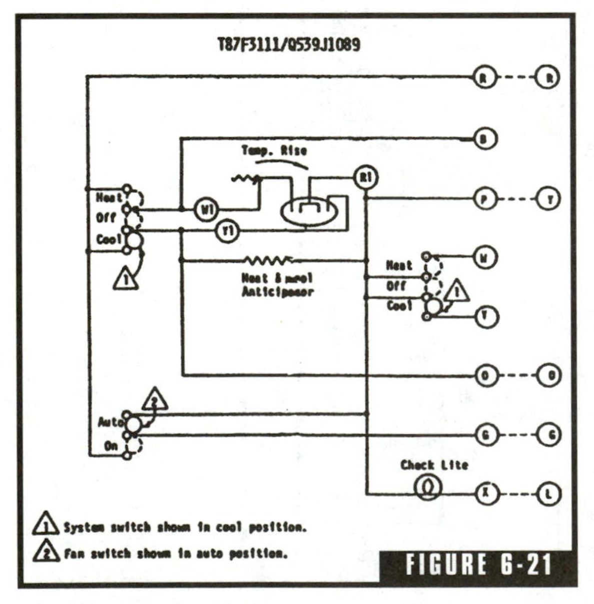

Wiring diagram. (*Courtesy of Bard.*)

Another simple check can be made on the air filter. If the filter is dirty, it could be the cause of your problem. Once you get past the registers, dampers, and filter, the process gets a little more complicated.

The next thing to check is the compressor. A compressor that runs but does not cool the evaporator coil points to either a defective compressor or a lack of refrigerant. Check the refrigerant to make sure it is at a normal operating charge. If necessary, add refrigerant to the system. If the problem still exists, you either have a bad compressor or a blocked water flow in the coil. It is also possible that there is not a sufficient water flow to maintain cool temperatures.

Investigate as much of the piping as you can to see if there are any visible signs of trouble, such as a kinked pipe. If there aren't, you must establish that there is an adequate water flow. You may want to increase the flow rate, depending upon what the specifications from the manufacturer of your equipment suggest.

Won't Cut On

Sometimes heat pumps won't cut on when there is a demand for heat. This problem is usually fairly simple to correct. To keep the troubleshooting as simple and effective as possible, check the air filter first. If it is dirty, replace it and try the unit. If it still refuses to cut on when heat is called for, move your troubleshooting to the thermostat.

Before you start digging around in the wiring of the thermostat, check to see that the controls are set in the proper position. You will feel silly if you tear the thermostat apart only to find that it was not set in the heating mode or that it is set below the existing room temperature.

Once you have determined that the thermostat is set properly, proceed to test it for defects and incorrect wiring. As always, refer to the recommendations made by the manufacturer during your troubleshooting.

The last component to check is the blower motor. It may be overheated or defective. Put the heat pump into the cooling mode and make the blower motor run. If the fan in the blower doesn't move, check for an open overload. Assuming that the motor is not overheated, you will have to replace it.

Ice

What would make the evaporator of your heat pump become covered in ice? A dirty air filter could be the cause of an ice buildup. Check the filter and replace it if necessary. Then, use the heat pump and see if the problem continues. In the meantime, check the air temperature in the home. If room temperature drops below 55°, icing can occur on the evaporator (Fig. 6-28).

The motor may be your problem. Check the motor to see that is set on the speed specified by the manufacturer. It may be necessary to set the motor to a higher speed. Also, check the blower motor to see that it is not overheated or tripping off on overload.

There is one other possible cause for the problem, but it is rare to encounter it. If the water in the system is too cold, the evaporator could ice up. This is highly unlikely, but it has been known to happen.

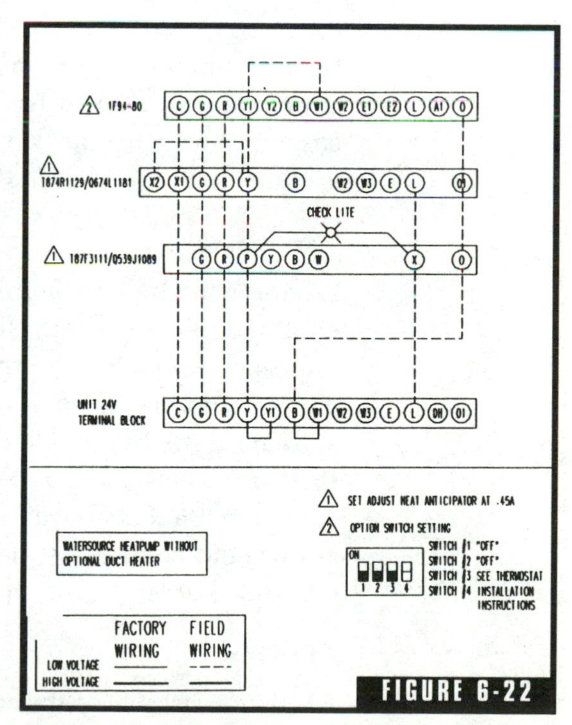

FIGURE 6-22

Wiring diagram. (*Courtesy of Bard.*)

Short Cycles

What should you look for if your heat pump short cycles? The thermostat is the most likely cause of the problem. Assuming

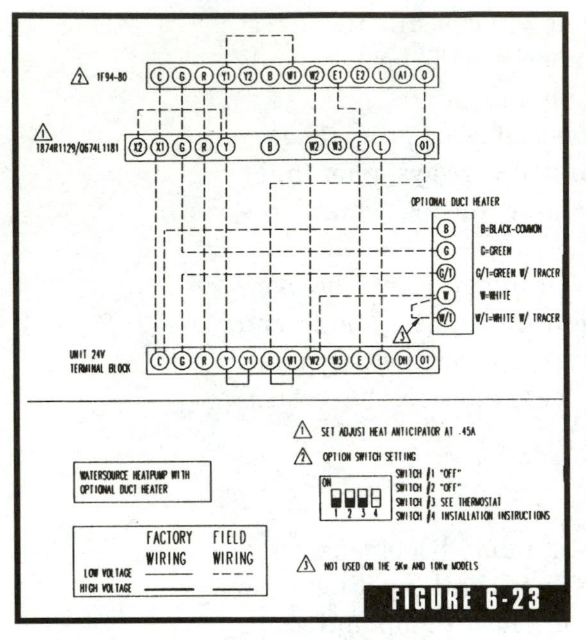

Wiring diagram. (*Courtesy of Bard.*)

that you have installed the thermostat in a recommended location, check to see if the differential is set too close in the control. If necessary, adjust the heat anticipator to balance out the thermostat. Inspect all wiring to locate any loose wires, and pay attention to the control contactor to see if it is working properly.

If none of the early troubleshooting steps reveal the cause of the problem, check out the compressor and refrigerant charge. If the refrigerant charge is low, the compressor may be running hot. It is also possible that the compressor overload is kicking in and causing the short cycles.

Poor Performance

Poor performance, in general, can be caused by a number of potential problems. If a heat pump is not producing enough warm or cool air, there will be many steps to take in finding the cause. Let's start with some of the easier aspects to check out and work our way through the many options that may be at fault for the poor performance.

Thermostat

Where is the indoor thermostat located? Is it on an outside wall? Is it mounted where direct sunlight shines on it through a window? Improperly locating the thermostat can easily affect the performance of a heat pump. If someone put the thermostat in a location where extreme shifts in temperature occur, such as from direct sunlight, you should relocate it. The thermostat should be positioned on an inside wall, about 5 feet above the floor, in a location that is not subjected to hot or cold conditions that are different from the rest of the climate-controlled area.

Air Flow

Air flow is critical to the successful operation of a heat pump. If return grilles are blocked, supply registers are closed, or dampers are malfunctioning, air flow will not be good and neither will the performance of the heat pump. Check for any reason that may cause insufficient air flow, like those just mentioned, and inspect all duct work to see that it is in place and not leaking conditioned air.

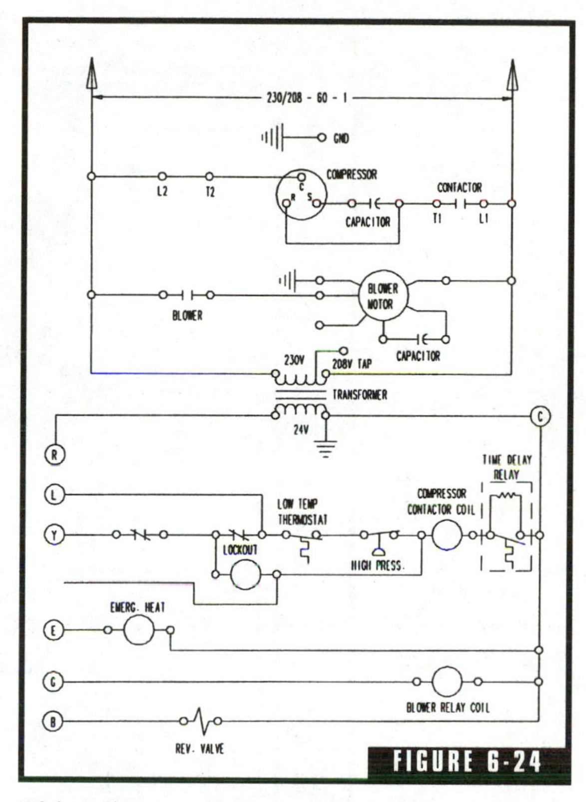

Wiring diagram. (*Courtesy of Bard.*)

Refrigerant

Refrigerant is needed to keep a heat-pump system running effectively. If there is too much or too little refrigerant in the system, problems can occur. Check the refrigerant level in the system to assure that it is at the level recommended by the manufacturer of your heat pump.

Water

Water is a key element in the operation of a water-source heat pump. If you are having problems with the performance of your heat pump, check the water pressure, temperature, and flow. If any of these aspects of the water are not what they should be, corrections will be necessary.

Reversing Valve

The reversing valve in your heat-pump system is another key element in the system. If the reversing valve is defective, it can allow refrigerant to move from the discharge side of the system to the suction side of the compressor.

Contaminants

Contaminants in the refrigerant system can wreak havoc with a heat pump. Moisture is all it takes to mess up the system. Solid particles can also create problems, such as clogged strainers or filters. If you suspect that the refrigerant system has become contaminated, you

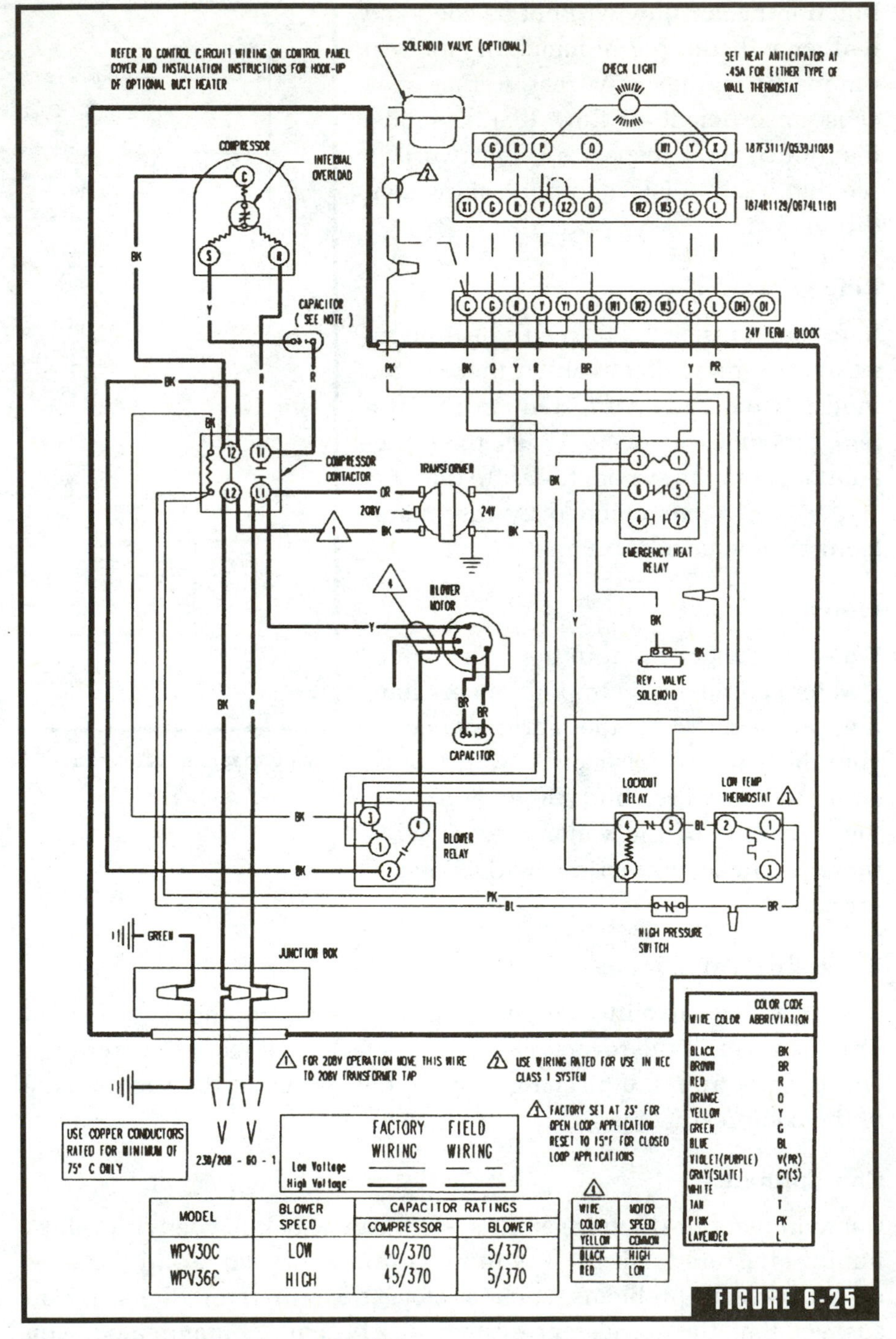

MODEL	BLOWER SPEED	CAPACITOR RATINGS	
		COMPRESSOR	BLOWER
WPV30C	LOW	40/370	5/370
WPV36C	HIGH	45/370	5/370

WIRE COLOR	MOTOR SPEED
YELLOW	COMMON
BLACK	HIGH
RED	LOW

COLOR CODE WIRE COLOR	ABBREVIATION
BLACK	BK
BROWN	BR
RED	R
ORANGE	O
YELLOW	Y
GREEN	G
BLUE	BL
VIOLET(PURPLE)	V(PR)
GRAY(SLATE)	GY(S)
WHITE	W
TAN	T
PINK	PK
LAVENDER	L

Wiring diagram. (*Courtesy of Bard.*)

should check strainers, filters, and capillary tubes for obstructions. It may very well be necessary to dehydrate and evacuate the whole system.

Operating Pressure

Heat pumps have specific operating pressures that they should work within. Check the specifications provided by the manufacturer of your equipment to make sure your unit is running at the proper operating pressure.

Blowers

Blowers for heat pumps are sometimes wired backward. If the capacitor leads are reversed in the motor, the fan can run backward. This, of course, is not right, and reversing the capacitor leads will be necessary.

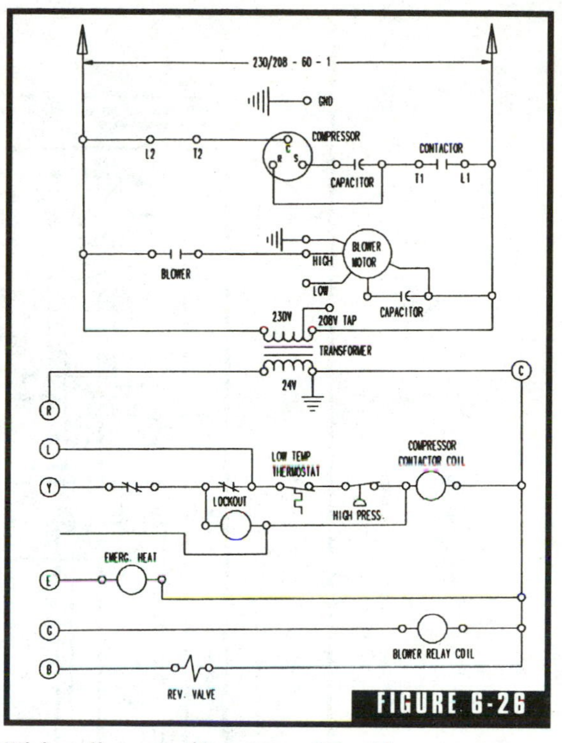

FIGURE 6-26

Wiring diagram. (*Courtesy of Bard.*)

Compressors

Compressors are a major part of a heat pump. If the compressor is bad, the heat pump cannot operate properly. To check your compressor, you will need to identify its discharge and suction pressures. If the discharge pressure is too low and the suction pressure is too high, there is a good chance the compressor is defective and will need to be replaced.

Size

The size of a heat pump is relevant to its ability to heat and cool appropriately. If a system is designed properly, there shouldn't be any problem related to the size of the heat pump. However, if a mistake was made in calculating loads and tonnage, the heat pump may be undersized. Minor mistakes in sizing can sometimes be made up for with insulation or other simple procedures. A major mistake in sizing could result in having to replace the heat pump with a larger unit.

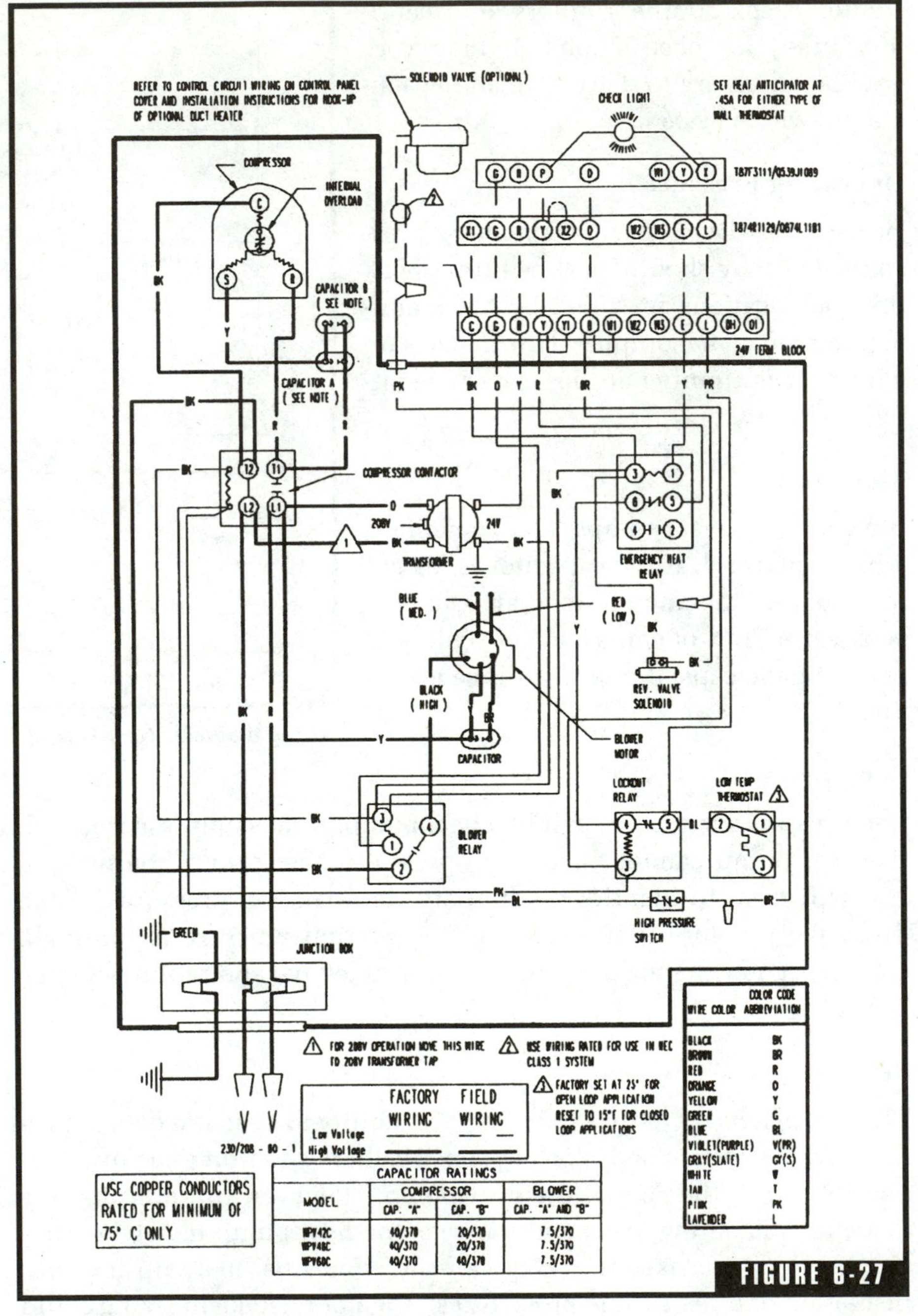

CAPACITOR RATINGS			
MODEL	COMPRESSOR		BLOWER
	CAP. "A"	CAP. "B"	CAP. "A" AND "B"
WPV42C	40/370	20/370	7.5/370
WPV48C	40/370	20/370	7.5/370
WPV60C	40/370	40/370	7.5/370

WIRE COLOR	COLOR CODE ABBREVIATION
BLACK	BK
BROWN	BR
RED	R
ORANGE	O
YELLOW	Y
GREEN	G
BLUE	BL
VIOLET(PURPLE)	V(PR)
GRAY(SLATE)	GY(S)
WHITE	W
TAN	T
PINK	PK
LAVENDER	L

FIGURE 6-27

Wiring diagram. (*Courtesy of Bard.*)

Low Pressure

If your heat pump is off because of the low-pressure cut-out control, the suction pressure may be too low during the cooling cycle. To verify this, you must inspect the blower, the water coil, and the filter.

If the heat pump is off because of the cut-out in a heating cycle, you should check the water flow and temperature in the system. Older heat pumps may have built up a scaling, usually due to hard water, in the water coil. This can contribute to restricted water flow and problems associated with it.

Two other possible causes for this problem are a defective low-pressure switch and a low refrigerant charge. In the case of the switch, you should check to see that it is not stuck in the open position. If the switch will not reset, it should be replaced. It is also possible that the switch is out of calibration.

In the case of a low refrigerant charge, you must add refrigerant to the system. There may be a leak in the system that caused the refrigerant level to drop. Inspect for any leaks and repair them if found. Remember to evacuate the system before you recharge it if leaks are found or the system has been opened.

High Pressure

Just as a heat pump can cut out due to low pressure, it can also cut out due to high pressure. The cause could be a defective high-pressure switch. Under such circumstances, you must test and evaluate the switch, just as was described for the low-pressure switch.

The refrigerant charge could be at fault, but this time it will be a matter of having too much refrigerant, rather than not enough. Check the refrigerant charge and bleed off any excess to bring the charge back within the system's designed level.

Other factors can contribute to a heat pump cutting off on the high-pressure control. If the system is shutting down during a cooling cycle, check the water

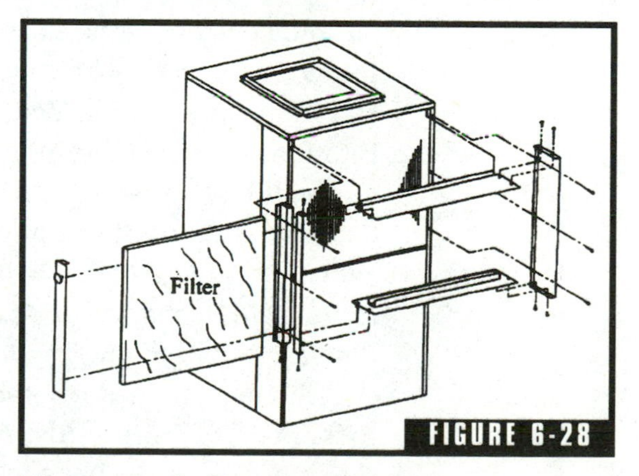

FIGURE 6-28

Filter removal. (*Courtesy of Bard.*)

flow and temperature. In older units, it is wise to inspect the water coil for a scale buildup.

When the unit is cutting off during a heating cycle, you should check air flow and temperature. It is also advisable to inspect the blower and filter. Again, if the system has some age on it, inspecting the coil for scale is a good idea.

Compressor Won't Run

When a compressor on a heat pump won't run, but the blower will, there are several potential problems to look for. Many of the checks involve working around electricity, so proceed cautiously, if at all.

The indoor thermostat for the system is the best place to begin your troubleshooting. Start by checking the thermostat settings. If they are all as they should be, check the wiring and calibration of the thermostat. Hopefully, you will identify the problem and be able to get on with more pleasant activities. If, however, the thermostat is not at fault, go over all of the wiring for the system and make sure the wiring has been done correctly and is not loose or broken. Also, check the voltage for the system to ensure that it is as it should be.

While we are talking about electrical checks, you should test the continuity of the compressor windings. This is done with an ohmmeter, but if you didn't already know that, you probably shouldn't be working around electricity. If your test proves the windings to be open, the compressor will have to be replaced.

Your unit could be on the blink due to the high- or low-pressure cut-out controls. This is easy to test for, if you are accustomed to working with controls. The first step in the process is to turn the thermostat to its off position. Wait 5 minutes and put the thermostat selector into the cooling mode. When this is done and the compressor cuts on, you will know the unit was off due to either the high- or low-pressure cut-offs. Should the compressor fail to start, one of the pressure switches may be bad. This can be tested by jumping the switches on an individual basis. By testing the switches one at a time, you can determine which one of them is faulty.

Moving on through the troubleshooting steps, you can test the lockout relay. To do this, cut the power off. Wait a few minutes and then cut the power back on. If the rely was stuck in an open position, cutting the power off should have closed it. If the relay did not reset itself, it will have to be replaced. While you are experimenting, check the capacitor. If it is defective, replace it.

The last few potential causes for your problem all deal with the compressor. Perhaps the compressor overload is tripped. To check for this, see if the compressor is hot. If it is, the overload will not reset itself until the compressor has cooled down. Assuming that the compressor is not hot, the overload may be defective. If the overload is external, it can be replaced. An internal overload that has gone bad will require the replacement of the compressor.

Sometimes a compressor will seize up. This can be tested by connecting an auxiliary capacitor in parallel with the run capacitor for a few moments. With this done, the compressor may start. If the compressor starts but falls back into the same problem, an auxiliary start kit might solve your problem. When a compressor will not start even with an auxiliary capacitor hookup, the compressor will normally have to be replaced.

The last thing to check is the grounding or burnout of the compressor. Check the internal windings to see if they are grounded to the compressor shell. If they are, expect to replace the compressor. When a compressor has burned out, a filter drier should be installed on the suction line.

We have now completed all the basic troubleshooting options for both air- and water-source heat pumps. When you buy a heat pump, the manufacturer will provide a list of instructions for handling problems with the specific unit. Always read and follow those instructions before attempting any troubleshooting or repair work.

General Maintenance and Repair of Heat Pumps

The maintenance and repair of heat pumps covers a lot of ground. The subject can be complex. A great deal of training and experience is required to produce top-quality service personnel. This, however, doesn't mean that average people can't handle some maintenance and repairs on their own. The last chapter provided a host of troubleshooting information. There's some more here, but most of what you are about to read has to do with the actual maintenance and repair of units. This is a large chapter, so let's get right down to work.

Control Boxes

All heat pumps have control boxes that house a majority of the electrical components used to keep the heat pump working. In split systems, which are used for most residential requirements, the control boxes are located in the outdoor unit. The exact location will vary from manufacturer to manufacturer, but the control box should be somewhere in the outside unit.

What's in the control box? When you look inside a control box, you are likely to find a number of components. For example, capacitors, defrost controls, contactors, start-assist devices, electrical circuits, and electrical components can all be found in control boxes.

Each heat pump will be supplied with a wiring diagram for the equipment inside the control box, and these diagrams should be followed closely.

Capacitors

Capacitors are used in conjunction with motors to provide increased torque for starting the motor. An average control box may contain up to three capacitors. One of them will help in starting the fan motor. Another will be used in conjunction with the motor for the compressor. If there is a third capacitor, it will be working as a start capacitor.

To check a capacitor, the first thing you should do is look at it closely. Is the case of the capacitor swollen? Are any leaks present? If the capacitor is not bulging or leaking, you should proceed to test the component in a more technical manner. To do this, you must first have an ohmmeter.

You should discharge the capacitor with a 15,000 ohm and 2-watt resistor. Then all wires should be removed. Your ohmmeter scale should be set to R x 100. At this time, place the ohmmeter probes across the terminals. What is the reading from your meter?

The meter should start at a low resistance value and move to a measurable resistance before it stops. When the indicator on the ohmmeter doesn't move at all, it indicates that the capacitor is open and must be replaced. An indicator that registers a zero reading and maintains it indicates the capacitor is shorted out and must be replaced.

There is another test that you can try on capacitors. This test is best done with the use of a capacitor tester. When such an instrument is not available, a substitution can be made with an ammeter and a voltmeter. To use an ammeter and a voltmeter, you must set the voltmeter in parallel with the charging circuit. The ammeter jaws should surround one leg of the charging circuit. With this done, energize the circuit long enough to get a meter reading.

Once you have your reading, you must do a little math to compute capacitance. The equation needed will have you multiply 2,650 times the amps. That number will then be divided by volts. The numerical answer to the equation should fall within 10 percent of the rating on the case value. It can be either 10 percent above or below the case rating and still be all right.

Protector Modules

You might find that your heat pump is equipped with a solid-state motor protector module. This module will be working in conjunction with the compressor. Since these modules work with the use of internal sensors that are buried in the motor windings of the compressor, you cannot condemn a module until both the module and the sensors are tested. This test begins with the module itself.

First, you must cut off the electrical power to the unit. Wait until the compressor is cool, and then place a jumper wire between the terminals of the module. Apply electrical power and bring power to the control circuit by setting the unit to call for cool air. If the compressor operates, you may be on the right track in blaming the module for your problem, but you will have to test the sensors to be sure. If the compressor doesn't operate, there may be a problem in the low-voltage electrical circuit, but you can rule out the protector module.

When you want to test the sensors, you must remove the leads from the module. Be careful not to mix up the leads in a way that will make putting them back in their proper place difficult. You might want to wrap some masking tape around the leads and mark them so that you will know exactly where to reinstall them.

You will need an ohmmeter to test the sensors. The sensors are sensitive, and no more than 5 volts should be applied to them. Any additional voltage could cause damage to the sensors. The ohmmeter used to test the sensors must have a range of 0 to 200,000 ohms.

When you test the sensor, the lack of a meter reading will indicate an open circuit. A zero reading on the meter points to a shorted or grounded circuit. Compare any other reading you obtain with the paperwork that came with your equipment. If the reading is within the acceptable range for the sensor, the protector module is bad and must be replaced. When you find that the sensors themselves are bad, the whole compressor will have to be replaced.

If your test indicates a need to replace the compressor, I recommend that you call in another professional to render a second opinion. It is possible that you made a mistake in your testing. The purchase and installation of a new compressor that is not needed is an expensive on-the-job lesson to learn.

Variable Resistors

Variable resistors are found in some heat pumps. These devices take the place of a relay and capacitor combination. The purpose of the variable resistor is to give a little added punch of voltage when it is needed to get a motor started. If this resistor is bad, the motor may not be able to start.

Testing a variable resistor is not difficult. You will need an ohmmeter to conduct the test. Cut off the electrical power to the unit and disconnect the wires from the starting device. Place the probes of the ohmmeter over the terminals and take a reading. If you don't get any reading, the device is open and must be replaced. Any good meter reading indicates that the circuit is complete and need not be replaced.

Compressors

Most compressors are hermetically sealed. This means that work on them in the field is very limited in scope. Due to their design and construction, compressors are one part of a heat pump that just will not allow a lot of on-site tinkering. Unfortunately, the fact that little can be done with a compressor in the field means that compressors usually must be replaced rather than repaired. There are, however, some onsite tests you can run, if you are equipped to do so, that will indicate if a compressor must be replaced.

An overcharged refrigerant system can damage a compressor. If the refrigerant is charged to a point where it returns to the compressor, problems are going to occur. One of the telltale signs of such a condition is a noisy compressor. There is another way to tell if a compressor is bad or if it is the valves working with the compressor that need attention. This test, however, requires the use of a gauge manifold set. Since this is a tool you are not likely to have, it will probably be best to call in a professional to run the test for you. However, since some readers will want to know how the test is done, let's look at the steps involved.

The valves that work in connection with compressors to provide seals between the low- and high-pressure sides of the system are very important. If these valves are damaged or malfunctioning, the compressor is in trouble. Before a troubleshooter can know if a compressor problem is in the compressor or in the valves, a test must be conducted with the use of a gauge manifold set.

The test will check to see if suction pressure will pull down and if the discharge will build up. Assuming that the system has a proper refrigerant charge, the test will reveal the condition of the compressor. If the suction pressure will not pull down and the discharge will not build up, the compressor is most likely defective. You should watch the gauge pressures when the system is shut down. If they equalize quickly, the problem may be in the valves, rather than with the compressor.

What would make a compressor motor fail? If it becomes grounded from wires touching the motor stator or crankcase, it could fail. When a winding becomes open, the compressor can fail. And, if windings short together, failure is likely.

Off on Overload

Some compressors are equipped with an internal line-break overload that can force the compressor to go off on overload. Not all compressors are equipped with these. Ones that are should be tested before you scrap the compressor. Check the paperwork from the heat pump manufacturer to determine if this is a check you should make.

The line-break overload reads current and winding temperature. Under extreme conditions, the device will shut down the compressor. If you have an overload protector, you will need an ohmmeter to check it out.

To test the overload device, you should first cut off all power to the unit. If necessary, wait until the compressor has cooled before proceeding with the test. When the compressor is cool and the power is off, remove the wires from the compressor terminal. Your ohmmeter should be set at R x 1.

When you look in the terminal box, you will see three terminals. One will be marked with the letter "C." This indicates the common terminal. Another terminal will be marked with the letter "S." This terminal is the start terminal. The last terminal will be the run terminal, and it will be marked with the letter "R."

Start your test with one probe of the ohmmeter on the common terminal and the other probe on the run terminal. Note the meter reading. Now, move the probe from the run terminal to the start terminal, and note the meter reading. A reading of low resistance means that you have ruled out the overload as being the problem. A reading of infinite resistance indicates that the overload may be open. To confirm this, position the probes of your meter on the run and start terminals. A reading of low resistance in this position verifies that the overload is open.

Check the Windings

You can also check the windings of the motor. Turn your ohmmeter to the R x 100 scale and then check to see if a winding is grounding out. To do this, place one of the ohmmeter probes on the compressor's discharge line. Move the other probe across the common, run, and start terminals. Keep a watchful eye on the meter as you touch the probe to the various terminals, and hope that you don't get a reading. If you do get a reading from any of the terminals, you've got a bad compressor.

There is another test that you should perform if the compressor has passed the previous tests. Set your ohmmeter to the R x 1 scale and connect one of the probes to the common terminal. Touch the other probe to the run terminal and notice if there is any reading. Then, move the probe from the run terminal to the start terminal and note the meter reading.

If you don't get any meter reading from the test, one of the windings is open. A zero reading or no reading indicates a short. Hopefully, you will get a low resistance reading, proving that the windings are okay. Check the specifications provided with your equipment to establish what readings you should obtain from this test.

Before you start the compressor, connect an ammeter to the common terminal. When you turn on the power, take note of the meter reading. Compare the reading with the Locked Rotor Amperage rating on the equipment information plate. The ammeter should draw at least the Locked Rotor Amperage amount when the motor starts, but then it should settle back down to a normal level. If the ammeter remains at a high reading, there is a problem with the compressor. It is also important to note that the hoses for the test gear must be connected to the proper refrigerant lines and that air from the hoses must be purged to prevent contamination of the refrigerant system.

Acid in a Compressor

Acid in a compressor is destructive, and if a compressor burns out, there is a good chance that acid will mix with the oil in the compressor. If the acid is left alone, it will gradually eat away at the compressor, causing long-term damage. To avoid this, an in-line drier must be installed when a compressor has gone through a burn-out. A burn-out can sometimes be smelled when the system is opened, but the only way to tell for sure if acid has gotten into the oil is to test the oil.

The first step will be the removal of the bad compressor. Then, an R-22 cylinder should be attached to the suction line and used to purge the system of as much oil and solid contamination as possible. Once this is done, a new compressor can be put in place.

A clean-up filter/drier should then be installed on the common suction line. If an existing filter is in place, it should be removed and discarded. A filter/drier that utilizes a removable, replaceable core is preferable, but any clean-up filter/drier can be used. The filter/drier must be rated for one size larger than the tonnage of the unit it is serving.

The next step is to connect the suction and liquid lines to the compressor and test the system for leaks. Then the entire system must be evacuated. When this is complete, the system can be charged with refrigerant.

Turn the unit on, but don't change the mode of operation. In other words, if the system went down in the heating mode, start it in the heating mode. This will minimize the risk of spreading contaminants. Run the system for a few hours to make sure all the contaminants will be trapped. When you feel sure all of the contaminants are captured, remove the drier from the unit and recharge the system.

Evacuating a Sealed System

Evacuating a sealed system will be necessary if the system contains moisture. Moisture can invade the system anytime the system is opened for repairs. To do this job professionally, a deep vacuum pump and a thermister vacuum gauge are needed. A vacuum of 300 micrometers is pulled on the system and the pump is valved off. The system vacuum should not rise at a rate of more than 100 micrometers per hour. Sometimes the rate of micrometer rise is greater than 100 micrometers per hour. This calls for more evacuation of the system until the systems vacuum will not rise faster than 100 micrometers per hour.

Sometimes it is necessary to install a filter/drier on the common suction line of the system. Any existing filter/drier that was present when the system failed should be removed and discarded so that a new one can be installed. A filter/drier with a replaceable core is the best type to install.

Just as we discussed earlier, the system should be operated for a few hours in the same mode that it was in when the failure occurred. The filter/drier should be checked after a couple of hours for a drop in pressure. After complete removal of moisture and contaminants has been accomplished, the system can be recharged with refrigerant and put back into normal operation.

There is one other thing that should be noted. If you are evacuating a sealed system when the temperature outside is below 40°, you must keep the outdoor unit warm. This can be done by placing a makeshift cover, like a tarp, over the outside unit and keeping a portable heater running near the compressor. Be careful, however, not to allow the portable heater to catch the tarp, or whatever cover is used, on fire.

Reversing Valves

Reversing valves are often considered the heart of a heat-pump system. Without these valves, the heat pump would be unable to provide both warm and cool air. These valves are extremely important to the satisfactory operation of a heat pump.

Reversing valves control the flow of refrigerant in a heat-pump system. They are electrically controlled and function on a pressure differential. Their sensitivity to pressure makes them rely on the system pressure to operate effectively. If the system pressure is out of kilter, the reversing valve will act up. This means that before you can pronounce a reversing valve to be defective, you must check not only the valve, but the system pressure as well. In fact, you should check the system pressure prior to testing the reversing valve. Refer to the manufacturer's specifications on operating pressures to determine if the system is running within the constraints of proper pressure loads.

Once you have established that the operating pressure of your system is at the proper level, you may begin to inspect the reversing valve. Start with a visual inspection of the valve body. Damage to the body of the valve can impede the slide assembly's movement on the inside of the valve. If you don't see any obvious problems, you will have to take your troubleshooting a step further.

Before you proceed with the reversing valve, you should first disconnect the fan motor of the condenser and remove the fan blade. When this is done, you must provide power to the reversing valve. Its solenoid coil must be energized to work. What you are about to do is easier for someone who has worked with reversing valves in the past, because the test requires a certain feel for what the valve should do. Without field experience, you may have difficulty in judging the reactions of the test.

In order to test the solenoid coil and needle valve of the reversing valve, you must remove the lock nut that holds the coil in place. Remember that electrical power for the solenoid must be on during this test. After the lock nut is gone, slowly slide the coil off its stem. Do you feel any resistance in the coil movement? You should, if the coil is in operating condition. As you continue to slide the coil along the stem, you should hear a clicking sound. This is made by the plunger needle, and the sound indicates that the needle is responding to the magnetic field of the solenoid coil.

Assuming that you have felt resistance and heard a clicking sound, slowly slide the coil back into its intended place on the stem. Did you hear another clinking sound? If you did, it meant that the plunger was triggered by the coil, and this is good. If you felt no resistance or didn't hear both of the clicking sounds, there is an electrical problem to be checked out.

There is a wide array of problems possible with a reversing valve. So far, you've only touched the tip of the iceberg in determining if the valve is bad or not. To go further in your troubleshooting, you should have a chart that gives instructions for doing what is known as the "touch test." Without such a chart, you will have to hunt and peck to find out if the valve is bad.

When you look at a reversing valve, you will notice six tubes. One of these tubes is a discharge tube from the compressor. A second tube is a suction tube going to the compressor. There is also a tube that goes to the inside coil. A fourth coil leads to the outside coil. Then there are two other tubes, one for the left pilot capillary tube and another for the right pilot capillary tube. With the help of a chart provided by the manufacturer, you can perform a troubleshooting sequence by taking note of the temperature of each tube. The temperature is not measured with a thermometer. All that you have to do is touch the tubes and determine if they are hot, cool, or the same temperature as the body of the reversing valve.

During normal operation in the cooling mode, valve number 1, the discharge tube, will be hot. Valve number 2, the suction tube, will be cool. The number 3 tube, the tube to the inside coil, will be cool. Tube number 4 will be hot. Tube number 5, the left pilot capillary tube, will feel the same as the reversing valve body. The right pilot capillary tube, tube number 6, will also be the same temperature as the valve body.

During normal operation in the heating mode, valve number 1, the discharge tube, will be hot. Valve number 2, the suction tube, will be cool. The number 3 tube, the tube to the inside coil, will be hot. Tube number 4 will be cool. Tube number 5, the left pilot capillary tube, will feel the same as the reversing valve body. The right pilot capillary tube, tube number 6, will also be the same temperature as the valve body.

Now that you know how the various tubes should feel in both the heating and cooling mode, you can assess the condition of your reversing valve. If the temperatures of the tubes are not as they are described above, you have a problem with the valve. By noting the differences in the tube temperatures, you can rule out many potential problems and concentrate on the ones that are able to occur under the given circumstances.

If you have a chart from the manufacturer of your heat pump, you will notice that different forms of malfunctions are listed on it. In conjunction with the malfunctions are temperatures for the six tubes. By touching and judging the temperature of each tube, you can move down your chart and eliminate some of the possible problems. Other problems will have to investigated to greater lengths. To expand on this, let's go through some of the troubleshooting and repair procedures you might face with a reversing valve.

Will Not Shift

If the reversing valve will not shift from cool to heat, there are six likely causes of the problem. You could spend time checking all six of these potential causes, or you can use the knowledge you are gaining here to minimize your work and maximize your efforts.

One of the first things to check when the valve will not shift from cool to heat is the coil. You learned how to do that just a few minutes ago. If the coil is bad, it must be replaced.

Checking the refrigerant charge is the next logical step. If the refrigerant charge is low, it must be recharged. You should also inspect for any leaks that may have caused the refrigerant charge to drop. A pressure differential in the charge could be too high, so check the pressure ratings as well. Once you have done this, you are ready to test the temperatures of the tubes. It will be helpful to have a note pad to write your findings in. Remembering the temperatures of each tube is critical in cutting down on the time it takes to troubleshoot the valve.

Let's start with worst possible outcome of your test. As we look at test results, you will notice the abbreviation of TVB. This stands for the same temperature as the valve body. Assume that you have just touched all of the tubes and came up with the following findings:

* Tube 1= warm
* Tube 2= cool
* Tube 3= cool
* Tube 4= warm
* Tube 5= TVB
* Tube 6= warm

If the tubes of your reversing valve have the temperatures indicated above, your compressor is probably bad. See how easy this is? Now, suppose you had slightly different temperature recordings. Let's say that the temperatures were as follows:

* Tube 1= hot
* Tube 2= cool
* Tube 3= cool
* Tube 4= hot
* Tube 5= hot
* Tube 6= hot

These temperature readings would indicate that both parts of the pilot are open. The back seat port did not close for some reason; it may be clogged. With this being the case, you should raise the head pressure and operate the solenoid in an attempt to clear the obstruction. If you are still unable to make the reversing valve shift, it should be replaced.

Assume you tested the temperature of the tubes and came up with the following results:

* Tube 1= hot
* Tube 2= cool
* Tube 3= cool
* Tube 4= hot
* Tube 5= TVB
* Tube 6= TVB

These readings would indicate clogged pilot tubes. To correct this problem, you would raise the head pressure and operate the solenoid to remove the obstructions. If this action was unsuccessful, you would have to replace the reversing valve. We'll talk about what is involved with replacing a reversing valve a little later on in the chapter.

Up to this point, each set of temperature readings has indicated a single problem. The troubleshooting process is not always that easy. Sometimes, there is more than one possibility for a problem associated with a given temperature reading. Let me give you an example of this.

You've just felt the six tubes associated with your reversing valve. The results are as follows:

* Tube 1= hot
* Tube 2= cool
* Tube 3= cool
* Tube 4= hot
* Tube 5= TVB
* Tube 6= hot

These readings indicate that you may have one of two problems. It will be up to you to determine which problem you have. Your readings prove that the pilot valve is all right, but there may be dirt in one of the bleeder holes. The other possibility is that a leak exists in the piston cup. The leak in the piston cup is easier to check for than dirt in the pilot is, so let's start with the piston cup.

Cut the heat pump off, and wait until the pressures equalize. Restart the unit, with the solenoid energized. If the valve shifts, try it again with the compressor running. If the valve won't shift with the compressor running, you must replace the reversing valve.

Assuming that the piston cup is not your problem, you must address the issue of dirt in the pilot. Kill the power to the solenoid, and raise the head pressure. Energize the solenoid in an attempt to break the dirt loose. If this doesn't work, you must remove the reversing valve and wash it out. Once the valve is clean, test it with air. If it won't work with air, the valve must be replaced. You should install a drier on the suction line and mount the valve horizontally.

Fails to Shift Completely

You might have a reversing valve that will start to shift but fails to shift completely. If you do, your temperature readings will look something like this:

* Tube 1= hot * Tube 2= warm * Tube 3= warm
* Tube 4= hot * Tube 5= TVB * Tube 6= hot

This type of reading means that either the body of the reversing valve is damaged, or there is not enough pressure differential at the start of the stroke or not enough flow to maintain the pressure differential. Obviously, you should inspect the valve body first. If it shows damage, replace it.

If the body of the valve is intact, you should check the operating pressure of the system. Also check the refrigerant charge. Raise the head pressure and see if the valve will shift. If it won't, install a valve with smaller ports.

Leak in Heating

If you feel you have a leak in heating, there are two possible causes for the problem. Testing the temperature of the tubes will guide you to the problem that deserves attention. In the temperature readings you are about to be given, I've used another abbreviation. The letters WVB stand for a temperature that is warmer than the valve body. Here are the readings:

* Tube 1= hot * Tube 2= cool * Tube 3= hot
* Tube 4= cool * Tube 5= TVB * Tube 6= WVB

These readings indicate that the piston needle on the end of the slide is leaking. Try operating the valve several times to see if the leak stops, stays the same, or grows worse. If it is leaking badly, you must replace the valve.

The other possible cause of this broken valve would be having both the pilot needle and the piston needle leaking. The temperature readings would, however, be different if this were the case. Those readings would be as follows:

* Tube 1= hot * Tube 2= cool * Tube 3= hot
* Tube 4= cool * Tube 5= WVB * Tube 6= WVB

If this were the case, you should operate the valve several times and check on the condition of the leaks. If they are bad, the valve would have to be replaced.

Reversal Not Completed

If reversal is not completed after the reversing valve has started to shift, there are three possible causes for the problem. Again, temperature readings from the tubes will allow you to narrow the field of possibilities considerably. Let's look at the three different readings you might get under these circumstances and see what they will tell you. Here is the first set of readings:

* Tube 1= hot
* Tube 2= warm
* Tube 3= warm
* Tube 4= hot
* Tube 5= hot
* Tube 6= hot

The readings above indicate that both ports of the pilot are open. To correct this, raise the system's head pressure and operate the solenoid. If the valve still will not shift, replace it. There is another combination of temperature readings that would indicate this same problem and repair procedure. Those readings are:

* Tube 1= hot
* Tube 2= hot
* Tube 3= hot
* Tube 4= hot
* Tube 5= hot
* Tube 6= hot

The remaining possibilities are that the body of the valve is damaged or the compressor is not pumping enough to make the valve operate properly. If either of these are the case, your temperature readings should look like this:

* Tube 1= hot
* Tube 2= hot
* Tube 3= hot
* Tube 4= hot
* Tube 5= TVB
* Tube 6= hot

If the valve body is damaged, it must be replaced. When the valve is hanging up in mid-stroke, due to the compressor not pumping enough, you can try raising the head pressure to see if the valve will work. If it doesn't, replace it with a valve that has smaller ports.

From Heat to Cool

The last type of malfunction you may have to work with is when the reversing valve will not shift from heat to cool. There are six reasons why this might happen, but temperature readings from the valve tubes will cut the list down to size quickly.

Let's say that your temperature readings look like this:

* Tube 1= hot * Tube 2= cool * Tube 3= hot
* Tube 4= cool * Tube 5= TVB * Tube 6= TVB

The above readings indicate that the pressure differential is too high or the pilot tube is obstructed. Start by raising the head pressure and operating the solenoid. This may free the obstruction from the pilot tube. If there is still no shift, and if the pressure differential is not too high, the valve must be replaced.

You can check to see if the pressure differential is a problem by shutting down the unit. The valve should reverse during the equalization period. If it doesn't, it's time to install a new reversing valve.

Now, let's see what might be wrong with the valve if the temperature readings are as follows:

* Tube 1= hot * Tube 2= cool * Tube 3= hot
* Tube 4= cool * Tube 5= hot * Tube 6= TVB

These readings point to dirt in a bleeder hole. Kill the power to the solenoid and raise the head pressure. Energize the solenoid in an attempt to break the dirt loose. If this doesn't work, you must remove the reversing valve and wash it out. Once the valve is clean, test it with air. If it won't work with air, the valve must be replaced. You should install a drier on the suction line and mount the valve horizontally.

There is also a possibility that these same readings would indicate a leak in the piston cup. Cut the heat pump off, and wait until the pressures equalize. Restart the unit, with the solenoid energized. If the valve shifts, try it again with the compressor running. If the valve won't shift with the compressor running, you must replace the reversing valve.

What would the following temperature readings indicate?

* Tube 1= hot * Tube 2= cool * Tube 3= hot
* Tube 4= cool * Tube 5= hot * Tube 6= hot

These readings mean that you have a defective pilot and must replace the reversing valve. There is one last set of readings that you might get when the reversing valve will not shift from heat to cool. Those readings are going to indicate that the compressor is bad and must be replaced. The readings are as follows:

* Tube 1= warm
* Tube 2= cool
* Tube 3= warm
* Tube 4= cool
* Tube 5= warm
* Tube 6= TVB

Replacing a Reversing Valve

Replacing a reversing valve is not very difficult if you are able to evacuate and recharge the refrigerant system. The first step toward replacing a reversing valve is the removal of the solenoid coil. Once that is out of the way, you must break the solder joints that hold the valve in place. To loosen these joints, you are going to have to heat the copper tubing with a torch. But, you must not allow the body of the valve to get too hot. Even though you are going to trash the existing valve, too much heat on its body could introduce contaminants into the system.

To keep the valve body from overheating, wrap it in wet rags. The water in the rags will keep the body cool enough that no damage should be caused to the rest of the system. It may be necessary to pour additional water on the rags as you heat the pipes, so have some handy.

For obvious reasons, you won't want to pull the valve out of place with your bare hands. The copper tubing will be hot enough to brand you if it touches your skin. Use either heavy gloves or a pair of pliers to free the valve from the piping. Once the solder joints are loose, the valve can be removed. You must keep the joints hot enough to maintain liquidity of the solder, and you may have to twist the valve somewhat to get it free. If you've never done this before, you will probably have to make a few attempts before you are able to remove the valve.

When the old valve is out of the way, you should heat the ends of the copper tubing and wipe of all excess solder with a cloth. Remember that the tubing will be very hot, so wear thick gloves during this procedure. Once the ends of the tubing cool down, they should be sanded with a light-grit sandpaper. Now you are ready to install the new valve.

Just as the old valve wasn't allowed to overheat, the new valve must also be protected. Once the new valve has been placed on the tubing, wrap it with wet rags. Remember to protect the system and the valve tubes from any moisture that may attempt to get in.

The valve should be installed horizontally. It is also advisable to handle the valve carefully, to avoid any damage to the body. Check the valve to make sure it falls within an acceptable rating for your system, and solder it into place. Test your new installation for leaks and then reinstall the solenoid coil.

Now it is time to add refrigerant to the system. Refer to the manufacturer's recommendations to determine the amount of refrigerant to add. With this done, you are ready to test the heat pump. Don't operate the reversing valve until you know the system is running properly. If the system is not charged properly, the reversing valve cannot work the way it should.

When you are sure the system is working correctly, test the reversing valve. Put it through its cycles several times to check its operation. Assuming that it works well after a dozen or so cycles, you can pat yourself on the back for a job well done.

Superheat Charging

When a professional technician is asked to charge a heat pump system with refrigerant, the method most often used is called the superheat-charging method. This manner of charging a system is only done with the system in a cooling mode. As long as the system being charged is of a capillary-fed type, the superheat charging is very accurate.

To charge a system with refrigerant, you will need a few tools and the charts that were packed with your heat pump. To complete this chore, a thermometer is attached to the suction line where it enters the condensing unit. A suction gauge is attached to the suction line port of the condensing unit. The heat pump is then started and allowed to stabilize.

Once the system has stabilized, readings for the pressure in the suction line will be taken and recorded. The temperature of the suction line will also be tested and recorded. The last piece of the puzzle is the temperature of the outdoor air. With these three numbers, you can refer to the charts provided with your heat pump to see where you stand with your refrigerant charge.

You can plot your readings on the suction-line-temperature chart provided with your heat pump. As an example, the outside air temperature will be marked on the left side of the chart. The suction pressure will be labeled on the top of the chart. As you move from left to right, horizontally, to connect these two numbers with a corresponding number on the chart, you will see what the suction line temperature should be. If the temperature in the suction line of your system matches the recommendation on the chart, you're all set.

Normally, if the temperature reading for your system is lower than what is recommended, you will have to inspect the system to see if it is overcharged. If so, you must purge the refrigerant until the proper temperature is arrived at. A temperature in your system that is higher than what the chart recommends indicates the system is low on refrigerant and must be charged.

In addition to the chart that allows you to confirm the desired temperature of your suction line, there will also be a chart that provides recommendations for heating performance. You will have to use these charts to check the suction pressure of your system. To do this, you must take a temperature reading of both the outdoor and the indoor air. With these two temperatures known, you can use the performance charts to establish desired discharge pressures and suction pressures. The paperwork that comes with your heat pump will explain how to use the charts.

Expansion Valves

Many heat pump manufacturers provide thermostatic expansion valves to be used with their equipment. The troubleshooting and repair procedures involved with these valves are extensive. The remainder of this chapter will explain to you how to diagnose and correct problems with expansion valves.

If you decide to work with expansion valves, you should know how to measure and adjust the operating superheat. People who do not possess this knowledge and ability will be severely hampered in their troubleshooting and repair work with expansion valves.

Measuring and adjusting the operating superheat of a heat pump requires a temperature-pressure chart (this should be available from the manufacturer of your heat pump), a thermometer, and a suction gauge. The first step in the process is determining the suction pressure. This is done by attaching a gauge to the system and taking a reading. Once you have an accurate pressure reading, you can determine the saturation temperature by referring to the temperature-pressure chart supplied by the manufacturer of the heat pump.

The next step in measuring operating superheat involves taking the temperature of the suction gas. To do this, clean part of the suction line near the remote bulb location. Professionals would use a potentiometer to take the temperature reading, but it is possible to get a decent reading with a common thermometer. Tape the thermometer to the cleaned section of the suction line. It is important to protect the thermometer from ambient air temperature. You want a reading of the gas temperature, not the open-air temperature.

Once you have gotten an accurate temperature of the suction gas temperature, you should subtract the saturation temperature that you determined from the temperature-pressure chart from the temperature of the suction gas. This will give you the superheat reading of the suction gas.

Assuming that your unit is equipped with an external adjustment valve, adjusting the operating superheat is no problem. If you remove the cap from the bottom of the valve, you should see an adjusting stem. This stem can be turned to increase or decrease the amount of refrigerant flowing through the valve. Turning the stem clockwise will decrease the flow. A counterclockwise rotation of the stem will increase the amount of refrigerant allowed to go by the valve. Increasing the flow of refrigerant will lower the superheat, and decreasing the flow of refrigerant will increase the superheat.

Not all expansion valves have external adjustments. If the adjustments on your valve are internal, the act of altering the superheat will be a bit more complicated. You will have to pump your unit down and disconnect the outlet line from the valve. Once you've done this, you can adjust the stem in the same way described in the last paragraph. Now that you know the basics for controlling operating superheat, let's look at some of the potential problems you may encounter that are involved with expansion valves.

You may have an occasion when you have low suction pressure and high superheat. There are many reasons why this may happen. Let's say, for example, that the problem is being caused by the inlet pressure being too low from an excessive vertical lift, undersized liquid line, or excessive low condensing temperature. To correct this problem, you would have to increase the head pressure. If the problem is that the liquid line is too small, it will have to be replaced with a larger-diameter line. This, however, is only one of many possible causes and solutions for the problem; let's look at the rest of them.

Valve Restriction

You might have a valve that is restricted by pressure drop through the evaporator. If this is the case, you should change to an expansion valve that has an external equalizer. But, let's say that your valve already has an external equalizer, and it is what's causing the problem. The line for the equalizer may be restricted. This would mean you'd have to replace the equalizer. It is also possible that the valve is equipped with an improper equalizer. Under these conditions, you would have to replace the equalizer with one that was designed for your needs.

Gas in the Line

You may get gas in your liquid line. This can happen due to a pressure drop in the line or an insufficient refrigerant charge. If this happens, you should check the refrigerant charge and add refrigerant if necessary. Make sure the size of the liquid line is what the manufacturer's specifications call for. You may have to clean the strainers and replace the filter drier. It is also a good idea to increase the head pressure or decrease the temperature to guarantee you have solid liquid refrigerant at the valve inlet. It may not be necessary to complete all of these steps, but one (or all) of them should solve your problem.

Clogged Filter

A clogged filter screen can give you low suction pressure and high superheat. To solve this problem, all you have to do is clean the filter.

Superheat Is Too High

If the operating superheat is too high, you could experience low suction pressure and high superheat. If this is the case, simply adjust the superheat as you were told earlier.

Oil

If you use the wrong type of oil, your heat pump may suffer from the symptoms of low suction pressure and high superheat. To correct this, all you have to do is purge the system and install the proper type of oil.

Too Small

It is possible that the orifice of the valve is too small. Replacing the valve with one that has a larger opening will correct this problem.

Plugged Orifice

Sometimes a plugged orifice will be encountered. This is usually a matter of wax and oil clogging the opening. The wax and oil indicate that the wrong type of oil was used in the system. Purge the system and replace the oil content with a proper oil. You may also want to install a filter/drier to prevent moisture and dirt from blocking the orifice of the valve.

Power Assembly

A faulty power assembly can inhibit the proper operation of your heat pump. When the power assembly goes on the blink, it is best to replace it.

Remote Bulb

The gas-charged remote bulb of a valve may lose control due to the tubing for the bulb being colder than the bulb. The best solution to this problem is the replacement of the existing bulb with a "W" cross-ambient power assembly.

Frost

If you have frost developing on your line, it is generally an indication of a restriction in the line. Frost or a temperature reading that is below normal indicates a good place to look for obstructions in the line.

Other Probable Causes

There are other probable causes of a low suction pressure and a high superheat. For example, an obstructed line could be the culprit. If the liquid line is too small, that could be the source of your problem. It is also conceivable that the suction line is too small. If the wrong type of oil was inducted into the system, it could be blocking the refrigerant flow. The solenoid valve may be too small or not operating properly. Valves that are too small or that are not opening fully could be at fault. There are, to be sure, many possible causes for a low suction pressure and a high superheat. Now, suppose you have low suction pressure and low superheat. How does that change your troubleshooting and repair procedures? Well, let's delve into it and see.

Compressor

If the compressor is oversized or running too fast, it could cause your system to have low suction pressure and low superheat. To correct this, you might have to install a different pulley on the compressor, one that would make it run at the proper speed. Another option might be to install a compressor capacity control.

Oil in the Evaporator

An excessive buildup of oil in the evaporator could also create low suction pressure and low superheat. This is usually a result of the suction piping not being installed properly, but it may be necessary to install an oil separator. Check the piping to see that it will return oil properly. If it will not, change the piping arrangement so that it will. Should the piping already be installed properly, defer to the installation of an oil separator.

Ice

An ice buildup will generally mean that the evaporator is too small. The solution to this problem is simple; replace the evaporator with one that is of the proper size.

Uneven Loading

Uneven loading of the evaporator, or inadequate loading, can be the reason why you are fighting with a low suction pressure and a low superheat. Inadequate distribution of air or brine flow is the cause in this case. To correct the problem, you must balance the evaporator's load distribution by providing a proper air or brine flow.

Poor Distribution

Poor distribution in the evaporator will cause liquid to divert to the path of least resistance, so to speak. This can lead to throttling the valve prior to all passages receiving an adequate flow of refrigerant. When you suspect this is your problem, you should clamp a power assembly remote bulb to the free-draining suction line. The line must be cleaned before the bulb is attached. Then you will install a distributor that controls the flow. Take some time to balance the evaporator's load distribution, and your problem should be solved.

High Suction

If you have a unit that is exhibiting a high suction pressure and a high superheat, there are only a few likely causes of the problem. The compressor may be too small. If it is, you must replace it with a larger compressor. The evaporator may be too large. This too requires a complete replacement with an evaporator of the correct size. There is a chance that the compressor's discharge valves are leaking. In this case, you must repair or replace the valves. The fourth possibility is an unbalanced system. If the load is in excess of the design conditions, you must balance the load.

Low Superheat

A low superheat with a high suction pressure can be caused by a number of things. The compressor may be too small, in which case you must replace the compressor with one of the proper size. Your problem may be solved by simply adjusting the superheat setting. If the discharge valves associated with the compressor are leaking, they will have to be repaired or replaced. The expansion valve may be of the wrong size. This can lead to gas in the liquid line, requiring the replacement of the valve. There are some other possible causes, so let's take a look at them.

Let's say you have low superheat and high suction pressure. What is likely to cause this combination? Well, the compressor could be too small. If it is, you must replace it with a larger unit. One of the first things to check out is the adjustment of the superheat. When this adjustment is too low, all you have to do is turn it up.

Compressor valves sometimes leak, as you've seen before. A leak in a compressor valve can cause low superheat and high suction pressure. Check the valves, and if they're leaking, repair or replace them. If your expansion valve is not sized properly, you may get gas in your liquid line. This, too, can cause a low superheat and a high suction pressure. To solve the problem, replace the valve with one that is of the proper size.

Moisture is a frequent enemy of heat pumps. If moisture freezes the expansion valve in an open position, low superheat and high suction pressure are likely. To solve this problem, you must thaw the valve. An easy way to do this involves wetting rags with hot water and applying them to the valve. Saturate the rags with hot water and continue wrapping the valve until the ice melts. Once you've solved the immediate problem, install a filter/drier to prevent the problem from reoccurring.

If the diaphragm of an automatic expansion valve breaks, it will result in a liquid feedback. This will call for replacement of the power assembly. Another cause of liquid feedback can be that the valve is sticking in an open position. You may have to replace the valve under these conditions, but it is also possible that a good cleaning will do the job. In any event, a filter should be installed to prevent future problems of this nature.

High Discharge Pressure

When a heat pump has a high discharge pressure, you may have too much refrigerant in the system. Check the refrigerant charge and purge it if necessary. If the condenser is dirty, you should clean it. Maybe the condenser or liquid receiver is too small. If this is the case, replace the equipment with a unit that is of the proper size.

You may find that air or noncondensables have entered the system. This will be cause to purge and recharge the entire system. A water valve that is out of adjustment can be causing your problem. If the cooling water is above the design temperature, you should increase the supply of water. It may even be necessary to install a larger valve.

Sometimes the cooling water is inadequate due to an inadequate supply of water or a bad water valve. If this is the case, start the pump and open the water valves. Should you find problems with the valves, repair or replace them.

The last likely cause of a high discharge pressure is the air-cooled condenser. It may not be getting enough air circulation. You should also check for worn belts and pulleys that may be slipping. If these conditions check out okay, inspect the blower motor to assure that it is sized properly.

Erratic Discharge Pressure

An erratic discharge pressure with your heat pump could be related to one of six causes. The controls could be bad on the fluctuating discharge pressure. If they are, you will have to adjust, repair, or replace them. It may be necessary to replace the condensing water regulating valve, if it is bad. Another possibility is an improper charge of refrigerant. Check your refrigerant levels and add refrigerant as needed.

If the cooling fan for the condenser cycle is out of whack, you must figure out why it is not operating properly and correct it. The cycling of the evaporative condenser is another consideration to take into account. Check the spray nozzles, coil surface, control circuits, and thermostat overloads. You may find that it will be necessary to clean the nozzles or coil surface. If any other defective equipment is detected, it should be replaced.

Erratic Suction Pressure

If you are experiencing an erratic suction pressure, check the superheat adjustment; it may need to be changed. If this isn't the cause of the problem, consider installing a P-trap on the suction line to provide a free-draining line. A restricted external equalizer line could also be causing your trouble. Check the line and replace it if necessary.

If the remote bulb location is not right, you could have erratic suction pressure. Check the location of the bulb, and change it if necessary. Remember to clean the suction line thoroughly before clamping the bulb into place at a new location.

If you are experiencing a change in pressure drop across the valve, you may have a faulty condensing water regulator. If so, replace it. A flood of backflow from the liquid refrigerant, caused by an uneven evaporator loading or an improper liquid distribution device, is another possibility. Even an improperly mounted evaporator could be at fault. These problems could require you to remount the evaporator lines at a proper angle, replace the distributor, or install the proper load distributor devices to balance the air velocity over the evaporator coils.

Inspect each valve in your system to assure that it has its own equalizer line going directly to a proper location on the evaporator outlet. If it doesn't, rework the system so that it does.

As a last check, inspect spray nozzles, coil surface, control circuits, and thermostat overloads. The nozzles or coil surface may require cleaning. Control circuits and thermostat overloads may require replacement.

You should now know that heat pumps can be complicated creatures. Some elements of them can be dealt with by people who have not been trained specifically to work with them. Know your limitations, though. Don't get in over your head. If you have doubts, call in a trained service technician.

CHAPTER

8

A Basic Overview of Modern Radiant Heat Sources

Radiant heat is popular in many regions and comes from a number of heating sources. People often have a particular type of heat that comes to mind when they think of radiant heat. To some people, it's hot-water baseboard heat. Some people get a vision of a cast-iron radiator. Others think of wood stoves. Basically, any heat source that provides heat without some direct fan or other form of circulation can be considered a radiant heat source. By definition, radiant is issuing from a source in rays. This term could be applied to many types of heat sources. There are, however, a few types of heating systems that are commonly known as radiant heat sources. Typically, the types of heat considered in the trades to be radiant sources use water or steam to operate.

Solar heating gives off heat from a variety of sources. For example, heat can be gathered in and stored in a heat mass, such as heavy tile floors. The heat is gained through windows or solar panels when there is adequate heat from the sun. As the sky darkens, the heat absorbed during the day in the heat mass is radiated throughout the darker hours. This type of heating system is best known as a solar heating system.

Wood stoves are fueled by wood. Other stoves use pellets or coal to generate heat. The fuel in the stove creates heat that is radiated from the stove. In many cases, the heat is emitted from the stove without manipulation. Some stoves use fan systems to disperse the heat.

Again, the heat from a stove can be considered to be radiant heat. But, in the heating business, heat from stoves is known as stove heat, or by some similar term.

When you talk with a heating expert and are interested in discussing radiant heat, it will usually involve a heating system that relies on hot water and some type of convector or radiator. These systems use water, or some water-based solution, to move thermal energy from one place to another. For example, a boiler in a basement may create the heat and copper tubing with aluminum convection fins might transport and deliver the heat to various parts of a home. A system such as this is the type that is commonly referred to as radiant heat. In place of the copper tubing with fins, cast-iron radiators could be used. Another way is to install plastic piping in flooring, often in concrete that stores heat, so that the in-floor piping radiates heat.

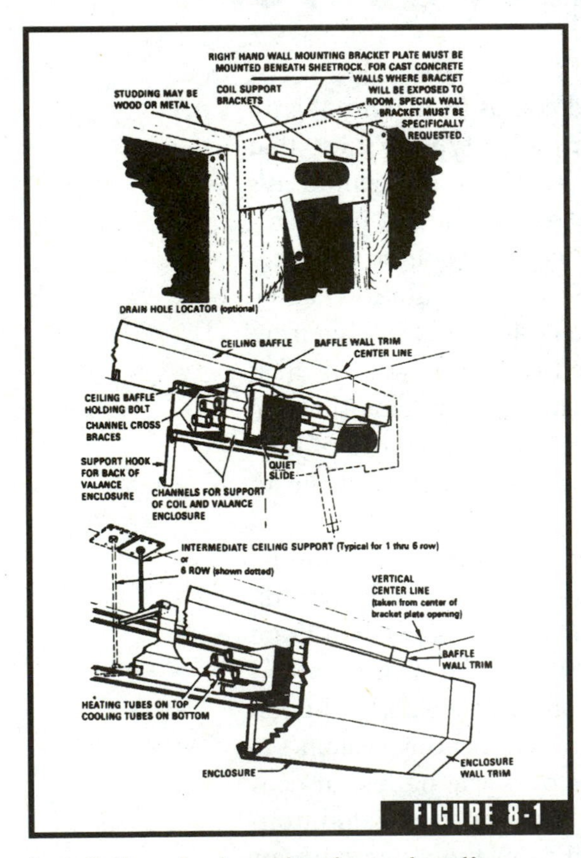

Installation of valance heating and cooling system. (*Courtesy of Edwards.*)

It is common for people to believe that the water used in a radiant system is the heat source. This is not the case. The truth is, water is the means of transporting heat. It is neither the heat source nor its destination. Thermal energy, which is the real heat, is generated at a heat source, such as a boiler, and transported with water to a point of dispersion. When the water reaches the desired point of distribution, heat is released by way of a heat emitter. The water doesn't leave the system; it merely allows the heat it is transporting to be released through the emitter. There are even valance systems which can be used (Figs. 8-1 to 8-5).

Why Water?

Water is used to convey thermal energy because it works very well in that role. In fact, water is practically an ideal choice for the transportation of thermal energy. Overall, water is readily available and

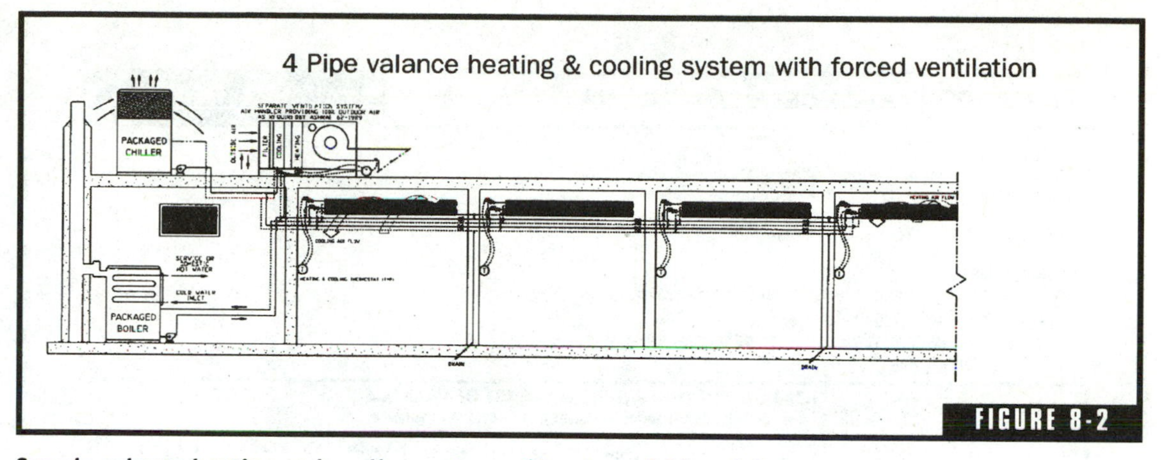

Sample valance heating and cooling systems. (*Courtesy of Edwards.*)

inexpensive. Being nontoxic and nonflammable adds to the benefits of using water as a means of transporting heat. To top it all off, water has one of the highest heat storage abilities known to science. Water can be used in all three of its forms, solid, liquid, and vapor, for heating and cooling purposes.

Hydronic systems use water in its liquid state. When used in a hydronic system, the temperature of water can range from 50° to 250°. Pressurization is required at the extreme temperature range to keep water in its liquid form. This prevents the water from turning to steam. Cooler temperatures, even those below 32°, can be used when antifreeze is introduced into the system. If this is done, the solution is known as brine. A brine system might be used for snowmelting or some similar special application.

Steam heat was used largely before hydronic heating. In modern heating systems, hydronic systems are much more common and popular than steam systems. In early stages of development, hydronic systems relied on buoyancy to circulate water. This was possible due to the nature of water properties at different temperatures. Hot water will rise from a heat source, such as a boiler (Fig. 8-6). When the water is hot, its natural properties will allow it to rise to heat emitters. When the water cools, it flows downward, back to the heat source. To make this type of system work, pipes have to be sized precisely and installation procedures must be performed perfectly (Fig. 8-7).

EDWARDS VALANCE HEATING: RATINGS

VALANCE HEATING

NOTE: The following heat output ratings apply on two-pipe systems where the cooling circuits also are used for heating.

HEAT OUTPUT: BTU/hr., per lineal foot of valance.

CONDITIONS:
- Water flow rate: 1 gpm thru ½" (nom.) finned tubing.
- For ratings at Flow Rates of 2, 3, and 4 gpm, see over.
- Room temperature of 70°F at 5 foot occupancy level, 24" from outside wall.

HEAT OUTPUT IN BTU/HR/LINEAL FOOT OF VALANCE
for various average water temperatures T (avg) thru valance

T (Avg): / Δ T*:	130° / 60°	140° / 70°	150° / 80°	160° / 90°	170° / 100°	180° / 110°	190° / 120°	200° / 130°	210° / 140°	220° / 150°	
Valance Tubes	BTUH	BTUH	BTUH	BTUH	BTUH	BTUH	BTUH	BTUH	BTUH	BTUH	
2	180	240	310	370	420	490	550	600	670	720	2
3	280	370	460	550	640	720	820	900	1000	1090	3
4	370	490	610	730	850	970	1090	1200	1330	1460	4
5	450	610	760	910	1060	1210	1360	1490	1660	1810	5
6	530	720	910	1090	1270	1460	1630	1790	2000	2180	6
8	710	960	1210	1450	1700	1930	2180	2390	2660	2910	8

NOTE: The above figures are heating ratings for valance and are based on 8 feet of free clear ceiling area per lineal foot of active finned valance tube. The free clear ceiling area must be thermally trapped to a depth of 1 ft. (See Fig. A) For example, for 1 foot length of 4 Row valance the effective radiant area is 4 x 8 or 32 square feet for the above rating. If the ceiling area available is only 6 square feet per foot of active finned tube length, multiply the rating by 0.94. If the ceiling area available is 12 square feet per foot of active finned tube length, multiply by 1.12. (See Table 1)

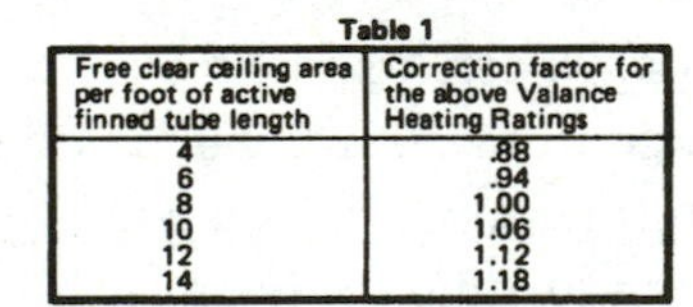

Table 1

Free clear ceiling area per foot of active finned tube length	Correction factor for the above Valance Heating Ratings
4	.88
6	.94
8	1.00
10	1.06
12	1.12
14	1.18

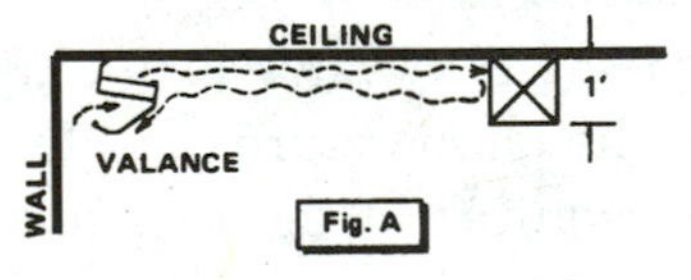

*Δ T is the difference in temperature between the average water temperature passing thru the valance coil and the room temperature of 70°F at 5 foot occupancy level. For actual measurements, the thermometer bulb shall be sheilded from all radiation effects.

If automatic changeover from cooling to heating is employed, the manufacturer recommends that an outdoor-indoor compensating water temperature control be used to prevent the rapid vaporization from the valance coils of condensate which has accumulated on the coils during the cooling cycle. If the vaporization of the condensate is excessively rapid the water vapor may condense on cold room surfaces adjacent to the valance.

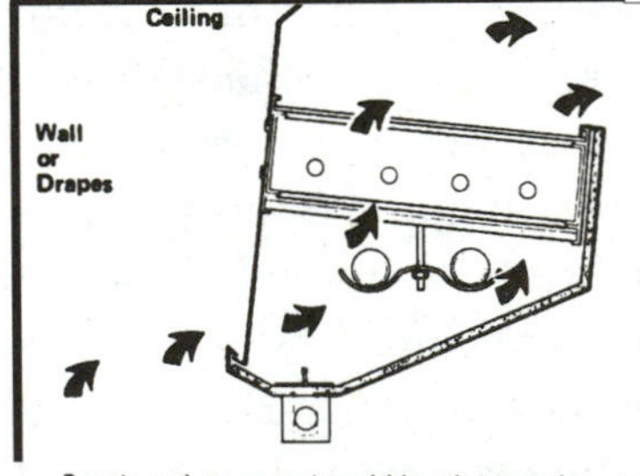

Supply and return mains within valance enclosure are to be of rust-proof tubing such as copper. Pipe should not be used.

FLOW OF AIR THROUGH VALANCE DURING HEATING CYCLE

FIGURE 8-3

Performance data for valance systems. (*Courtesy of Edwards.*)

EDWARDS VALANCE HEATING: RATINGS

VALANCE HEATING

NOTE: The following heat output ratings apply on two-pipe systems where the cooling circuits also are used for heating.

CONDITIONS:

– Water flow rate: 2 gpm thru ½" (nom.) finned tubing.
– Room temperature of 70°F at 5 foot occupancy level, 24" from outside wall.

HEAT OUTPUT IN BTU/HR/LINEAL FOOT OF VALANCE
for various average water temperatures T (avg) thru valance

T (Avg): Δ T*:	130° 60°	140° 70°	150° 80°	160° 90°	170° 100°	180° 110°	190° 120°	200° 130°	210° 140°	220° 150°	
Valance Tubes	BTUH	BTUH	BTUH	BTUH	BTUH	BTUH	BTUH	BTUH	BTUH	BTUH	
2	190	250	320	380	440	510	560	610	680	740	2
3	290	380	480	560	650	740	840	920	1030	1120	3
4	380	510	620	750	870	1000	1120	1230	1370	1500	4
5	470	620	780	930	1090	1250	1400	1540	1710	1860	5
6	550	740	930	1120	1310	1500	1670	1040	2050	2240	6
8	730	990	1240	1490	1740	1980	2240	2460	2740	2990	8

CONDITIONS:

– Water flow rate: 3 gpm thru ½" (nom.) finned tubing.
– Room temperature of 70°F at 5 foot occupancy level, 24" from outside wall.

HEAT OUTPUT IN BTU/HR/LINEAL FOOT OF VALANCE
for various average water temperatures T (avg) thru valance

T (Avg): Δ T*:	130° 60°	140° 70°	150° 80°	160° 90°	170° 100°	180° 110°	190° 120°	200° 130°	210° 140°	220° 150°	
Valance Tubes	BTUH	BTUH	BTUH	BTUH	BTUH	BTUH	BTUH	BTUH	BTUH	BTUH	
2	190	250	320	380	440	510	570	620	690	750	2
3	290	380	480	570	660	750	850	930	1035	1130	3
4	380	510	630	760	880	1010	1130	1240	1380	1510	4
5	470	630	790	940	1100	1260	1410	1550	1725	1880	5
6	550	750	940	1130	1320	1510	1690	1860	2070	2260	6
8	740	1000	1250	1500	1760	2000	2260	2480	2760	3020	8

CONDITIONS:

– Water flow rate: 4 gpm thru ½" (nom.) finned tubing.
– Room temperature of 70°F at 5 foot occupancy level, 24" from outside wall.

HEAT OUTPUT IN BTU/HR/LINEAL FOOT OF VALANCE
for various average water temperatures T (avg) thru valance

T (Avg): Δ T*:	130° 60°	140° 70°	150° 80°	160° 90°	170° 100°	180° 110°	190° 120°	200° 130°	210° 140°	220° 150°	
Valance Tubes	BTUH	BTUH	BTUH	BTUH	BTUH	BTUH	BTUH	BTUH	BTUH	BTUH	
2	190	250	330	390	450	520	580	630	700	760	2
3	300	390	490	580	670	760	870	950	1050	1150	3
4	390	520	640	770	900	1030	1160	1260	1410	1540	4
5	480	640	810	960	1120	1280	1440	1580	1760	1920	5
6	560	760	960	1150	1350	1540	1720	1900	2110	2300	6
8	750	1020	1270	1530	1790	2040	2300	2530	2810	3080	8

NOTE: The above figures are heating ratings for valance and are based on 8 feet of free clear ceiling area per lineal foot of active finned valance tube. The free clear ceiling area must be thermally trapped to a depth of 1 ft. (See Fig. A) For example, for 1 foot length of 4 Row valance the effective radiant area is 4 x 8 or 32 square feet for the above rating. If the ceiling area available is only 6 square feet per foot of active finned tube length, multiply the rating by 0.94. If the ceiling area available is 12 square feet per foot of active finned tube length, multiply by 1.12. (See Table 1)

FIGURE 8-4

Performance data for valance systems. (*Courtesy of Edwards.*)

EDWARDS DUAL-A-MATIC VALANCE HEATING: RATINGS

(DUAL-A-MATIC) VALANCE HEATING

NOTE: On two-pipe systems, for valance heating ratings where the cooling circuits are also used for heating, see Form 7-VH-E.

HEAT OUTPUT: BTU/Hr., per lineal foot of valance.

CONDITIONS:
- Water flow rate: 3 gpm thru ½" (nom.) finned tubing.
- Room temperature of 70°F at 5 foot occupancy level, 24" from outside wall.

VALANCE HEATING RATINGS, BTUH per LINEAL FOOT of VALANCE for 4-PIPE (DUAL-A-MATIC) UNITS at various average water temperatures, °F, T (avg)

T (avg):		130° avg	140° avg	150° avg	160° avg	170° avg	180° avg	190° avg	200° avg	210° avg	220° avg
* Δ T:		60°	70°	80°	90°	100°	110°	120°	130°	140°	150°
Valance Tubes											
Cooling†	Heating	BTUH	BTUH	BTUH	BTUH	BTUH	BTUH	BTUH	BTUH	BTUH	BTUH
2	1	140	190	240	290	330	380	430	470	520	560
3	2	240	320	400	480	550	620	710	780	870	940
4	3	330	450	550	670	770	880	990	1090	1210	1320
5	4	420	570	710	850	990	1130	1270	1400	1550	1690
6	5	500	690	860	1040	1210	1380	1550	1710	1900	2070
8	6	690	940	1170	1410	1650	1870	2120	2330	2590	2830

NOTE: The above figures are heating ratings for valance and are based on 8 feet of free clear ceiling area per lineal foot of active finned valance tube. The free clear ceiling area must be thermally trapped to a depth of 1 ft. (See Fig. A) For example, for 1 foot length of 4 Row valance the effective radiant area is 4 x 8 or 32 square feet for the above rating. If the ceiling area available is only 6 square feet per foot of active finned tube length, multiply the rating by 0.94. If the ceiling area available is 12 square feet per foot of active finned tube length, multiply by 1.12.(See Table 1)

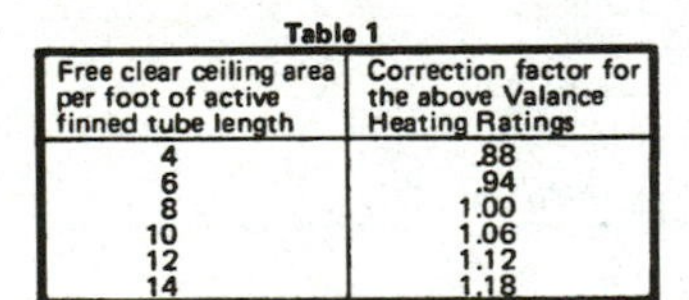

Table 1

Free clear ceiling area per foot of active finned tube length	Correction factor for the above Valance Heating Ratings
4	.88
6	.94
8	1.00
10	1.06
12	1.12
14	1.18

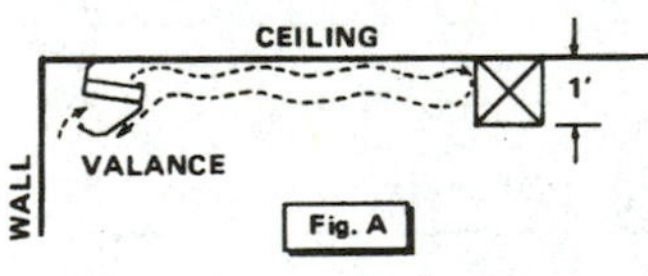

†For valance cooling ratings, see Data Forms EDF-7-102F,G,H,J.

* Δ T is the difference in temperature between the average water temperature passing thru the valance coil and the room temperature of 70°F at 5 foot occupancy level. For actual measurements, the thermometer bulb shall be shielded from all radiant effects.

If automatic changeover from cooling to heating is employed, the manufacturer recommends that an outdoor-indoor compensating water temperature controller be used on the circulating hot water circuit to prevent too rapid vaporization from the valance coils of condensate which has accumulated on the fins during the cooling cycle. If the vaporization of the condensate on the coil is excessively rapid the water vapor evaporated from the coil may condense on cold room surfaces.

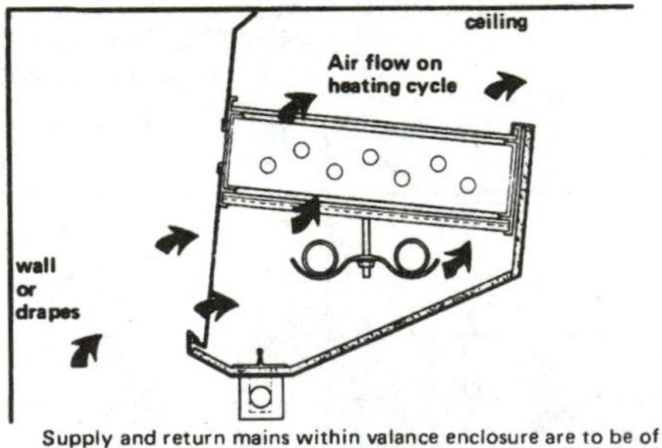

Supply and return mains within valance enclosure are to be of rust-proof tubing such as copper. Pipe should not be used.

FIGURE 8-5

Performance data for valance systems. (*Courtesy of Edwards.*)

Modern hydronic systems utilize electrically powered circulating pumps to push and pull water to higher levels in greater quantities for better performance. The use of pumps also increases the functions available from a hydronic system. For example, a single heat source, such as a boiler, can provide both heat for a building and domestic (potable) hot water. A hydronic heating system that is designed and installed properly can produce superior performance in fuel efficiency, comfort, and the creation of hot water for bathing and cooking. All in all, hydronic heating systems are extremely good and are, in many regions, the most efficient heating system available.

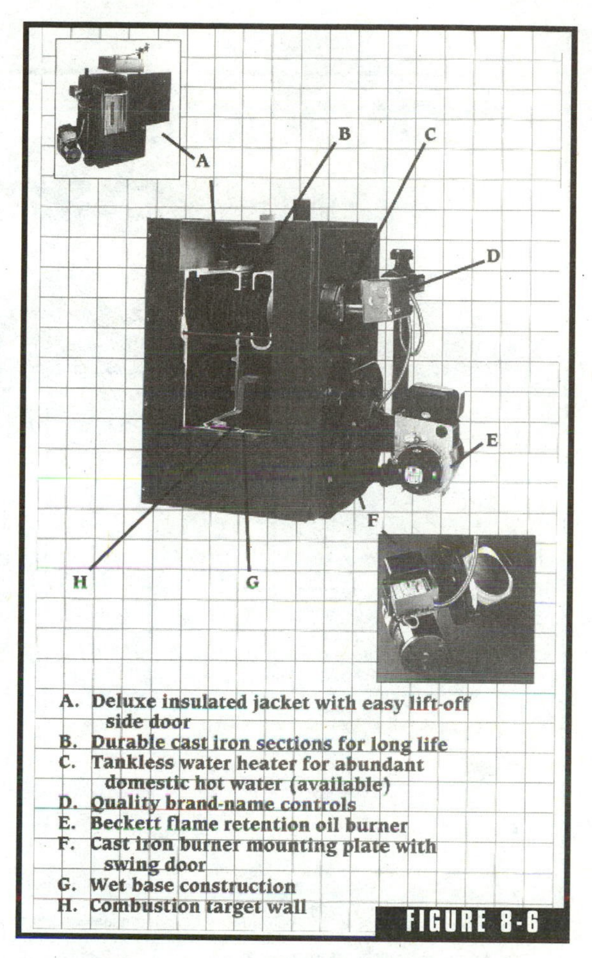

FIGURE 8-6

Typical boiler. (*Courtesy of Burnham.*)

Why Hydronic Systems?

Why do so many people prefer hydronic heating systems? The systems are cost-effective, efficient, affordable, and they can provide superior comfort. There are a number of criteria that ideal heating systems must meet to be considered suitable. Comfort is certainly one of them, and it is an often misunderstood concept. Ask average people what comfort from a heating systems means to them, and you will likely hear that comfort means being warm during cold days and nights. This is a part of the comfort needed, but it is not nearly the whole story. Got your attention? Read on.

Heating comfort is not as simple as producing heat. True human comfort is affected by the way in which the human body losses heat. Statistics indicate that normal adults who are engaged in light activity produce heat at a rate of about 400 British thermal units (Btus) per hour. Nearly half of this heat is released by thermal radiation to cold-

SPECIFICATIONS

Burnham
AMERICA'S BOILER COMPANY

Energy EPA DOE

V7 RATINGS

BOILER MODEL (1)(2)(3)	BURNER CAPACITY GPH (4)	DOE HEATING CAPACITY MBH (5)	I=B=R NET RATINGS			CHIMNEY REQUIREMENTS SIZE IN. x IN. x FT.	AFUE %	
			WATER MBH (6)	STEAM MBH (7)	STEAM SQ. FT.		STEAM	WATER
V-72	0.60	68	57	N/A	N/A	8x8x15	N/A	80.0
V-713	0.75	89	77	N/A	N/A	8x8x15	N/A	83.5
V-73	1.05	121	105	91	379	8x8x15	81.4	82.2
V-714	1.20	140	121	N/A	N/A	8x8x15	N/A	81.7
V-74	1.35	156	136	117	488	8x8x15	81.3	82.0
V-75	1.65	191	166	143	596	8x8x15	81.2	81.9
V-76	1.90	221	192	166	692	8x8x15	81.5	82.0
V-77	2.10	245	213	184	767	8x12x15	81.8	82.1
V-78S	2.35	266*	–	200	833	8x12x15	N/A	N/A
V-78W	2.35	275*	239	–	–	8x12x15	N/A	N/A
V-79S	2.60	298*	–	224	933	8x12x15	N/A	N/A
V-79W	2.60	299*	260	–	–	8x12x15	N/A	N/A
V-73WM	0.75	89	77	N/A	N/A	8x8x15	N/A	84.1
V-74WM	1.05	125	109	N/A	N/A	8x8x15	N/A	83.9
V-75WM	1.35	160	139	N/A	N/A	8x8x15	N/A	83.8
V-76WM	1.65	195	170	N/A	N/A	8x8x15	N/A	83.6
V-73WR (8)	0.60	73	63	N/A	N/A	8x8x15	N/A	86.2
V-74WR (8)	0.80	98	85	N/A	N/A	8x8x15	N/A	86.2
V-75WR (8)	0.90	110	96	N/A	N/A	8x8x15	N/A	86.2
V-76WR (8)	1.15	141	123	N/A	N/A	8x8x15	N/A	86.1
V-73SR	0.60	72	N/A	54	225	8x8x15	84.0	N/A
V-74SR	0.80	96	N/A	72	300	8x8x15	84.0	N/A
V-75SR	0.90	108	N/A	81	338	8x8x15	84.0	N/A
V-76SR	1.15	138	N/A	104	433	8x8x15	84.0	N/A

*I=B=R Gross Outputs

NOTES:

1. Add suffix "T" for boiler with tankless heater. V72 not available with tankless heater.
2. V-72, V-713 and V-714 available as a packaged water boiler only.
3. WM, WR and SR boilers require efficiency conversion kits.
4. The I=B=R Burner Capacity is based on oil having a heating value of 140,000 BTU/GAL.
5. DOE Heating Capacity and Annual Fuel Utilization Efficiency (AFUE) are based on U.S. Government tests at 13.0% CO_2 with No. 1 maximum smoke and -0.02 inches water column over fire draft on standard equipment package.
6. Net I=B=R Ratings are based upon a pick-up allowance of 1.15. Consult manual for unusual piping and pickup requirements.
7. Net I=B=R Ratings are based upon a pick-up allowance of 1.333. Consult manual for unusual piping and pickup requirements.
8. As an ENERGY STAR partner, Burnham has determined that this product meets the ENERGY STAR guidelines for energy efficiency.

Maximum Working Pressure: 15 PSI steam; 30 PSI water.

FIGURE 8-7

Boiler ratings. ***(Courtesy of Burnham.)***

er surfaces. Some 30 percent of the heat production is lost to surrounding air by way of convection. Over 20 percent of the heat created is lost through the skin via evaporation. This is all well and good, but what does it have to do with heating systems? A lot! A heated area that is not designed properly can cause discomfort for some people. If body heat is allowed to leave a body too quickly or too slowly, a person can experience discomfort.

The design of a heating system must take into account the components of the space being heated. For example, the number of windows affects the design of a heating system. You can stand in a room where the temperature is 72° and still feel chilly if there are a lot of windows in the room during a cold spell outside. For maximum comfort, a heated space must be balanced properly. This means controlling air temperature, average surface temperature, and relative humidity. When a hydronic heating system is installed properly, it can control both air and surface temperatures. A humidifier can take care of balancing the humidity. Combining the hydronic system with a humidifier, you can create a very comfortable environment.

Comfort is not the only advantage of a hydronic heating system. If you have ever lived in a house with a noisy heating system, you know how distracting the routine noise of a heating system can be. When a hot-water heating system is installed and purged properly, it can operate with almost no noise. This is another nice advantage of a hydronic system, but there are still more reasons to choose a hydronic system.

It's common knowledge that some types of heating systems can produce dust and dirt. Fan-forced systems are sometimes blamed for moving all sorts of airborne particles around heated space. For people with allergies, airborne dust, mold, and other particles in the air can produce great discomfort. Hot-water heating systems don't produce airborne particles through duct work. The result is a clean heating system that does not inflict problems with airborne particles.

The installation options for hydronic heating systems are broad. With the proper design, a hydronic system can provide radiant floor heating (Fig. 8-8), baseboard heating, domestic water heating, spa heating, and so forth. A single heating source can be used to branch out into many forms of heat dispersion. Combine this with the energy savings possible with hydronic systems and you have a very good heating system.

FIGURE 8-8

Radiant floor heating piping. (*Courtesy of Wirsbo.*)

Ease of Installation

The ease of installation is another quality that has made hydronic heating popular. Heating systems that require ducts are not particularly difficult to install in new construction, but they can be a pain in the neck when remodeling a building or home. Radiant heat is delivered mostly by pipes, generally pipes with a diameter of 1 inch or less. It's much easier to route a small pipe than a large duct. With the use of soft, rolled copper tubing or plastic tubing, heating pipes can be installed in places that might seem very unlikely locations for heating distribution. Even rigid tubing or piping is much easier, in most cases, to install than ducts.

There are definite advantages to the space-saving installation of heating pipes in remodeling. Ducts can be installed without undue problems in new construction, but the ducts still require a substantial amount of space. If a hydronic system is installed, the space saved can be used for plumbing chases or other needs. It can reduce or eliminate the need for false ceilings and boxed corners. These are factors that must be considered when deliberating on the best type of system to install.

Heat Transfer

Heat transfer is a critical element of a heating system. There are three types of heat transfer. Conduction, thermal radiation, and convection are the three options to consider when planning heat transfer in a system. Old-fashioned radiators use conduction as a heat transfer. Simply put, conduction is the movement of heat through a solid object. How well heat conducts through an item is related to the temperature difference across the solid material and its thermal conductivity. Knowing how to design the thickness of a conductor in compliance with the temperature of the heat being transferred is essential to a suitable design.

Convection is the type of heat transfer found in most modern hydronic systems. This is accomplished most often with the use of finned baseboard heating elements. These elements consist of thin-wall copper tubing that is surrounded by numerous aluminum fins.

Heat transfers from the copper tubing to the fins and is then dispersed into the heated area. This type of system can be termed a natural convection system.

Another type of convection system is the forced convection system. This type of system involves the use of a blower. In effect, a blower produces forced air over a heating coil and, therefore, pushes heat from the coil into the heated area. Any of the three types of heat transfer can be effective. They can even be used in connection with each other. All aspects of a design criteria must be considered when arriving at a suitable heating system.

Basic Components

A typical hydronic heating system is made up of four basic component areas. The first major component is the heat source. In most cases, this is a boiler. Usually, the boiler will be fueled by either gas or oil (Figs. 8-9 and 8-10). The boiler may be made of cast iron or steel. It is possible for the heat source to be a heat pump or a solar collector, but boilers are far and away the more common type of heat sources. The heat source is what generates heat for a heating system. Once the heat is generated, it must be distributed.

The distribution system for common hydronic heating systems is normally piping or tubing. Valves and circulating pumps are also part of a normal hydronic distribution system. To distribute heat in a hydronic system, water is typically moved through pipes or tubing to heat emitters and returned to the boiler for reheating. The water used for heating is retained in the system. It has to be replenished periodically at small levels, but this is done automatically by the heating system.

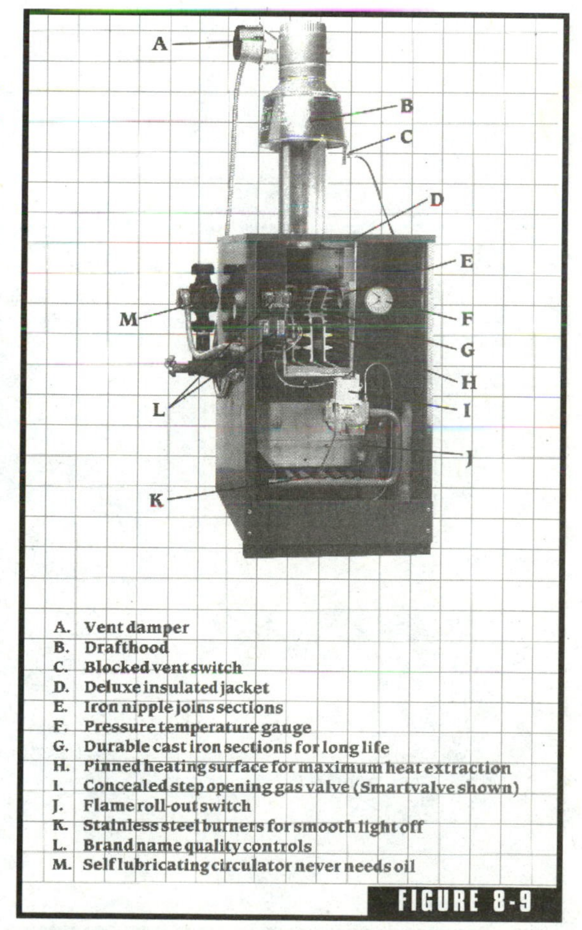

FIGURE 8-9

Gas-fired boiler. ***(Courtesy of Burnham)***

SPECIFICATIONS

Burnham
AMERICA'S BOILER COMPANY

SERIES 2 RATINGS Natural Gas or LP Gas

BOILER NUMBER (1)(2)	AGA INPUT MBH (3)	DOE HEATING CAPACITY MBH	I=B=R NET RATING MBH (4)	AFUE %		APPROX. SHIPPING WEIGHT (LBS.)	MINIMUM RECOMMENDED CHIMNEY SIZE ROUND Dia. (In.) x Ht. (Ft.)
				STANDING PILOT AND VENT DAMPER	EI AND VENT DAMPER		
202BWN(V)(S)	37.5	31	27	80.1	82.3	222	4 x 15
202BXWN(V)(S)	50	41	37	80.5	83.5	272	4 x 15
203BWN(V)(S)	62	52	45	80.0	82.6	272	4 x 15
204BWN(V)(S)	96	80	70	80.1	82.3	316	5 x 15
205BWN(V)(S)	130	108	94	80.2	82.0	364	6 x 15
206BWN(V)(S)	164	136	118	80.3	81.7	424	6 x 15
207BWN(V)(S)	198	163	142	80.4	81.4	468	7 x 15
208BWN(V)(S)	232	191	166	80.0	81.2	524	7 x 15
209BWN(V)(S)	266	218	190	80.0	80.9	560	8 x 15
210BWN(V)(S)	299	244	212	80.0	80.6	618	8 x 15
203BhWNS	62	52	45	N/A	84.0	274	4 x 15
204BhWNS	96	80	70	N/A	84.0	319	5 x 15
205BhWNS	130	108	94	N/A	84.0	368	6 x 15
206BhWNS	164	136	118	N/A	84.0	429	6 x 15

Water Only - 30 PSI working pressure
DOE heating capacity and annual efficiency are based on U.S. Government standard tests.

1. When ordering, suffix 'N' indicates natural gas. Substitute 'P' when ordering propane. Add 'V' to indicate 24 volt standing pilot. Add 'S' to indicate Smartvalve intermittent ignition. Smartvalve is standard on 2Bh.
2. Series 2h high efficiency boiler not available with LP gas.
3. Inputs shown are for installations at sea level and elevations up to 2,000 ft. For elevations above 2,000 ft., ratings should be reduced at the rate of four percent (4%) for each 1,000 ft. above sea level.
4. Net I=B=R ratings shown are based on normal I=B=R piping and pick up allowance of 1.15. Consult Burnham for installations having unusual piping and pick up requirements such as intermittent system of operation, extensive piping systems, etc.

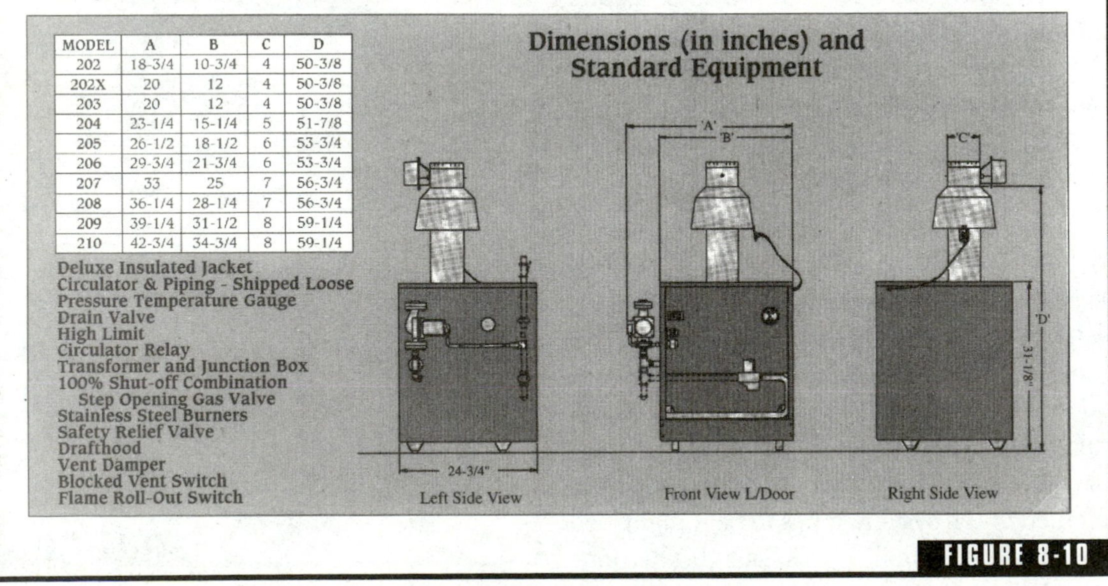

MODEL	A	B	C	D
202	18-3/4	10-3/4	4	50-3/8
202X	20	12	4	50-3/8
203	20	12	4	50-3/8
204	23-1/4	15-1/4	5	51-7/8
205	26-1/2	18-1/2	6	53-3/4
206	29-3/4	21-3/4	6	53-3/4
207	33	25	7	56-3/4
208	36-1/4	28-1/4	7	56-3/4
209	39-1/4	31-1/2	8	59-1/4
210	42-3/4	34-3/4	8	59-1/4

FIGURE 8-10

Specifications for a gas-fired boiler. (*Courtesy of Burnham.*)

Once a system has a heat source and a distribution system, it needs a heat emitter. This is most often either baseboard heating units or tubing installed in a floor. But, it can be a radiator or wall panels. The use of wall panels is growing, while the use of old-style radiators is declining. The heat emitters used, and there may be more than one type of emitter in a system, are what provide heat to a heated area.

The final component category is the control system. This is made up of thermostats, aquastats, and switches. When building a functional system, the four component categories must be matched properly. This is not a complicated process, but it is an important one. Designers and engineers at wholesalers are generally more than willing to help contractors design effective systems. Using basic design principles is all that is usually needed to obtain a satisfactory system.

Cold Climates

Hydronic heating systems are most popular in cold climates. For example, when I was building homes in Virginia, the most common type of heating system was a heat pump. This was due to the fact that heating and air conditioning could be produced from a single system. The climate in Virginia was fairly moderate, so air-source heat pumps worked well. During remodeling jobs, I ran into some hydronic heat-

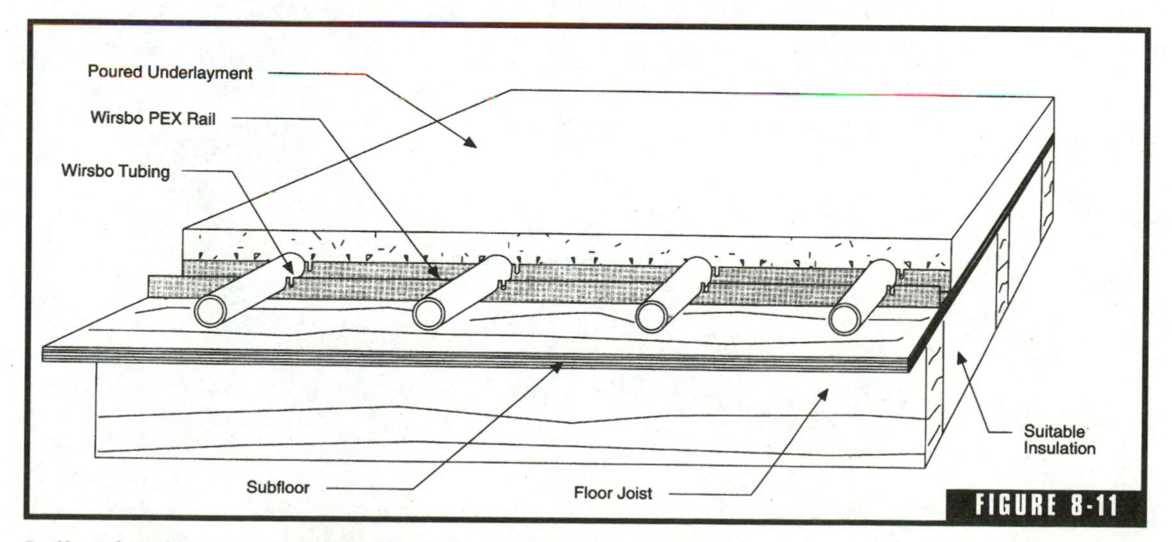

FIGURE 8-11

Radiant heating system installed in a floor. (*Courtesy of Wirsbo.*)

ing systems, but they were rare in Virginia. I moved to Maine about 11 years ago, and the roles reversed.

Maine has extremely harsh winters. I've lived in central Maine, near the coast, for many years and moved to far northern Maine about 2 years ago. Throughout the state, hot-water baseboard heat is the king of heating systems. Radiant heat (Fig. 8-11) in floors is also popular and growing in demand. Commercial buildings use heat pumps, but I have yet to build, remodel, plumb, or see a house in Maine that has a heat pump as a heat source. Water-source heat pumps would work well in Maine, but the cost of these systems can be prohibitive. Far and away, hydronic heating systems are the system of choice in Maine. This is due to the extreme cold. The same holds true in other cold regions.

If you work in warm climates, you will probably not use many hydronic systems. Heat pumps will most likely be where your demand is. Even electric baseboard heat is preferred in very warm climates, since the heat is used rarely and the equipment is inexpensive to purchase and install. There are a number of types of heating systems. Each type has it advantages and disadvantages. Contractors must learn how to match system attributes to defined needs. This book provides all the basic knowledge needed to make good decisions in most heating applications. Let's move on now. Turn to the next chapter, and we will explore the topic of oil-fired boilers.

CHAPTER 9

Boilers

Oil-fired boilers are the most common type used where I live, in the state of Maine. Why is this? Part of the reason is that natural gas is not available in many parts of the state. If it were, I suspect that gas-fired boilers would be much more popular. With the exception of some of the southern cities in Maine, natural gas is mostly unavailable. If a gas-fired unit is wanted, it is likely to have to be a liquified petroleum (LP) gas unit. Fuel oil is, without a doubt, the prime source of fuel for boilers in Maine. This is not the case in all areas of the country. Gas-fired boilers are very efficient and quite popular where natural gas is available. However, remote areas don't always enjoy gas service. It is often the colder climates where the most remote regions exist; such is the case in Maine. Therefore, oil-fired boilers are frequently used in hydronic systems.

Are oil-fired boilers good heat sources (Figs. 9-1 and 9-2)? Yes, they are fine heat sources. When I lived in Virginia, I installed heat pumps in my personal homes and in the homes that I built for most of my customers. Over the last decade, I've installed oil-fired boilers almost exclusively. This is because I relocated to Maine about 11 years ago. I used an oil-fired boiler in the home I built for myself and in all the homes I've built for customers. In addition to my work as a builder, my plumbing and heating company has installed numerous oil-fired

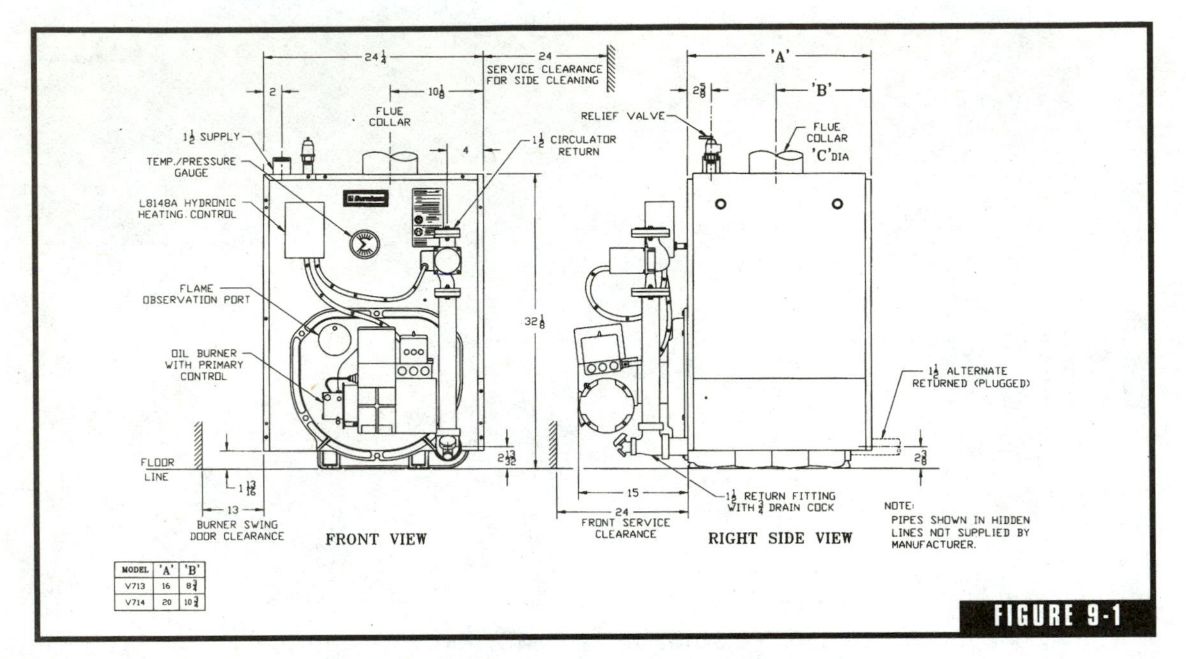

FIGURE 9-1

Packaged water boiler WITHOUT a tankless heater. (*Courtesy of Burnham.*)

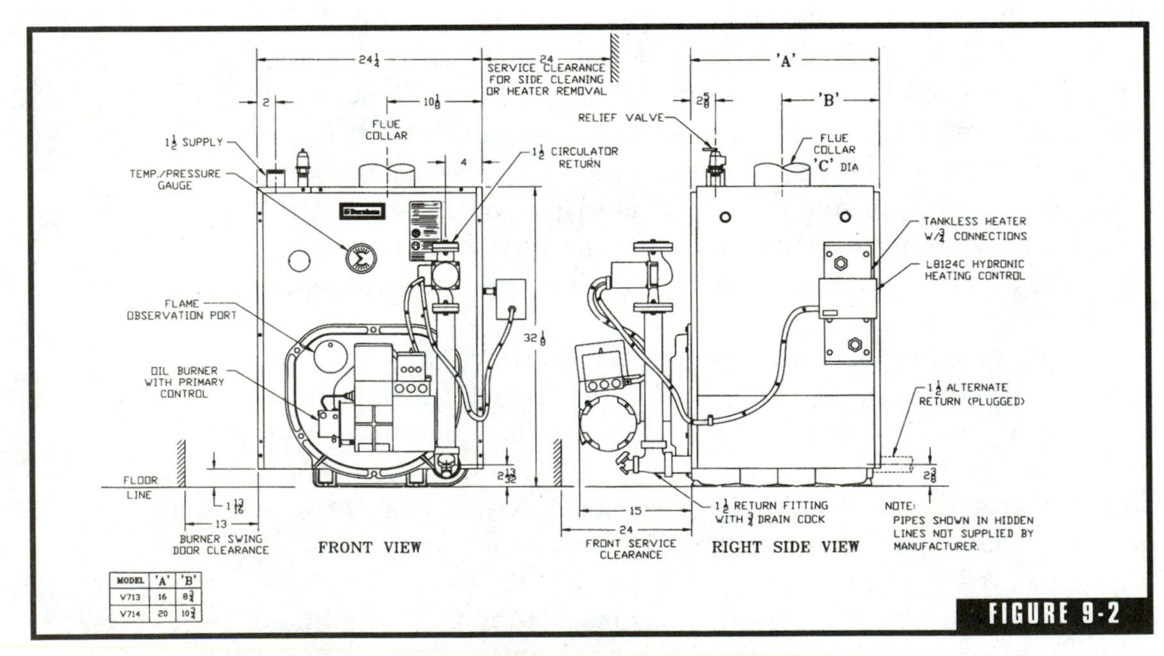

FIGURE 9-2

Packaged water boiler WITH a tankless heater. (*Courtesy of Burnham.*)

hydronic heating systems. Simply put, oil-fired hydronic heating systems rule the roost in most of Maine.

What makes the oil-fired system so desirable? Well, in Maine, it is partially due to the fact that natural gas is not readily available in most areas. Plus, fuel oil is a common source of heating combustion, and Maine likes to do things the way they've "always" been done. By this, I mean that Mainers are not fast to accept change. It's sort of like the saying, If it ain't broke, don't fix it. Certainly, gas-fired boilers (Fig. 9-3) are great choices when the fuel of choice is available. Personally, given a choice, I probably would opt for natural gas over fuel oil, but this is mostly a personal opinion.

There are some distinct differences between oil-fired boilers and gas-fired boilers (Figs. 9-4 and 9-5). We are going to talk about gas-fired boilers later in the chapter, but this section is dedicated to oil-fired units. When you talk of oil-fired boilers, you must consider both cast-iron boilers and steel boilers. This is the basic difference between the two types of oil-fired boilers, but there are many, many other components involved in such heating systems. This chapter is going to explore the major elements of such systems. We will talk about con-

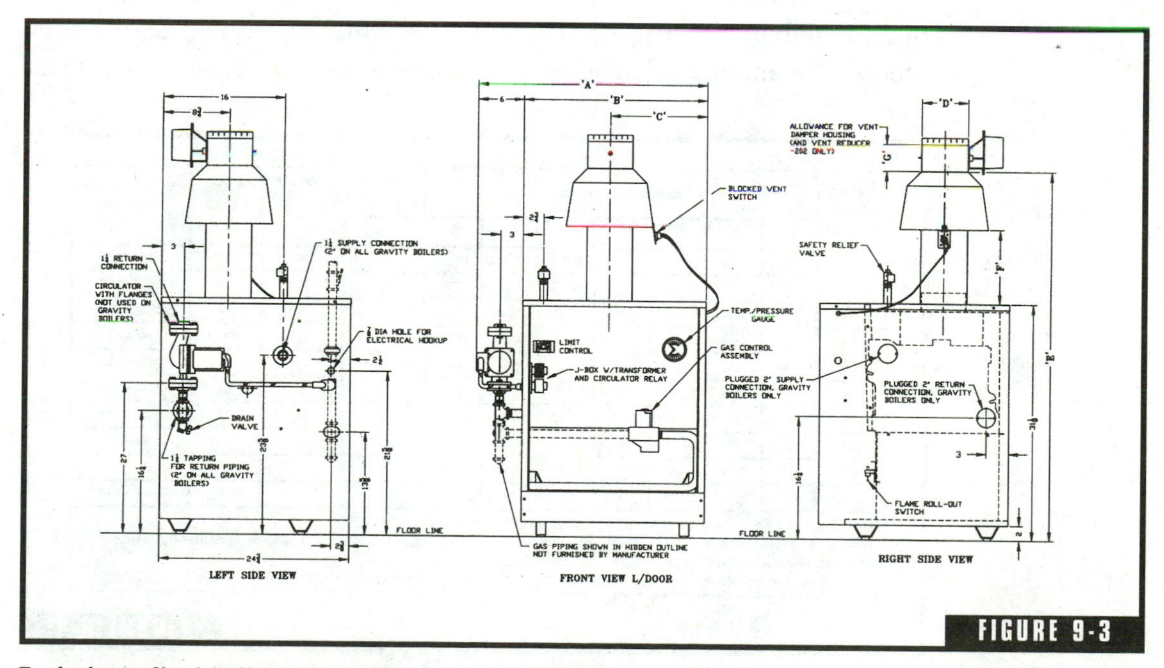

FIGURE 9-3

Typical gas-fired boiler setup. (*Courtesy of Burnham.*)

trols, valves, flue requirements (Fig. 9-6), installation requirements, nozzles, operating fundamentals, and much more. So, get ready for a crash course in the installation and use of oil-fired boilers.

Cast-Iron Boilers

Cast-iron boilers are the most common type of heat source used in residential hydronic applications. Known as sectional boilers, these boilers can be purchased as individual components and put together on a job. However, most of the boilers sold are already assembled. The boilers are heavy. They can easily weigh 500 pounds. This makes getting the boilers set in place difficult at times. When conditions allow the use of an appliance dolly, the job of moving a cast-iron boiler is not nearly as intimidating as it is when the boilers must be carried by workers (Figs. 9-7 to 9-10).

Modern boilers are much smaller than their predecessors. This is due to increased efficiency and new designs. Even though the boilers look small, don't underestimate their weight. Even little boilers are heavy. Wet-based, sectional boilers are assembled by bolting the sections together. When assembled, the sections are known as a boiler block. The amount of heat output available from a cast-iron boiler is

Boiler Model Number	Dimensions [inches]							Gas Connection for Automatic Gas Valve		Water Content [gallons]	Recommended Vent Size [1] [2]
	A	B	C	D	E	F	G	Natural	LP		
202	18¾	10¾	6-3/8	4	45-5/8	8½	10 [3]	½	½	2.5	3" dia. x 15 ft.
202X	20	12	6	4	45-5/8	8½	4¾	½	½	3.2	4" dia. x 15 ft.
203	20	12	6	4	45-5/8	8½	4¾	½	½	3.2	4" dia. x 15 ft.
204	23¼	15¼	7-5/8	5	47-1/8	9-1/8	4¾	½	½	4	5" dia. x 15 ft.
205	26½	18½	9¼	6	48½	9¾	5¼	½	½	4.7	6" dia. x 15 ft.
206	29¾	21¾	10-7/8	6	48½	9¾	5¼	½	½	5.5	6" dia. x 15 ft.
207	33	25	12½	7	50-1/8	10-3/8	6-5/8	¾	¾	6.2	7" dia. x 15 ft.
208	36¼	28¼	14-1/8	7	50-1/8	10-3/8	6-5/8	¾	¾	7	7" dia. x 15 ft.
209	39½	31½	15¾	8	52	11	7¼	¾	¾	7.7	8" dia. x 15 ft.
210	42¾	34¾	17-3/8	8	52	11	7¼	¾	¾	8.5	8" dia. x 15 ft.

[1] 15' chimney height is from bottom of draft hood opening to top of chimney.
[2] Refer to the National Fuel Gas Code, Appendix G for equivalent areas of circular and rectangular flue linings.
Maximum Allowable Working Pressure - 30 PSI (Water Only)
[3] 202 only. Dimension 'G' includes allowance for 4" x 3" reducer furnished with boiler. See Figure 8.

FIGURE 9-4

Vent sizing chart. (*Courtesy of Burnham.*)

directly proportional to the number of sections used to create the boiler block. The more sections there are, the more heat potential there is.

Some manufacturers offer their boilers as fully packaged units. This means that the heating unit is shipped complete with the boiler block, a burner assembly, a circulating pump, and controls. Buying a packaged boiler is convenient. It can also save time during an installation. However, packaged boilers are not customized, and this can be a problem. It is sometimes better to match components to a boiler on a job-by-job basis. For example, a prepackaged circulator that will work well for a baseboard heating system might be too small to use with radiant floor heating loops. If packaged systems are to be used, they must be assessed for proper sizing.

Cast-iron sectional boilers can be used only with closed-loop systems. When installed properly, closed-loop systems rid themselves of excess air after just a few

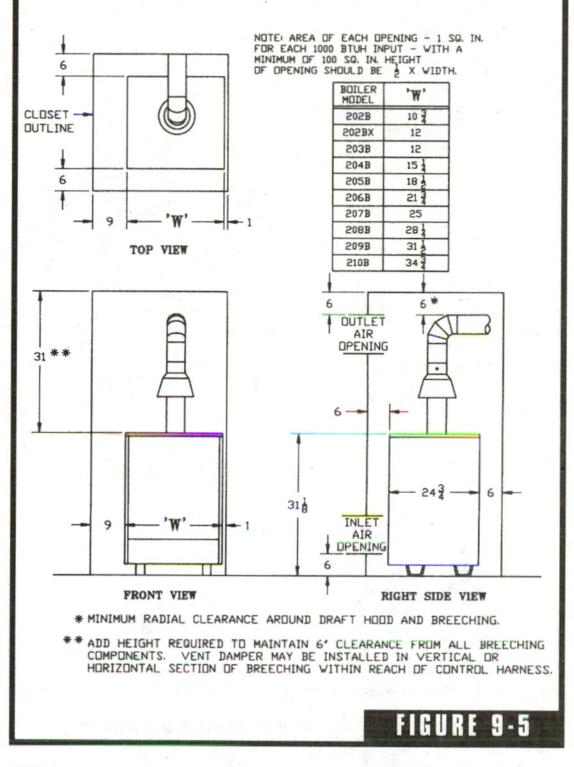

Minimum clearances for a gas-fired boiler. (*Courtesy of Burnham.*)

Maximum Capacity of Schedule 40 Pipe in CFH For Gas Pressures of ½ psig or Less

Length [Feet]	0.3 inch w.c. Pressure Drop				0.5 inch w.c. Pressure Drop			
	½	¾	1	1¼	½	¾	1	1¼
10	132	278	520	1,050	175	360	680	1,400
20	92	190	350	730	120	250	465	950
30	73	152	285	590	97	200	375	770
40	63	130	245	500	82	170	320	660
50	56	115	215	440	73	151	285	580
60	50	105	195	400	66	138	260	530
70	46	96	180	370	61	125	240	490
80	43	90	170	350	57	118	220	460
90	40	84	160	320	53	110	205	430
100	38	79	150	305	50	103	195	400

FIGURE 9-6

Capacities for gas pressure. (*Courtesy of Burnham.*)

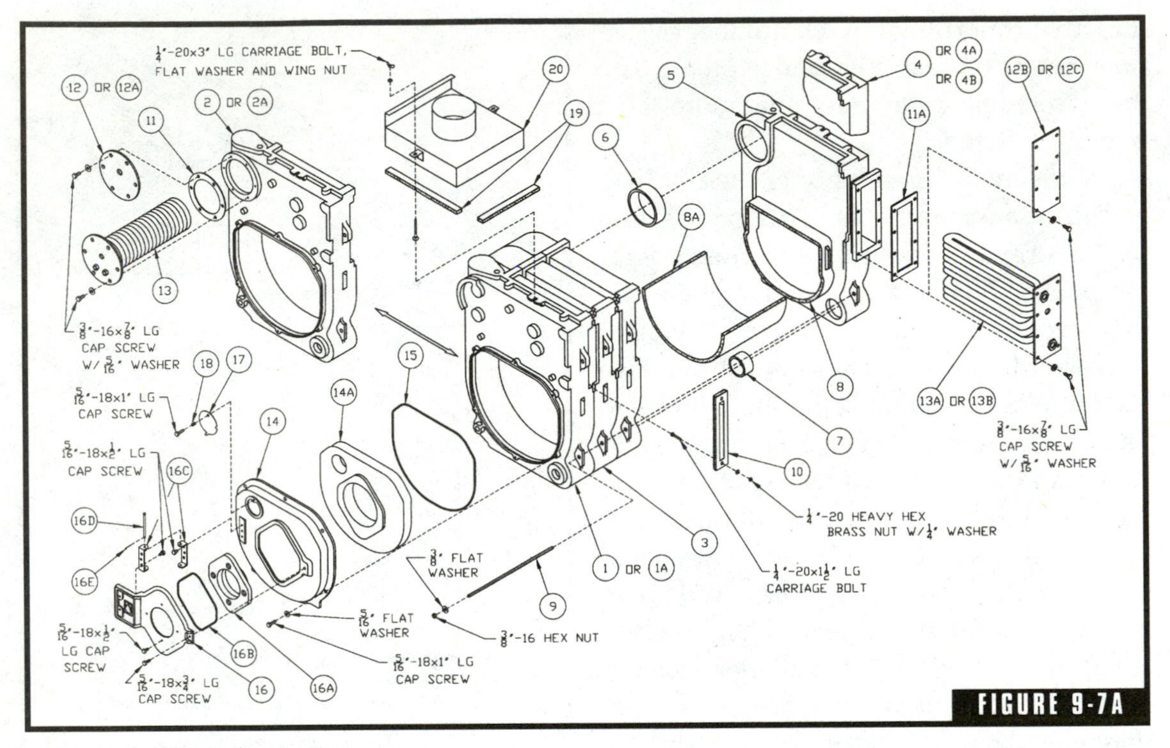

Bare boiler assembly. (*Courtesy of Burnham.*)

days of operation. This is important, because air in a system where a cast-iron boiler is used can result in rust and corrosion. Assuming that a cast-iron boiler is installed professionally, it should last for decades. It's very possible for a cast-iron boiler to continue working after 30 years of service.

Steel Boilers

Steel boilers are often sold with a 20-year warranty. This is the same warranty that is typically offered with cast-iron boilers. Even though both boilers usually come with the same warranty, there is market resistance in many regions against steel boilers. Many old-time installers don't feel that steel boilers can compete with cast-iron boilers in durability. Steel boilers are lighter than cast-iron boilers. The price of a steel boiler is usually several hundred dollars less than a comparable cast-iron boiler. With the many advantages of a steel boiler, it's surprising that more contractors are not using them.

Steel fire-tube boilers surround steel tubes with water and allow hot combustion gases to pass through the tubes. Spiral baffles are inserted in the fire tubes to increase heat transfer. This is done by inducing turbulence and slowing the passage of the exhaust gases. The baffles are called turbulators. The fire tubes may be installed vertical-

Item No.	Description	Part No.	V72	V713	V73	V714	V74	V75	V76	V77	V78	V79
BARE BOILER ASSEMBLY (See Page 32 for Illustration)												
1	Front Section (Non-Htr.), Machined Water	71727019	1	—	1	—	1	1	1	1	1	1
	Front Section (Non-Htr.), Machined Water	71727015	—	1	—	1	—	—	—	—	—	—
1A	Front Section (Non-Htr.), Machined Steam	71727018	—	—	1	—	1	1	1	1	1	1
2	Front Section w/Htr. Opening, Machined Water	71727016	—	—	1	—	1	1	1	1	1	1
2A	Front Section w/Htr. Opening, Machined Steam	71727017	—	—	1	—	1	1	1	1	1	1
3	Center Section	7172202	—	1	—	2	—	—	—	—	—	—
	Center Section	71727021	—	—	1	—	2	3	4	5	6	7
4	Narrow Back Section (Non-Htr.), Machined	7172204	—	1	—	1	—	—	—	—	—	—
4A	Wide Back Section (Non-Htr.), Machined Water	71727033	1	—	1	—	1	1	1	1	1	1
4B	Wide Back Section (Non-Htr.), Machined Steam	71727034	—	—	1	—	1	1	1	1	1	1
5	Rear Section w/Htr. Opening, Machined	7172203	—	1	—	1	—	—	—	—	—	—
	Rear Section w/Htr. Opening, Machined	71727036	—	—	1	—	1	1	1	1	1	1
6	5" Cast Iron Slip Nipple	7066003	1	—	2	—	3	4	5	6	7	8
7	2½" Cast Iron Slip Nipple	7066001	1	4	2	6	3	4	5	6	7	8
8	Rear Target Wall Insulation	8202701	1	1	1	1	1	1	1	1	1	1
8A	Combustion Chamber Liner, ½" x 23" x 4"	82027023	1	—	—	—	—	—	—	—	—	—
	Combustion Chamber Liner, ½" x 24" x 5"	82022033	—	1	—	—	—	—	—	—	—	—
	Combustion Chamber Liner, ½" x 23" x 8"	82027031	—	—	1	—	—	—	—	—	—	—
	Combustion Chamber Liner, ½" x 24" x 9"	82022043	—	—	—	1	—	—	—	—	—	—
	Combustion Chamber Liner, ½" x 23" x 12"	82027041	—	—	—	—	1	—	—	—	—	—
	Combustion Chamber Liner, ½" x 23" x 16"	82027051	—	—	—	—	—	1	—	—	—	—
	Combustion Chamber Liner, ½" x 23" x 20"	82027061	—	—	—	—	—	—	1	—	—	—
	Combustion Chamber Liner, ½" x 23" x 24"	82027071	—	—	—	—	—	—	—	1	—	—
	Combustion Chamber Liner, ½" x 23" x 28"	82027081	—	—	—	—	—	—	—	—	1	—
	Combustion Chamber Liner, ½" x 23" x 32"	82027091	—	—	—	—	—	—	—	—	—	1
9	Tie Rod, 3/8"-16 x 7¾" Lg.	80861072	2	—	—	—	—	—	—	—	—	—
	Tie Rod, 3/8"-16 x 12½" Lg.	80861010	—	2	2	—	—	—	—	—	—	—
	Tie Rod, 3/8"-16 x 17" Lg.	80861011	—	—	—	2	2	—	—	—	—	—
	Tie Rod, 3/8"-16 x 20¾" Lg.	80861012	—	—	—	—	—	2	—	—	—	—
	Tie Rod, 3/8"-16 x 25¼" Lg.	80861013	—	—	—	—	—	—	2	—	—	—
	Tie Rod, 3/8"-16 x 27½" Lg.	80861014	—	—	—	—	—	—	—	2	—	—
	Tie Rod, 3/8"-16 x 31½" Lg.	80861015	—	—	—	—	—	—	—	—	2	—
	Tie Rod, 3/8"-16 x 36½" Lg.	80861036	—	—	—	—	—	—	—	—	—	2

FIGURE 9-7B

Boiler parts list. (*Courtesy of Burnham.*)

ly or horizontally, depending on the brand of the boiler. Some boilers are built to have combustion gases pass through multiple fire tubes to increase heat. Generally speaking, a steel boiler that has horizontal fire tubes and passes gases through multiple tubes will be more efficient than a boiler that is equipped with single-pass vertical tubes.

Just as cast-iron boilers are suitable only for closed-loop systems, so are steel boilers. Corrosion can be a problem with steel boilers if dissolved air is present in the hydronic system. Since closed-loop systems rid themselves of air quickly, if they are designed and installed properly, corrosion is not usually a problem with steel boilers.

Fuel Oil

The choice of a fuel oil for oil-burning boilers is not complicated. There are two common choices. A number 1 fuel oil is a suitable choice for an oil burner. A sometimes more expensive, but better choice, is a number 2 fuel oil. The number 2 fuel gives more heat per

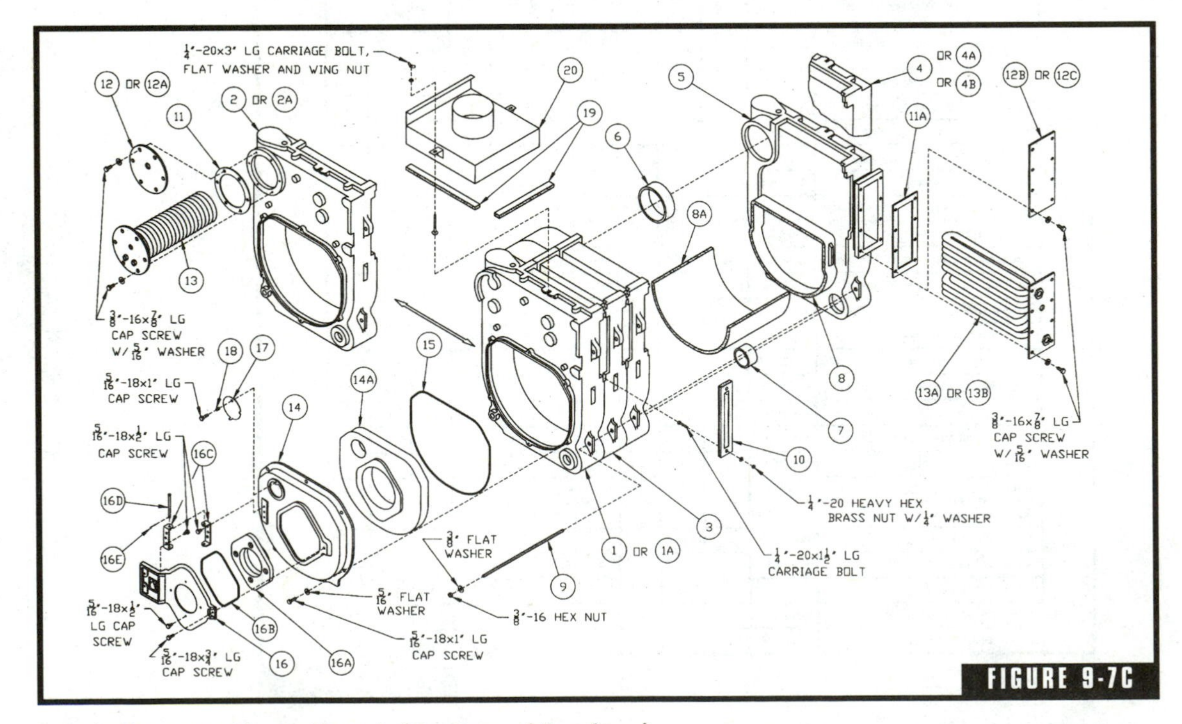

Bare boiler assembly continued. (*Courtesy of Burnham.*)

CONTINUED FROM PAGE 33

Item No.	Description	Part No.	V72	V713	V73	V714	V74	V75	V76	V77	V78	V79
BARE BOILER ASSEMBLY (See Page 34 for Illustration)												
10	Flue Cover Plate Assembly	6112214	1	2	2	3	3	4	5	6	7	8
11	Heater Cover Plate Gasket - Front Heater	8036068	—	—	1	—	1	1	1	1	1	1
11A	Heater Cover Plate Gasket - Rear Heater	8036058	—	1	1	1	1	1	1	1	1	1
12	Blank Heater Cover Plate (WB) - Front Heater	7036002	—	—	1	—	1	1	1	1	1	1
12A	Tapped Heater Cover Plate (WBTL) - Front Heater	7036001	—	—	1	—	1	1	1	1	1	1
12B	Blank Heater Cover Plate (SB & WB) - Rear Heater	7036020	—	—	1	—	1	1	1	1	1	1
12C	Tapped Heater Cover Plate (WBTL) - Rear Heater	7036021	—	1	1	1	1	1	1	1	1	1
13	222 Heater Carton Assy. - (WBT) Front Heater —OR—	6036007	—	—	1	—	1	1	1	1	1	1
	A54 Heater Carton Assy. - (WBT) Front Heater	6031204	—	—	—	—	—	—	—	1	1	1
13A	V1-2 Heater Assy. - (SBT & WBT) - Rear Heater	6036031	—	1	1	—	1	1	1	1	1	1
13B	V1-1 Heater Carton Assy. - (WBT) Rear Heater	6036030	—	—	—	1	—	—	—	—	—	—
14	Burner Swing Door Mounting Plate	7172705	1	1	1	1	1	1	1	1	1	1
14A	Burner Swing Door Mounting Plate Insulation	82027021	1	1	1	1	1	1	1	1	1	1
15	Rope Gasket - Burner Mounting Plate - 53" Lg.	72027014	1	1	1	1	1	1	1	1	1	1
16	Burner Swing Door	7172706	1	1	1	1	1	1	1	1	1	1
16A	Burner Swing Door Insulation	82027022	1	1	1	1	1	1	1	1	1	1
16B	Rope Gasket (Burner Swing Door) - 28" Lg.	72027015	1	1	1	1	1	1	1	1	1	1
16C	Hinge	7022701	2	2	2	2	2	2	2	2	2	2
16D	Hinge Pin	80861614	1	1	1	1	1	1	1	1	1	1
16E	Hairpin Cotter	80861667	1	1	1	1	1	1	1	1	1	1
17	Observation Port Cover	7026001	1	1	1	1	1	1	1	1	1	1
18	Spring 3/8" I.D. x ½" Lg.	8026015	1	1	1	1	1	1	1	1	1	1
19	Canopy Sealing Strip (½" x 1" x 98")	7202710	1	1	1	1	1	1	1	1	1	1
20	V72 Canopy Assembly, Type II	611270211	1	—	—	—	—	—	—	—	—	—
	V713 Canopy Assembly	6112203	—	1	—	—	—	—	—	—	—	—
	V73 Canopy Assembly, Type II	611270311	—	—	1	—	—	—	—	—	—	—
	V714 Canopy Assembly	6112204	—	—	—	1	—	—	—	—	—	—
	V74 Canopy Assembly, Type II	611270411	—	—	—	—	1	—	—	—	—	—
	V75 Canopy Assembly, Type II	611270511	—	—	—	—	—	1	—	—	—	—
	V76 Canopy Assembly, Type II	611270611	—	—	—	—	—	—	1	—	—	—
	V77 Canopy Assembly, Type II	611270711	—	—	—	—	—	—	—	1	—	—
	V78 Canopy Assembly, Type II	611270811	—	—	—	—	—	—	—	—	1	—
	V79 Canopy Assembly, Type II	611270911	—	—	—	—	—	—	—	—	—	1

FIGURE 9-7D

Boiler parts list. (*Courtesy of Burnham.*)

gallon than number 1 fuel does. It is common for number 1 fuel to be used in vaporizing or pot-type oil burners. Number 2 fuel is a more common choice for standard oil burners. Check the manufacturer's specifications to determine which type of fuel should be used in the oil burner that you are working with.

The storage of fuel oil is usually handled with standard metal oil tanks (Fig. 9-11). The tanks are available in both vertical and horizon-

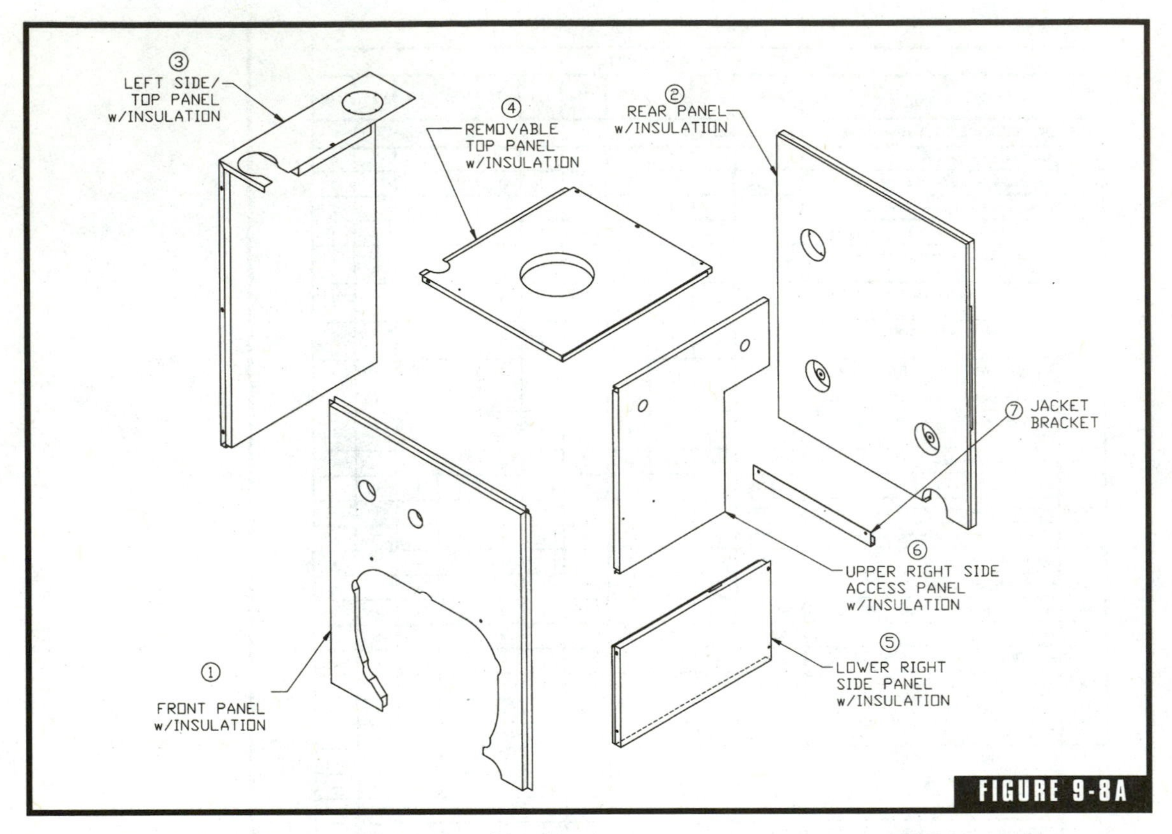

Flush boiler jacket. (***Courtesy of Burnham.***)

Item No.	Description	Part No.	V713	V714
V713 and V714 Flush Jacket Components - Items 1 thru 6 Include Insulation As Part of Assembly				
1	Jacket Front Panel Assembly	60427097	1	1
2	Jacket Rear Panel Assembly	60427096	1	1
3	Jacket Left Side/Top Panel Assembly	604270314	1	—
		604270414	—	1
4	Jacket Top Panel Assembly	604270322	1	—
		604270422	—	1
5	Jacket Lower Right Side Panel Assembly	604270337	1	—
		604270437	—	1
6	Jacket Upper Right Side Access Panel Assembly without Heater Opening (WB Only)	604270336	1	—
		604270436	—	1
	Jacket Upper Right Side Access Panel Assembly with Heater Opening (WBT Only)	604270335	1	—
		604270435	—	1
7	Jacket Bracket (WB Only)	7042715	1	1
	Jacket Bracket (WBT Only)	7042714		

FIGURE 9-8B

Flush boiler jacket parts list. (***Courtesy of Burnham.***)

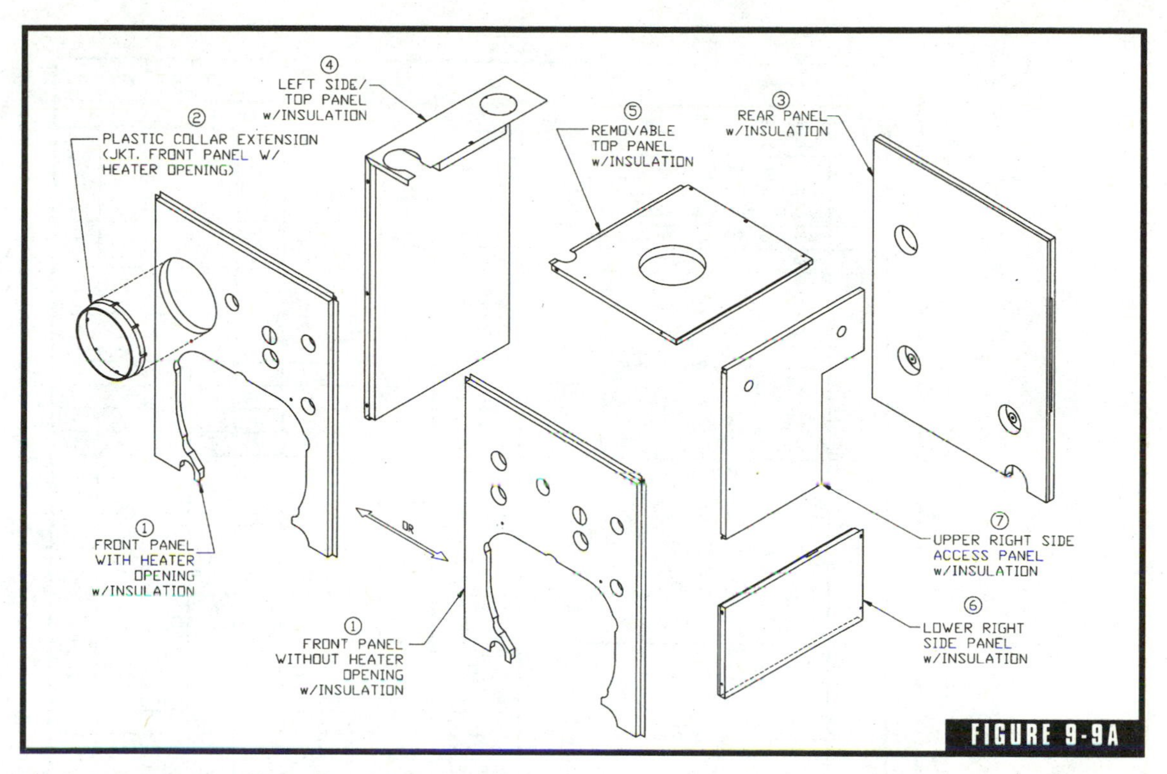

Flush boiler jacket. (*Courtesy of Burnham.*)

tal models. Vertical tanks are the most common. It's best when the tanks can be installed in a basement or cellar. Tanks installed outdoors in extremely cold climates can be affected by condensation. Water can build up in a tank and dilute the fuel oil. The tanks must be equipped with both a fill pipe and a vent pipe. When tanks are installed inside a home or building, both the fill and vent pipes are piped to the outside.

Fill pipes should have a minimum diameter of 2 inches. Vent pipes should have a diameter of 1¼ inches. Fill pipes must be equipped with watertight metal caps. Vent pipes have to be covered with a weatherproof hood or cap. It's common for vents to terminate with whistler-type caps so that people filling systems can tell when the tank is getting full. Vent pipes should be installed so that they are at least 2 feet from the outside wall of a building, to allow good air circulation (Fig. 9-12).

Item No.	Description	Part No.	V72	V73	V74	V75	V76	V77	V78	V79
V72 Thru V79 Flush Jacket Components - Items 1 thru 7 Include Insulation As Part of Ass'y										
1	Jacket Front Panel Ass'y w/o Htr. Opg. (Wtr. Blr.)	60427090	1	1	1	1	1	1	1	1
	Jacket Front Panel Ass'y w/o Htr. Opg. (Stm. Blr.)	60427091	—							
	Jacket Front Panel Ass'y w/Htr. Opg. (Wtr. Blr.)	60427092	—							
	Jacket Front Panel Ass'y w/Htr. Opg. (Stm. Blr.)	60427094	—							
2	Plastic Collar Extension-Jkt. Frt. Panel w/Htr. Opg.	8032704	—		1	1	1	1	1	1
3	Jacket Rear Panel Assembly	60427093	1	1	1	1	1	1	1	1
4	Jacket Left Side/Top Panel Assembly	604270241	1	—	—	—	—	—	—	—
		604270341	—	1	—	—	—	—	—	—
		604270441	—	—	1	—	—	—	—	—
		604270541	—	—	—	1	—	—	—	—
		604270641	—	—	—	—	1	—	—	—
		604270741	—	—	—	—	—	1	—	—
		604270841	—	—	—	—	—	—	1	—
		604270941	—	—	—	—	—	—	—	1
5	Jacket Top Panel Assembly	604270221	1	—	—	—	—	—	—	—
		604270321	—	1	—	—	—	—	—	—
		604270421	—	—	1	—	—	—	—	—
		604270521	—	—	—	1	—	—	—	—
		604270621	—	—	—	—	1	—	—	—
		604270721	—	—	—	—	—	1	—	—
		604270821	—	—	—	—	—	—	1	—
		604270921	—	—	—	—	—	—	—	1
6	Jacket Lower Right Side Panel Assembly	604270234	1	—	—	—	—	—	—	—
		604270334	—	1	—	—	—	—	—	—
		604270434	—	—	1	—	—	—	—	—
		604270534	—	—	—	1	—	—	—	—
		604270634	—	—	—	—	1	—	—	—
		604270734	—	—	—	—	—	1	—	—
		604270834	—	—	—	—	—	—	1	—
		604270934	—	—	—	—	—	—	—	1
7	Jacket Upper Right Side Access Panel Assembly without Heater Opening	604270233	1	—	—	—	—	—	—	—
		604270333	—	1	—	—	—	—	—	—
		604270433	—	—	1	—	—	—	—	—
		604270533	—	—	—	1	—	—	—	—
		604270633	—	—	—	—	1	—	—	—
		604270733	—	—	—	—	—	1	—	—
		604270833	—	—	—	—	—	—	1	—
		604270933	—	—	—	—	—	—	—	1
	Jacket Upper Right Side Access Panel Assembly with Heater Opening	604270332	—	1	—	—	—	—	—	—
		604270432	—	—	1	—	—	—	—	—
		604270532	—	—	—	1	—	—	—	—
		604270632	—	—	—	—	1	—	—	—
		604270732	—	—	—	—	—	1	—	—
		604270832	—	—	—	—	—	—	1	—
		604270932	—	—	—	—	—	—	—	1

FIGURE 9-9B

Flush boiler jacket parts list. (*Courtesy of Burnham.*)

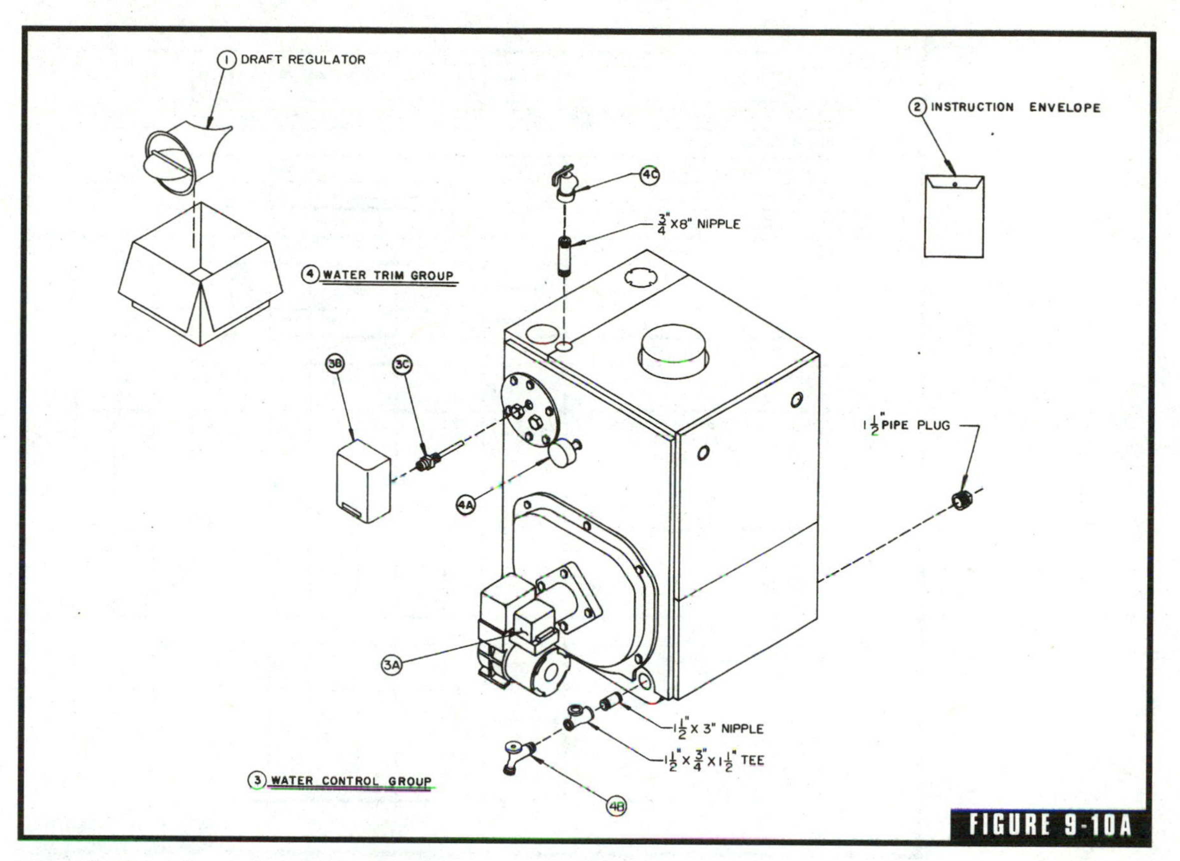

Water boiler trim and controls. (*Courtesy of Burnham.*)

Operating Fundamentals

The operating fundamentals of oil burners are not very difficult to understand. High-pressure atomizing oil burners contain nine major parts. They are a nozzle, a nozzle tube, a nozzle strainer, ignition electrodes, an electrodes bracket, an air entrance, an air adjustment collar, a fan, and rotary turbulator vanes. This type of burner is sometimes called a sprayer. This is because fuel oil is sprayed rather than vaporized. The burner unit is often called a gun. Basically, oil is forced under pressure through a special gun-like atomizing nozzle. The fuel is broken into very small liquid particles to form a fuel spray (Figs. 9-13 to 9-15).

ITEM NO.	DESCRIPTION	Part No.	V72	V713	V73	V714	V74	V75	V76	V77	V78	V79
V72 Thru V79, V713 and V714 WATER BOILERS - TRIM AND CONTROLS (See Page 42 for Illustration)												
1. DRAFT REGULATOR												
1A	DR-6 Draft Regulator	8116029	1	1	1	1	1	1	1	----	----	----
1B	DR-7 Draft Regulator	8116001	----	----	----	----	----	----	----	1	1	1
2. INSTRUCTION ENVELOPE CONTAINING:												
2A	Installation and Operating Instructions	8142711	1	1	1	1	1	1	1	1	1	1
2B	Limited Warranty Mailer (Water Boilers)	81460135	1	1	1	1	1	1	1	1	1	1
2C	I=B=R Pamphlet	81460061	1	1	1	1	1	1	1	1	1	1
3. WATER CONTROL GROUP												
3A	Honeywell R4184D (1027/1001) Protectorelay	80160473	1	1	1	1	1	1	1	1	1	1
3B	Honeywell L8148A1090 Hi Limit Circ. Relay (WB)	80160449	1	1	1	1	1	1	1	1	1	1
	--OR -- Honeywell L8124C1102 Hi & Lo Limit, Circ. Relay (WBT)	80160406	----	1	1	1	1	1	1	1	1	1
3C	Honeywell #123870A Immersion Well, ¾NPT x 1½" Insulation (WB)	80160426	1	1	1	1	1	1	1	1	1	1
	-- OR -- Honeywell #123871A Immersion Well, ½NPT x 3" Insulation (WBT)	80160497	----	1	1	1	1	1	1	1	1	1
4. WATER TRIM GROUP												
4A	2½" Dia. Temp/Pressure Gauge, ENFM #41042.5210	8056169	1	1	1	1	1	1	1	1	1	1
4B	¾NPT Drain Cock, Short Shank (WB and WBT), Conbraco #31-606-02	806603011	1	1	1	1	1	1	1	1	1	1
4C	¾NPT, F/F 30 LB. Relief Valve, Conbraco 10-408-05	81660319	1	1	1	1	1	1	1	1	1	1

FIGURE 9-10B

Water boiler parts list. (*Courtesy of Burnham.*)

High-pressure atomizing oil burners require pressure to operate. Residential applications usually need 80 to 125 pounds per square inch (psi) of pressure. A commercial burner may operate at pressures ranging from 100 to 300 psi. The burner process starts with oil being delivered to and through a nozzle. The oil is broken into small particles and sprayed into the ignition area. An air supply is brought in through a case opening and forced through a draft tube with the help of a fan. The air mixes with the fuel mist. A turbulator turns to mix the air and fuel completely. This mixture is ignited with a spark that is provided by a transformer, which converts electrical current and feeds it to electrodes, where a spark is produced (Figs. 9-16 to 9-18).

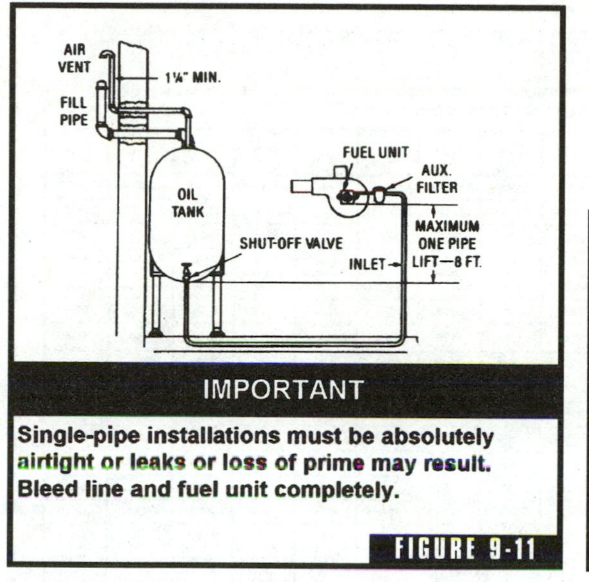

Typical oil storage tank setup. (*Courtesy of Burnham.*)

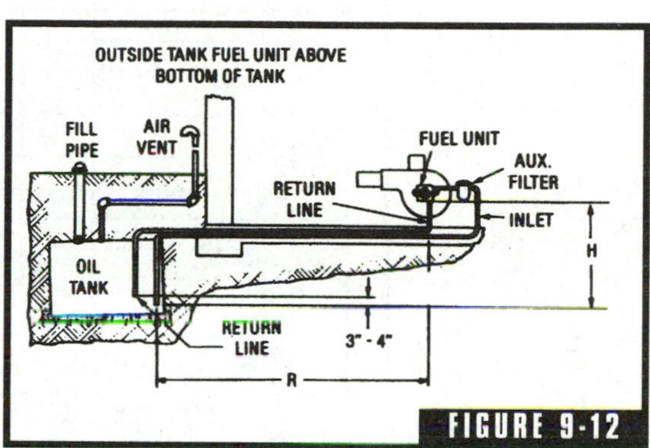

Outside fuel tank. (*Courtesy of Burnham.*)

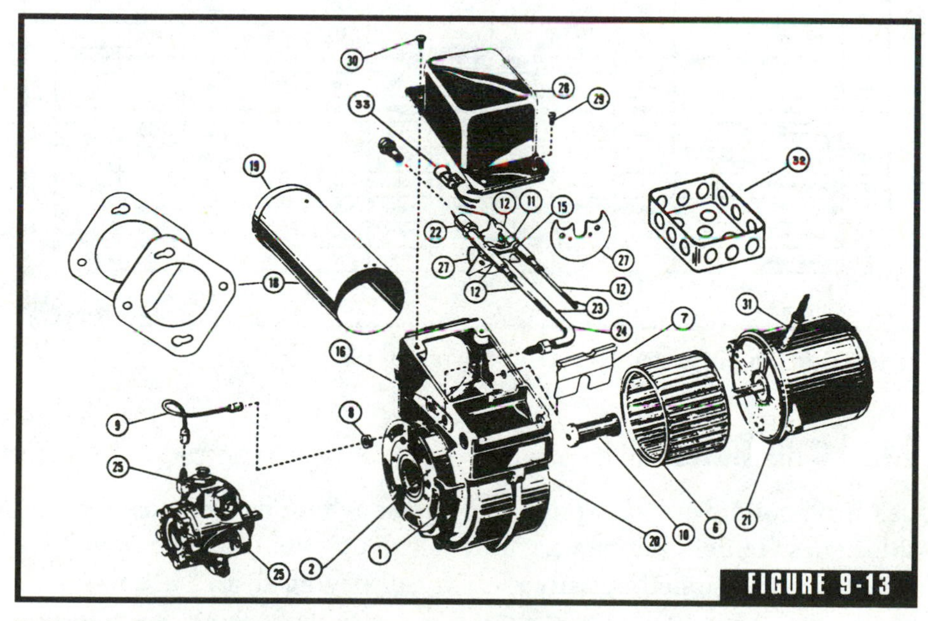

Exploded view of an oil burner. (*Courtesy of Burnham.*)

BECKETT PART NOS. FOR V7 SERIES BOILERS

NOTE: When ordering parts always give the serial and model numbers shown on the boiler and burner. Also provide the name of the part(s) and part number as listed below.

Item No.	Description	V72	V713	V73	V714	V74	V75	V76	V77	V78	V79
Air Tube Combination		AF72BN	AF72YH	AF44YH	AF72YY	AF44WP	AF44YB	AF44XO	AF72WK	AFG50MD	AFG50MD
Spec No.		BCB3230	BCB3226	BCB3213	BCB3236	BCB3214	BCB3215	BCB3216	BCB3217	BCB3218	BCB3234
1	Air Band	3492	3492	3492	3492	3492	3492	3492	3492	3492	3492
	Air Band Nut	4150	4150	4150	4150	4150	4150	4150	4150	4150	4150
	Air Band Screw	4198	4198	4198	4198	4198	4198	4198	4198	4198	4198
2	Air Shutter	3709	3709	3709	3709	3709	3709	3709	3709	3709	3709
	Air Shutter Screw	4198	4198	4198	4198	4198	4198	4198	4198	4198	4198
6	Blower	2999	2999	2999	2999	2999	2999	2999	2999	2999	2999
7	Low Firing Rate Baffle	3708P	3708P	3708P	---	---	---	---	---	---	---
8	Bulkhead Fitting Locknut	3666	3666	3666	3666	3666	3666	3666	3666	3666	3666
9	Connector Tube Assembly	5636	5636	5636	5636	5636	5636	5636	5636	5636	5636
10	Coupling	2454	2454	2454	2454	2454	2454	2454	2454	2454	2454
11	Electrode Clamp	149	149	149	149	149	149	149	149	149	149
	Electrode Clamp Screw	4219	4219	4219	4219	4219	4219	4219	4219	4219	4219
12	Electrode Insulator Assembly	5780	5780	5780	5780	5780	5780	5780	5780	5780	5780
15	Spider Spacer Assembly	5503	5503	5503	5503	5503	5503	5503	5503	5503	5503
16	Escutcheon Plate	3493	3493	3493	3493	3493	3493	3493	3493	3493	3493
18	Gasket	31498	31498	31498	31498	31498	31498	31498	31498	31498	31498
19	Head	360003	360003	360003	360006	360004	360006	360012	360016	5913	5913
	Head Screws	4221	4221	4221	4221	4221	4221	4221	4221	4221	4221
	Hole Plug	2139	2139	2139	2139	2139	2139	2139	2139	2139	2139
20	Housing Ass'y w/Inlet Bell	5874	5874	5874	5874	5874	5874	5874	5874	5874	5874
21	Motor	2456	2456	2456	2456	2456	2456	2456	2456	2456	2456
22	Nozzle Adapter	213	213	213	213	213	213	213	213	213	213
23	Nozzle Line Electrode Assembly	NL72BN	NL72YH	NL44YH	NL72YY	NL44WP	NL44YB	NL44X0	NL72WK	NL50MD	NL50MD
25	Pump	2460	2460	2460	2460	2460	2460	2460	2460	2460	2460
27	Static Plate	31646	3384	3384	3384	3384	3383	3383	None	3383	3383
28	Transformer	2442	2442	2442	2442	2442	2442	2442	2442	2442	2442
29	Transformer Hinge Screw	4217	4217	4217	4217	4217	4217	4217	4217	4217	4217
30	Transformer Holding Screw	4292	4292	4292	4292	4292	4292	4292	4292	4292	4292
31	Wire Guard	10251	10251	10251	10251	10251	10251	10251	10251	10251	10251
32	Junction Box	3741	3741	3741	3741	3741	3741	3741	3741	3741	3741
33	Flame Detector	7006	7006	7006	7006	7006	7006	7006	7006	7006	7006

FIGURE 9-14

Parts list for an oil burner. (*Courtesy of Burnham.*)

Gas-Fired Boilers

Gas-fired boilers are very popular in areas where natural gas is available. These boilers can be very efficient to operate. Keeping operating costs down is something anyone who is paying to run a boiler can appreciate. Basic boiler construction is the same for both gas- and oil-fired boilers. There are, however, some differences, so let's discuss them now (Fig. 9-19).

BURNER SPECIFICATIONS

		Beckett AFG						Riello R40				Carlin EZ-1HP / 102CRD-3*		
	Firing			Settings			Delavan Nozzle		Settings		Delavan Nozzle	Settings		Delavan Nozzle
Boiler Model	Rate GPH	Head	Static Disc	Head	Air Shutter	Air Band	GPH x Angle Type	Burner Model	Head	Air (Approx.)	GPH x Angle Type	Head Bar	Air Band	GPH x Angle Type
V72	.60	F3[*]	3-3/8"	---	5	0	0.50 x 80° B	---	---	---	---	---	---	---
V73 R	.60	F0[*]	3-5/8"	---	5	0	0.50 x 80° B	F3	0.5	3.0	0.50 x 60° W	.50	0.40	0.50 x 60° B
V73 M	.75	F3	3-3/8"	---	7	0	0.60 x 80° A	F5	0.5	2.1	0.60 x 60° W	.60	0.55	0.60 x 70° A
V73	1.05	F3[*]	3-3/8"	---	10	2	0.85 x 80° A	F5	2.0	2.8	0.85 x 60° W	.85/1.00	0.85	0.85 x 60° A
V713	.75	F3[*]	3-3/8"	---	8	0	0.65 x 80° A	---	---	---	---	.85/1.00	0.85	0.85 x 60° A
V714	1.20	F6	3-3/8"	---	10	1	1.00 x 80° A	---	---	---	---	.60/.65	0.55	0.65 x 70° A
V74 R	.80	F3	3-3/8"	---	5	0	0.75 x 80° A	---	---	---	---	.60/.65	0.55	0.65 x 70° A
V74 M	1.05	F4	3-3/8"	---	7	0	0.85 x 80° B	F5	2.0	2.8	0.85 x 60° W	.85/1.00	0.65	0.85 x 60° B
V74	1.35	F4	3-3/8"	---	8	1	1.10 x 80° B	F5	3.5	3.6	1.10 x 60° W	1.10	1.10	1.10 x 60° B
V75 R	.90	F3	3-3/8"	---	5	0	0.90 x 80° B	---	---	---	---	.75	0.60	0.75 x 70° A
V75 M	1.35	F6	2-3/4"	---	10	0	1.10 x 80° B	F5	3.5	3.4	1.10 x 60° W	1.10	0.85	1.10 x 60° B
V75	1.65	F6	2-3/4"	---	9	2	1.35 x 80° B	F10	2.0	2.7	1.35 x 60° W	1.35/1.50	1.15	1.35 x 60° B
V76 R	1.15	F6	2-3/4"	---	7	0	1.10 x 80° B	---	---	---	---	.85/1.00	1.00	1.00 x 60° B
V76 M	1.65	F12	2-3/4"	---	8	1	1.35 x 80° B	F10	2.0	2.6	1.35 x 60° W	---	---	---
V76	1.90	F12	2-3/4"	---	9	2	1.50 x 80° B	F10	2.5	3.0	1.50 x 60° W	* 2	40.	1.65 x 80° B
V77	2.10	F16	---	---	10	2	1.75 x 80° B	F10	2.0	3.5	1.75 x 60° B	* 3	30.	1.75 x 60° B
V78	2.35	V1	2-3/4"	4	10	6	2.00 x 60° B	F10	2.5	4.2	2.00 x 60° B	* 5	40.	2.00 x 60° B
V79	2.6	V1	2-3/4"	6	10	5	2.25 x 60° B	---	---	---	---	* 9	100	2.25 x 70° B

Beckett AFG: Oil Pump Pressure - V74R, V75R, & V76R: 100. PSI - All Other Models: 140. PSI
[*]Low Firing Rate Baffle on V72, V73R, V73, and V713

Riello R40: Oil Pump Pressure: 145. PSI
Air Tube Combination: SBT
Insertion Length: 1-11/16"

Carlin EZ-1HP / 102CRD-3: Oil Pump Pressure: 150.PSI (EZ-1HP) 140.PSI (102CRD-3)
*102CRD-3 ON V76, V77, V78 and V79
Tube Insertion: 1-7/8"

FIGURE 9-15

Specifications for an oil burner. (*Courtesy of Burnham.*)

Gas burners can operate with either an atmospheric, yellow flame or with a power burner. There are gas burners designed for use with natural gas and others that are designed for use with bottled gas. Most of the bottled gas used is liquified petroleum gas (LPG). It's very important to match the burner unit to the exact type of gas to be used for ignition.

A majority of all gas burners used for residential purposes are designed for use as atmospheric, yellow flame burners (Fig. 9-20) . The atmospheric injection type of burner works on a principle similar to a Bunsen burner. The burner is nothing more than a small tube. This is placed inside of a larger tube. The bigger tube has holes in it that are positioned slightly below the top of the smaller, inner tube. Gas coming out of the small tube pulls air through the holes of the larger tube. This produces an induced current of air in the larger tube. The air enters through the holes and is mixed with gas in the tube. The mixture is burned at the top of the larger tube. This type of flame produces little light but extensive heat.

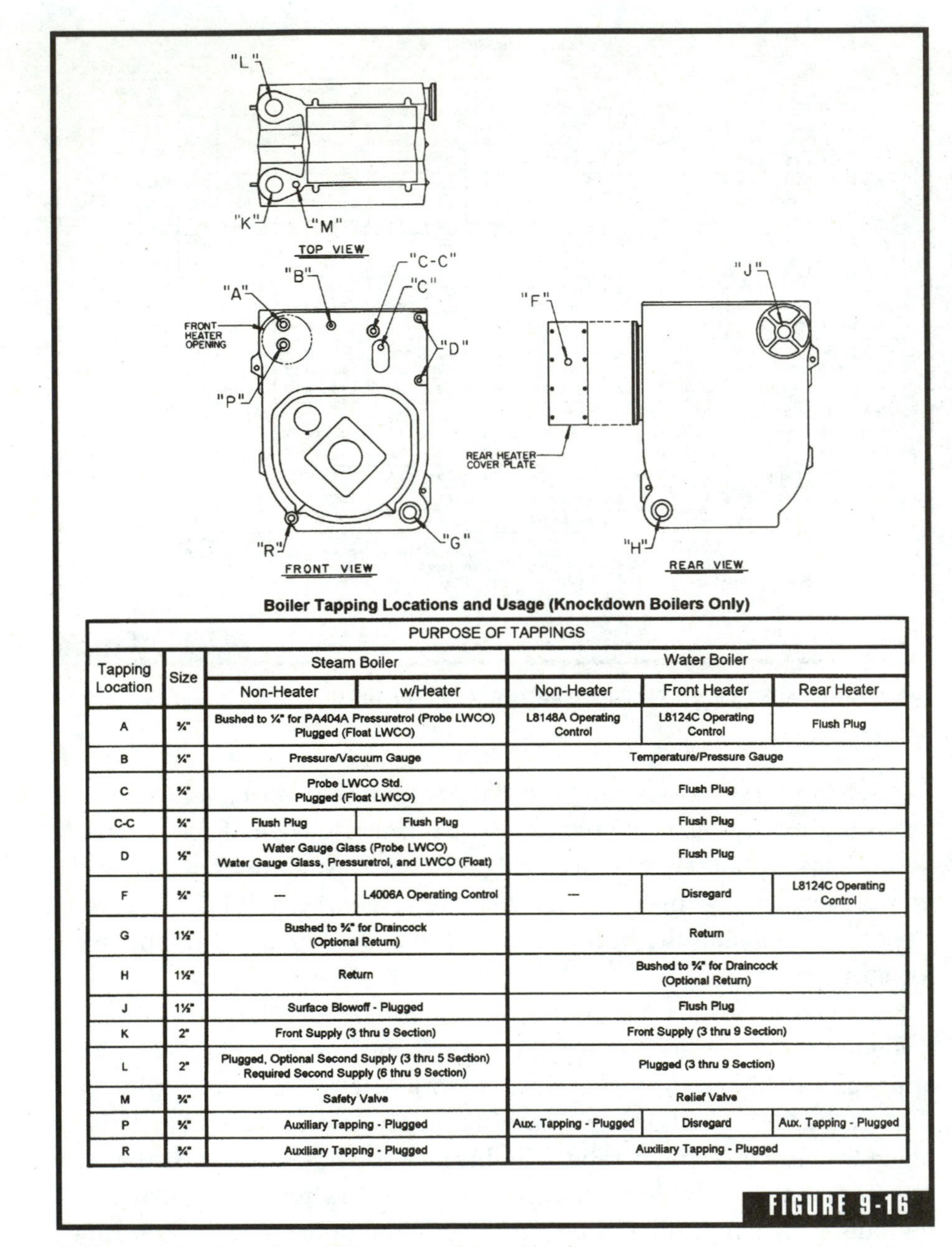

Boiler Tapping Locations and Usage (Knockdown Boilers Only)

Tapping Location	Size	PURPOSE OF TAPPINGS				
		Steam Boiler		Water Boiler		
		Non-Heater	w/Heater	Non-Heater	Front Heater	Rear Heater
A	¾"	Bushed to ¼" for PA404A Pressuretrol (Probe LWCO) Plugged (Float LWCO)		L8148A Operating Control	L8124C Operating Control	Flush Plug
B	¼"	Pressure/Vacuum Gauge		Temperature/Pressure Gauge		
C	¾"	Probe LWCO Std. Plugged (Float LWCO)		Flush Plug		
C-C	¾"	Flush Plug	Flush Plug	Flush Plug		
D	½"	Water Gauge Glass (Probe LWCO) Water Gauge Glass, Pressuretrol, and LWCO (Float)		Flush Plug		
F	¾"	—	L4006A Operating Control	—	Disregard	L8124C Operating Control
G	1½"	Bushed to ¾" for Draincock (Optional Return)		Return		
H	1½"	Return		Bushed to ¾" for Draincock (Optional Return)		
J	1½"	Surface Blowoff - Plugged		Flush Plug		
K	2"	Front Supply (3 thru 9 Section)		Front Supply (3 thru 9 Section)		
L	2"	Plugged, Optional Second Supply (3 thru 5 Section) Required Second Supply (6 thru 9 Section)		Plugged (3 thru 9 Section)		
M	¾"	Safety Valve		Relief Valve		
P	¾"	Auxiliary Tapping - Plugged		Aux. Tapping - Plugged	Disregard	Aux. Tapping - Plugged
R	¾"	Auxiliary Tapping - Plugged		Auxiliary Tapping - Plugged		

Boiler tapping locations. (*Courtesy of Burnham.*)

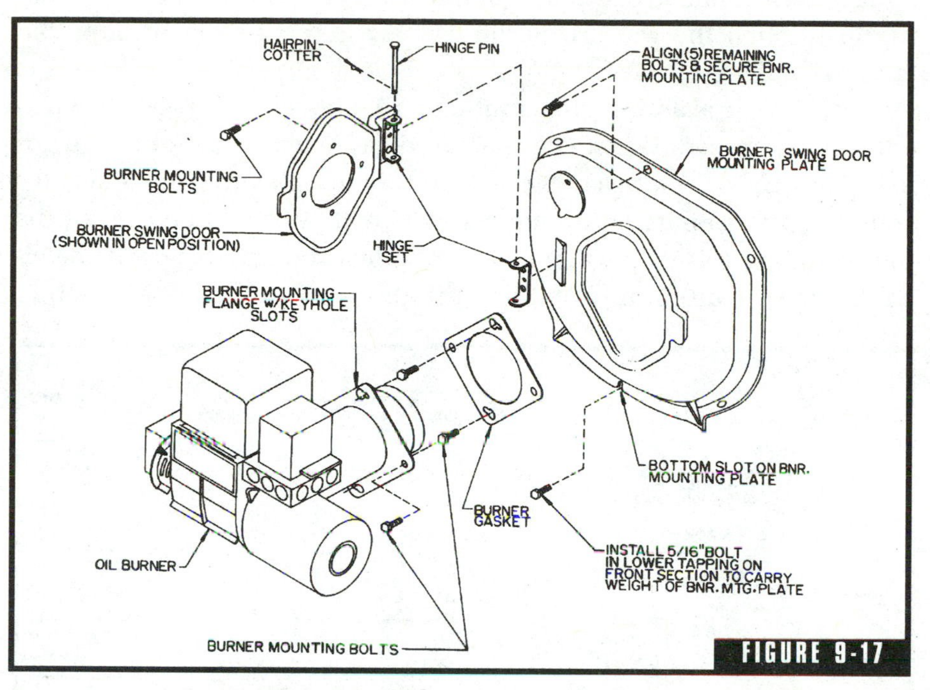

Burner mounting. (*Courtesy of Burnham.*)

The air supply for an atmospheric injection burner is either primary or secondary air. It is commonly introduced and mixed with the gas in the throat of a mixing tube. Gas passes through a small orifice in a mixer head, which is shaped to produce a straight-flowing jet moving at high velocity. The throat of the mixing tube can be called a venturi. Gas flowing through the venturi spreads and induces air in through an opening at an adjustable air shutter. Due to force from the energy of the gas stream, the mixture of gas and air is pushed into the burner manifold casting. At this point, the mixture goes through ports where additional air is

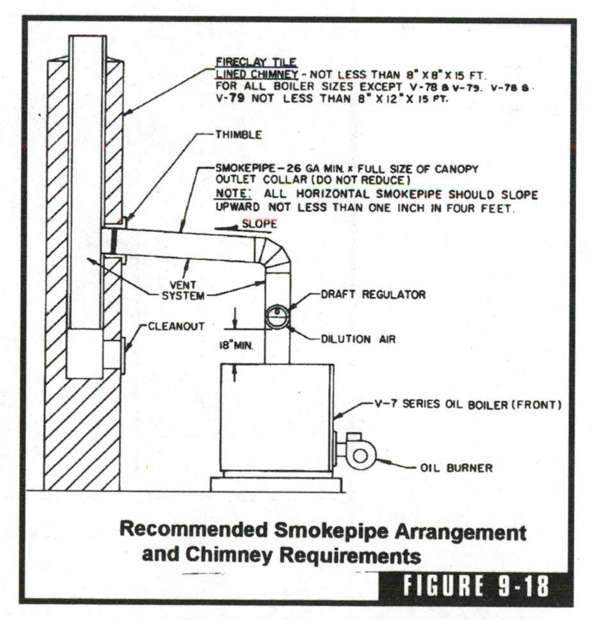

Suggested smokepump. (*Courtesy of Burnham.*)

added to the flame to complete combustion. Primary air is the air coming in through the venturi. Secondary air is the air that is supplied around the flame.

Primary air is admitted at a ratio of 10 parts air to 1 part gas when natural gas is used. When manufactured gas is used, the ratio changes to 5 parts air to 1 part gas. These ratios are just theoretical values. It's common for gas burners to operate well with 40 to 60 percent of the theoretical values. Some elements that affect the amount of gas needed include the uniformity of air distribution and mixing, the direction

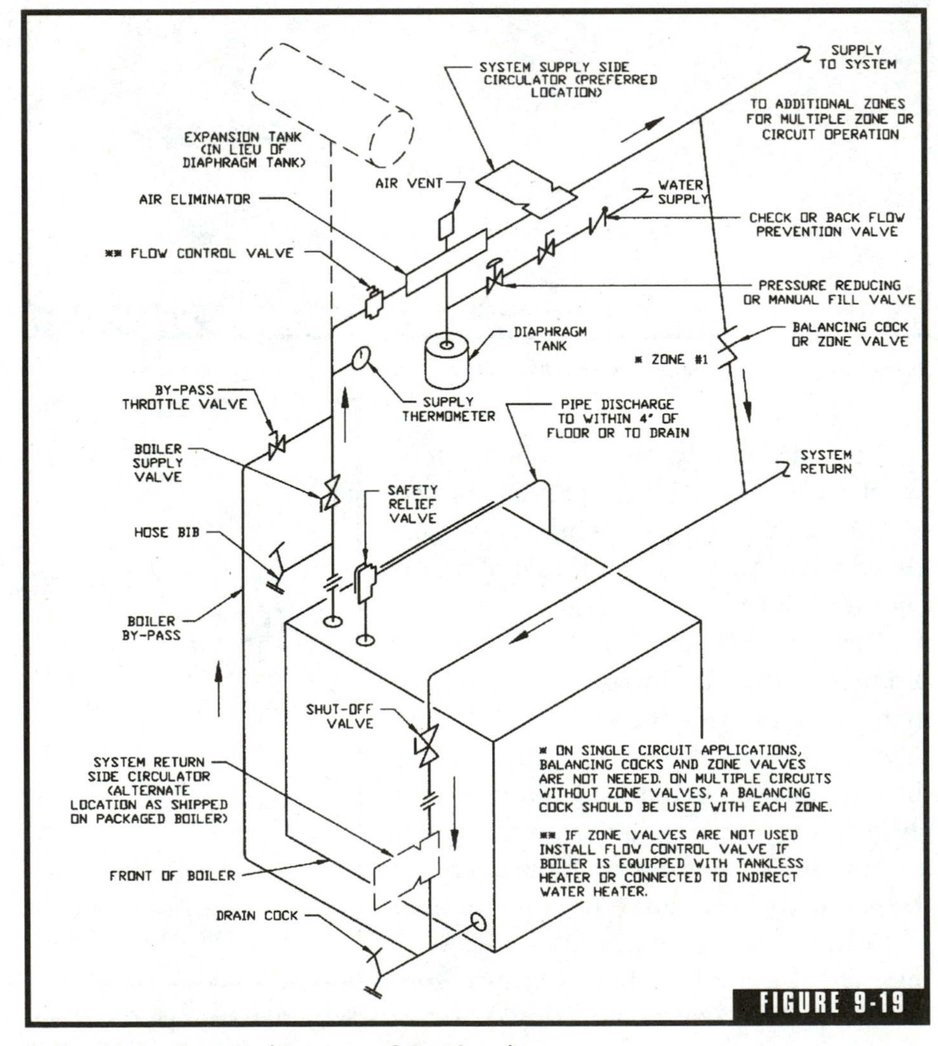

Boiler piping layout. (*Courtesy of Burnham.*)

of travel for gas from the burner, and the height and temperature of the combustion chamber (Fig. 9-21).

Natural drafts provide secondary air. This air should not exceed 35 percent. Draft hoods can be used to control secondary air. The hood can also control back drafts, which might blow out a flame. Ideally, the flame of a gas burner should burn as blue as possible. This occurs when the mixture of primary and secondary air is controlled properly.

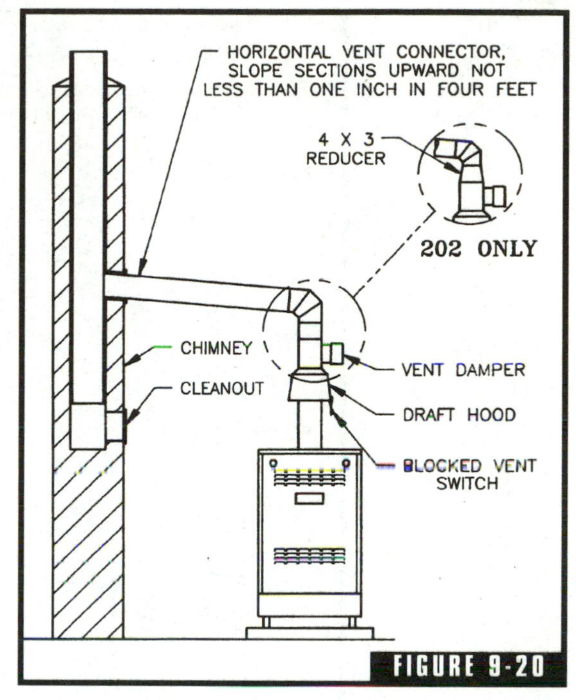

FIGURE 9-20

Boiler venting. (*Courtesy of Burnham.*)

Boiler Selection

Boiler selection depends on many factors. The first consideration is the type of fuel to be used to operate the boiler. It is possible to obtain a boiler that runs on electricity, but gas-fired and oil-fired boilers are, by far, much more common and more popular. Boilers that are designed for operation with natural gas can only be used where natural gas is offered to customers of utility providers. When available, natural gas is an excellent choice as a fuel.

Manufactured gas is usually available in any location. Some people dislike manufactured gas, due the large storage bottles used to hold the gas and the perceived risk of explosion. Oil-fired boilers require the use of large tanks, but fuel oil is not explosive in nature. It burns, of course, but it does not explode. Bottled gas can explode. Such accidents rarely happen, but it is a risk that some people are unwilling to accept.

Fuel oil is available in nearly every region and it's a good choice. Oil can be purchased in the off season, during warm weather, and stockpiled for winter use. Most codes allow the installation of two oil tanks for residential use. With two large tanks, homeowners can store summer oil at lower prices for winter use. Oil burners are dependable and make fine heat sources.

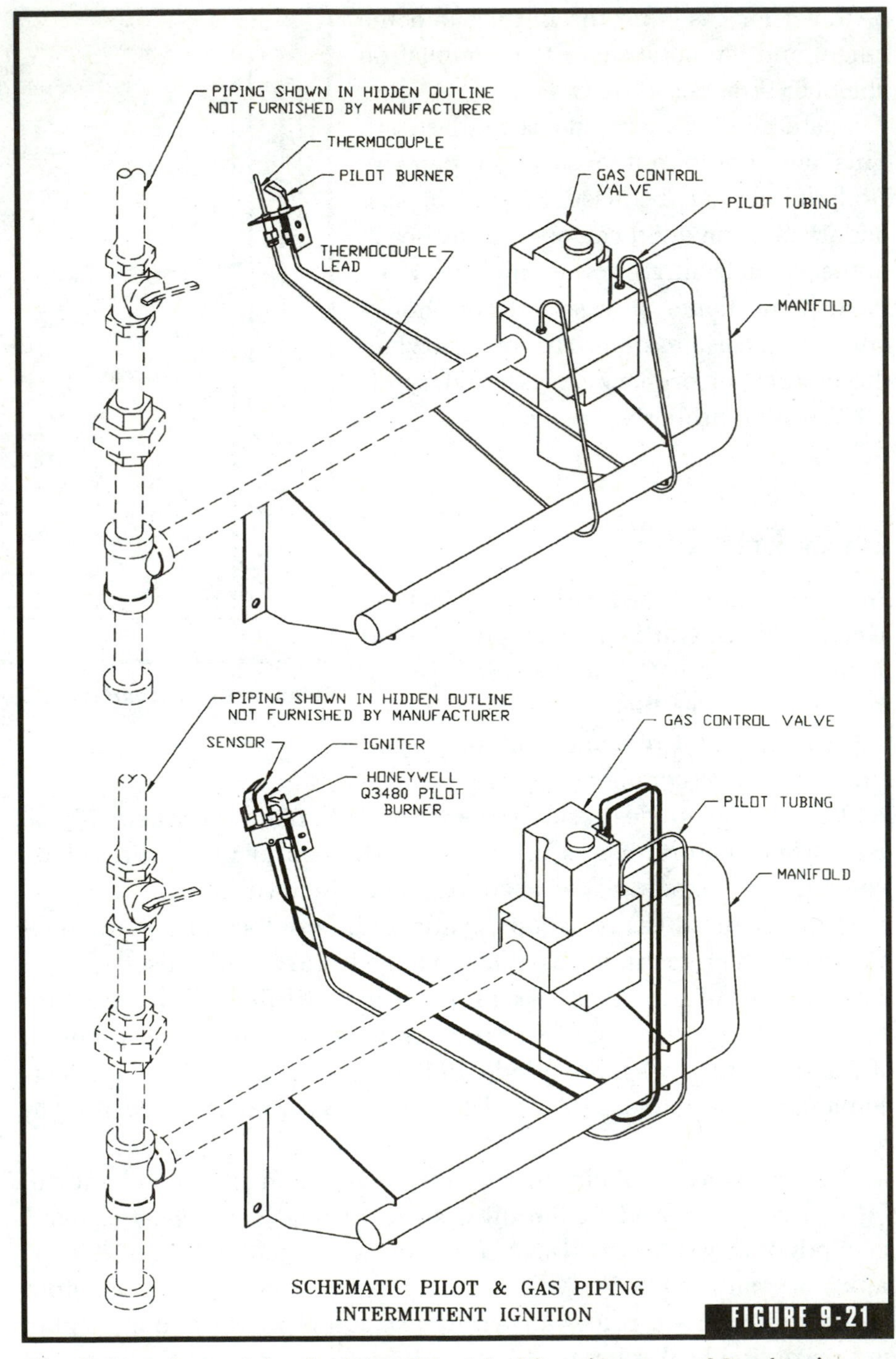

FIGURE 9-21

Pilot light and gas piping for intermittent ignition. (*Courtesy of Burnham.*)

The debate of cast-iron versus steel is one that continues to rage onward. There are advantages to both, and both are suitable for any residential application. Both types of boilers share similar warranties. Steel boilers tend to be less expensive. Cast-iron boilers may be a bit more efficient, but this is a debatable point. The ultimate decision between the two is up to the person paying for the unit. A quality boiler of either type will serve well for years to come.

Deciding on a boiler is only part of the process of building a hydronic heating system. There are many components that must be installed in conjunction with boilers. Circulators, relief valves, zone valves, expansion tanks, and other materials are all needed to make a working system. Let's turn to the next chapter and see what the main components are.

Components for Heating Systems

The components that go into a hydronic heating system are both numerous and important. There are many pieces of equipment needed to make a safe, functional hydronic heating system. People generally think of a boiler and some type of heat emitter, such as baseboard heating units. Rarely is much thought given to circulating pumps, expansion tanks, zone valves, relief valves, and other essential elements of a safe system. The components of a hydronic heating system that may seem inconsequential are not. They are key elements of the system. Failing to install an expansion tank or relief valve could prove disastrous. Without the use of a circulating pump, a modern hydronic system will not function very well. Zone valves are an inexpensive alternative to multiple circulating pumps when creating multiple heating zones with a system. A number of factors come into play when designing and installing a hydronic heating system. Learning what is needed, where it is needed, and how to install it is essential if you wish to install effective hydronic heating systems.

Modern Piping Materials

Modern piping materials in hydronic heating systems include, most often, copper tubing and cross-linked polyethylene tubing (PEX).

Another type of tubing sometimes used is polybutylene (PB) tubing. Copper tubing is the most popular type of piping used for general convection heating, such as systems utilizing baseboard heating elements, kick-space heaters, and space heaters. PEX tubing is most often used for radiant floor heating systems. PB tubing was being used before PEX tubing, and it is still in use, but PEX is replacing it quickly as a prime choice. This is due in no small part to recent problems with PB tubing and pending lawsuits arising from the problems.

The copper tubing used most in heating systems is type-M copper in a hard-drawn (rigid) form. Type-L copper is sometimes used in tight spots when a flexible, rolled copper tubing is more feasible. Distribution tubing typically has a diameter of 3/4 inch. Since copper expands and contracts with temperature variations, the tubing must be supported properly to maintain a quiet heating system.

PEX tubing is a polymer (plastic) material. It is extremely flexible, sold in long coils, and suitable for many hydronic applications. One of the most effective applications for PEX tubing is found in radiant floor heating. Standard PEX tubing can handle water with a temperature of 180° at 100 psi. If the pressure is reduced to 80 psi, the tubing can handle water temperatures up to 200°.

Both copper and PEX tubing have their places in heating systems. Matching the proper tubing to the job is important. A general rule of the thumb is to use copper tubing for general heating applications and to use PEX for radiant floor heating. There are exceptions to this, of course, but in general, the formula works.

Fittings

The fittings used in heating systems are often the same as those used in plumbing jobs. There are, however, some special fittings that are used most frequently with heating systems. Typical, generic fittings include couplings, slip couplings, reducing couplings, 45° elbows, 90° elbows, male adapters, female adapters, unions, and tees of both full size and reducing sizes. All of these are basic plumbing fittings. Now, let's look at the fittings that are primarily used with heating systems.

Baseboard tees, also called baseboard ells, are fittings that are shaped like a standard 90° elbow. However, the fitting has a threaded fitting in the bend of the elbow. The threads are there to accept the installation of an air vent/purger. These fittings are often used when copper tubing rises vertically through a floor and is turning on a 90° angle into a section of baseboard heat. The fittings are usually made of wrought copper or cast brass.

Diverter tees (Fig. 10-1) create a flow through a branch piping path that passes through one or more heat emitters before reconnecting to a primary piping circuit. The use of diverter tees can involve pushing or pulling water through a piping path. It can be difficult to tell a diverter tee from a regular tee when relying only on outward appearance. However, a peek inside will reveal the diversion section of the tee. When diverter tees are used, they must be installed in the proper location and direction. Diverter tees are equipped with arrows to indicate proper positioning for water flow. There are some installers in the trade who call the diverter tees venturi tees. Most commonly, they are known by the registered trademark name of MonoFlo tees, which is a product and trademark of the Bell & Gossett Company.

Dielectric unions are often used in plumbing applications. They are installed often on electric water heaters. These same unions are sometimes used with heating systems. A dielectric union mates together with two dissimilar metals. This breaks the continuity of a conductive reaction. In turn, it avoids galvanic corrosion. The unions are used to fit copper materials to steel materials.

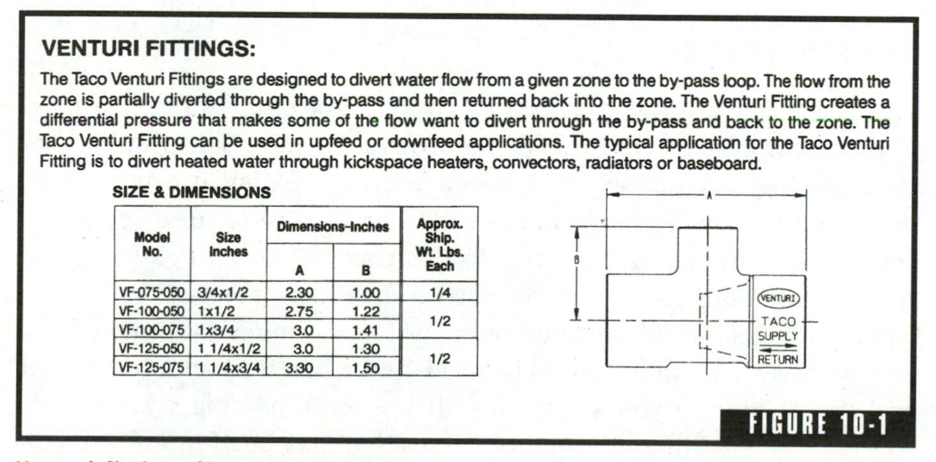

VENTURI FITTINGS:

The Taco Venturi Fittings are designed to divert water flow from a given zone to the by-pass loop. The flow from the zone is partially diverted through the by-pass and then returned back into the zone. The Venturi Fitting creates a differential pressure that makes some of the flow want to divert through the by-pass and back to the zone. The Taco Venturi Fitting can be used in upfeed or downfeed applications. The typical application for the Taco Venturi Fitting is to divert heated water through kickspace heaters, convectors, radiators or baseboard.

SIZE & DIMENSIONS

Model No.	Size Inches	Dimensions-Inches		Approx. Ship. Wt. Lbs. Each
		A	B	
VF-075-050	3/4x1/2	2.30	1.00	1/4
VF-100-050	1x1/2	2.75	1.22	1/2
VF-100-075	1x3/4	3.0	1.41	
VF-125-050	1 1/4x1/2	3.0	1.30	1/2
VF-125-075	1 1/4x3/4	3.30	1.50	

FIGURE 10-1

Venturi fitting. (*Courtesy of Taco.*)

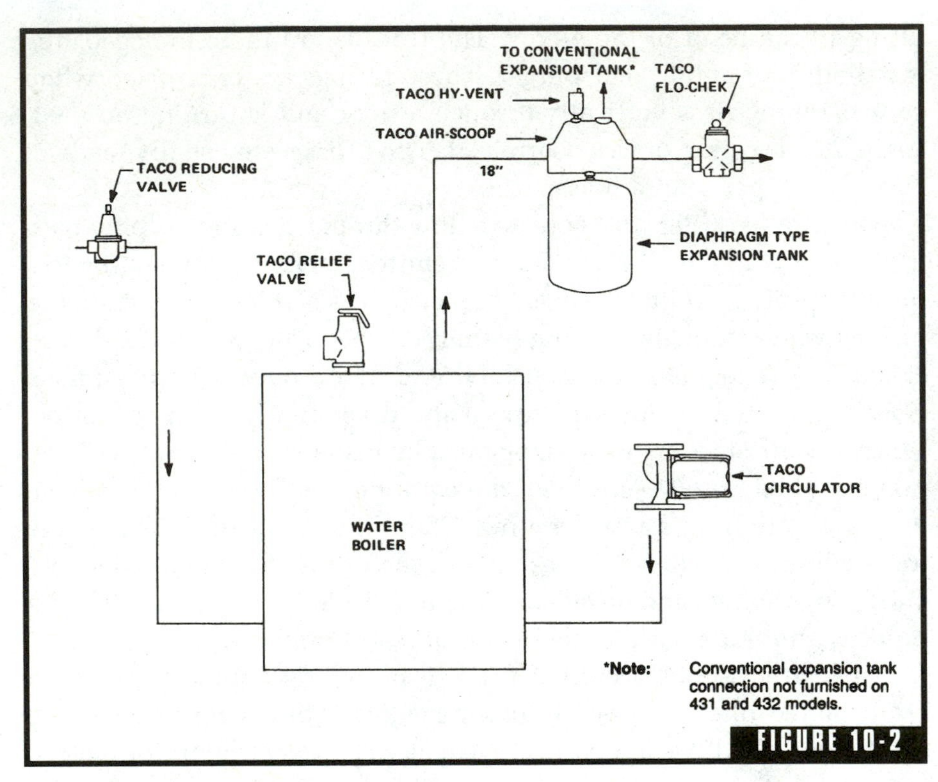

Air scoop. (*Courtesy of Taco.*)

Heating Valves

Valves used in heating systems (Fig. 10-2) can be of the same type used in plumbing systems, but this is not always the case. Selecting a valve for a heating system may seem like a simple task, but it can be more important than you might think. Valves are typically used for either component isolation or flow regulation. Gate valves and globe valves are two common types of valves used in heating systems. These same valves are also used in plumbing systems. Understanding which valves to use, why they should be used, and when they should be used will be of value to you as you install hydronic heating systems. So, let's explore the different types of valves that may be of interest to you in your heating jobs.

Gate Valves

Gate valves are used frequently in both plumbing and heating. These valves are intended for use as isolation valves. They are not meant to be used as flow regulators. This means that gate valves should either be fully open or completely closed. When open, a gate valve does not affect flow velocity very much. Don't install gate valves where flow regulation is required. If the gate is partially down in a gate valve, to regulate flow, there is a strong likelihood that vibration or chatter will be the result (Fig. 10-3).

Globe Valves

Globe valves can be used to regulate flow. In fact, that's what they are intended to do. There is a right and a wrong way to install a globe valve. Make sure that any globe valve you install is positioned so that

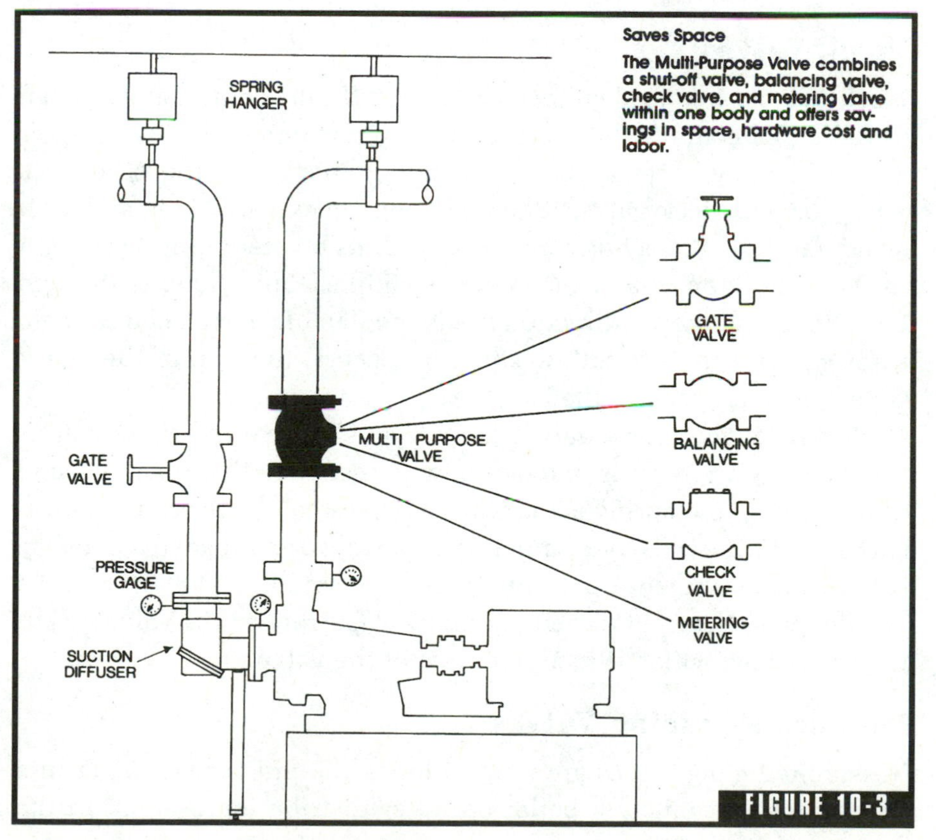

Multipurpose valve. ***(Courtesy of Taco.)***

water enters the lower body chamber. If water comes in from the other end, noise in the system is a strong possibility. Globe valves should not be used to isolate equipment. The design of a globe valve does not make it ideal for isolation. While globe valves can be used for isolation, they shouldn't be, since they are not the most efficient choice.

Ball Valves

Ball valves are probably one of the most used valves in the heating industry. These valves can be used for isolating components or regulating flow. The positive closing action of a ball valve makes it a fine choice as an isolation valve. While ball valves can be used for flow regulation, it is not considered wise to use ball valves to control flow when the flow must be reduced by more than 25 percent. There is no hard and fast rule on this. It is a recommendation to maintain valve condition.

Check Valves

Check valves are needed to ensure that fluids do not flow in an unwanted direction. There are two common types of check valves used in heating systems. One is a swing check and the other is a spring-loaded check valve. When a swing check valve is installed, it must be installed on a horizontal line with its bonnet pointing straight up. The operation of a swing check is simple. Fluid flowing through the valve in the proper direction holds the flap of a swing check open. If, for any reason, a backflow situation occurs, the flap of the check valve closes, preventing the backflow.

Spring-loaded check valves are not as sensitive to orientation as swing checks are. This is one reason why installers like spring-loaded valves. Due to the spring action, these valves can be used in any orientation. While it is not important that a spring-loaded check be installed in a straight-up position, it is essential that the valve be installed with the right direction of flow. An arrow on the valve makes it easy to know which direction to install the valve in.

Pressure-Reducing Valves

Pressure-reducing valves are used to lower the pressure of water in a system before it enters a boiler or water distribution system. In the case of heating, the pressure-reducing valve, also known as a boiler feed valve, is used to lower pressure from the water distribution system prior to it entering a boiler. As with check valves, pressure-reducing valves must be installed in the proper direction. The valves have arrows to point them in the direction of flow. There is usually a lever on the top of a pressure-reducing valve that can be lifted manually to speed up the filling process for a boiler during a setup and start-up procedure.

Pressure-Relief Valves

Pressure-relief valves are important safety valves. When temperature or pressure reaches a risky level, these valves will open to release water that might otherwise cause damage or even an explosion. These valves are required by code on all hydronic heating systems. It's common for new boilers to be shipped with a relief valve already installed. A typical rating for a relief valve in a small boiler is 30 psi. Never install a heating system without a relief valve. Relief valves are equipped with threads to accept a discharge tube. This tube is very important. If a relief valve blows off without a proper safety tube installed, people could be seriously injured from hot water or steam. I see far too many relief valves that are not equipped with discharge tubes. The tube should run from the valve to a point about 6 inches above a floor drain. If a drain is not available, at least extend the pipe to within approximately 6 inches above the finished floor level.

Backflow Prevention

Backflow prevention is an important part of a hydronic heating system. It simply would not do to have boiler water mixing with potable water. Local codes require backflow prevention, and there are various types of devices available to achieve code requirements. Most systems utilize in-line devices. These backflow preventers must be installed in the proper direction. They have arrows on the valve bodies to indicate the direction of flow. It is common for this type of device to have a threaded opening about midway on the valve that will act as a vent. Again, a safety tube should be installed on the threaded vent opening to prevent spraying or splashing if the valve discharges.

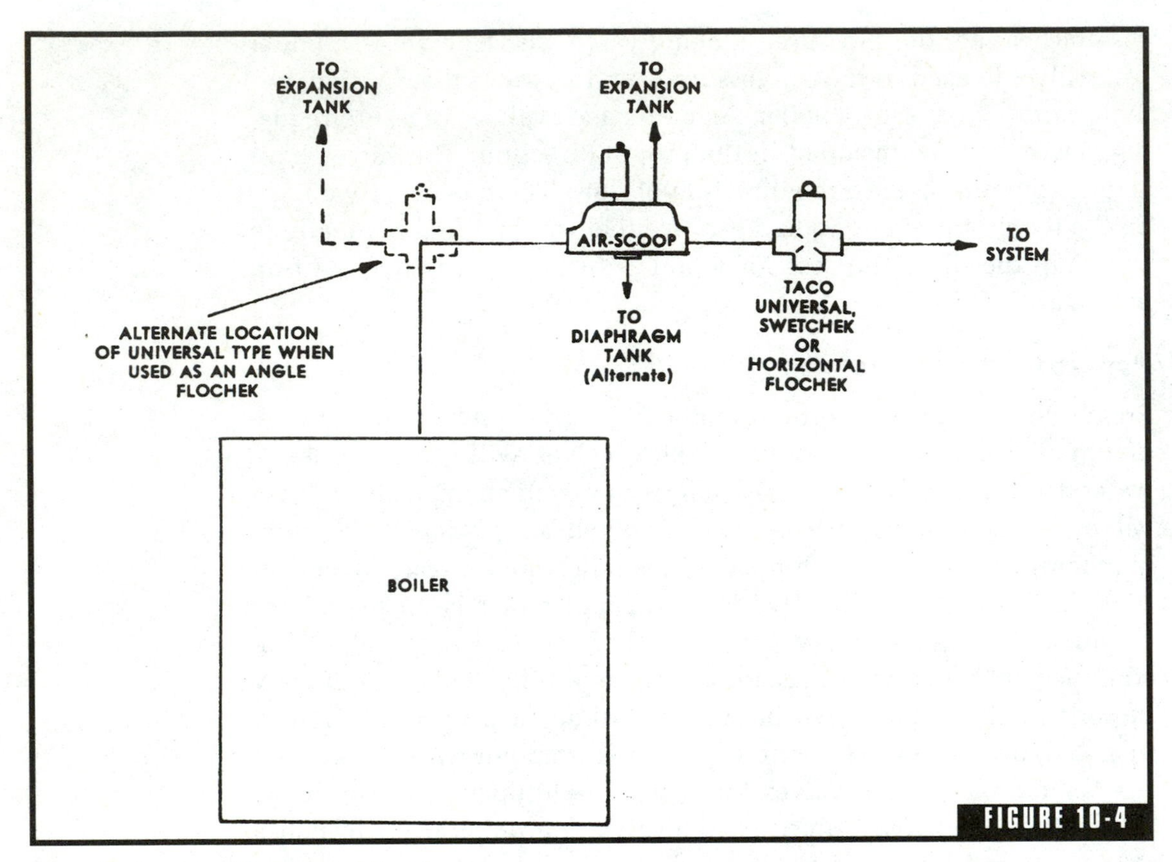

Flow-check valve. (*Courtesy of Taco.*)

Flow Checks

Flow-check valves (Fig. 10-4) are another type of valve used in heating systems. These valves have weighted internal plugs that are heavy enough to stop thermosiphoning or gravity flow when a system's circulating pumps are not running. Some flow checks have two ports, while others have three. Valves with two ports are intended for use in horizontal piping. When three ports exist, one can be plugged and the valve can be used in a vertical application. A small lever on top of the valves allows the valve to be opened manually in the event of a circulator failure. The lever, however, cannot be used to control flow rate. Again, the direction of flow is important when installing a flow check. An arrow on the body of the valve will indicate the direction of flow as it should be used with the valve (Fig. 10-5).

Mixing Valves

Mixing valves (Figs. 10-6 and 10-7) are used to mix cold water with hot water to create a regulated temperature in water being delivered from the mixing valve. This might be the case in a hydronic system that uses both baseboard heat emitter and radiant floor heating. The temperature of water for the floor heating would need to be lower than that of the water used for the baseboard system. This is possible with the use of a mixing valve. Using knobs or levers on the outside body of a mixing valve is all that is required to regulate the temperature of water being delivered from the valve.

Zone Valves

Zone valves can be used in place of additional circulators when a hydronic heating system is being zoned off into different zones. Circulating pumps cost more than zone valves, so the zone valves are often used in place of additional circulators. There are old-school installers who don't like zone valves. A preference between circulators

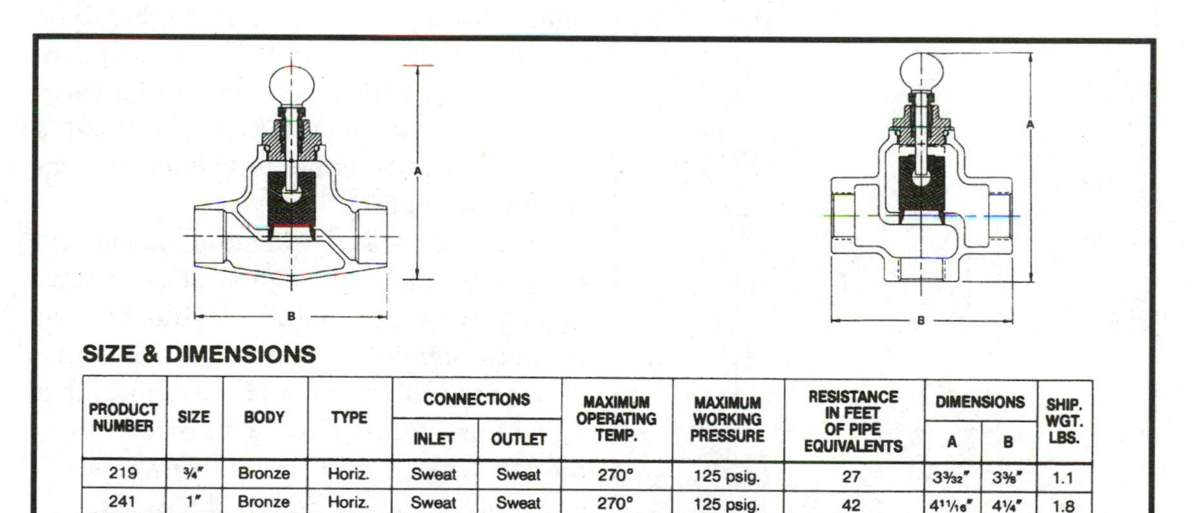

SIZE & DIMENSIONS

PRODUCT NUMBER	SIZE	BODY	TYPE	CONNECTIONS		MAXIMUM OPERATING TEMP.	MAXIMUM WORKING PRESSURE	RESISTANCE IN FEET OF PIPE EQUIVALENTS	DIMENSIONS		SHIP. WGT. LBS.
				INLET	OUTLET				A	B	
219	¾"	Bronze	Horiz.	Sweat	Sweat	270°	125 psig.	27	3 3/32"	3⅜"	1.1
241	1"	Bronze	Horiz.	Sweat	Sweat	270°	125 psig.	42	4 11/16"	4¼"	1.8
218	¾"	C.I.	Univ.	NPT	NPT	270°	125 psig.	27	4 29/32"	3 29/32"	1.1
220	1"	C.I.	Univ.	NPT	NPT	270°	125 psig.	42	5"	4¼"	3
221	1¼"	C.I.	Univ.	NPT	NPT	270°	125 psig.	60	5 11/16"	4¾"	4.8
222	1½"	C.I.	Univ.	NPT	NPT	270°	125 psig.	63	6 13/32"	6"	7.8
223	2"	C.I.	Univ.	NPT	NPT	270°	125 psig.	83	7"	6¾"	10.3

FIGURE 10-5

Flow-check data. ***(Courtesy of Taco.)***

TEMPERING VALVES:

Taco Tempering Valves are available in ½-inch and ¾-inch sweat connections. The basic construction is brass and stainless steel. The Tempering Valve saves energy by reducing the outgoing water temperature at the hot water source.

SIZE & DIMENSIONS

PRODUCT NUMBER	SIZE CONNS.	TYPE CONNS.	TYPE ADJUST.	RATINGS BATH	RATINGS GPM	TEMP. RANGE	MAXIMUM TEMP.	MAXIMUM PRESSURE	LENGTH	SHIP. WT. LB.
508	½"	Sweat	External	1-2	6	120-160	200°F	125 psig.	3¾"	.5
526	¾"	Sweat	External	1-3	12	120-160	200°F	125 psig.	3¾"	1.0

508, 526

FIGURE 10-6

Tempering valves. (*Courtesy of Taco.*)

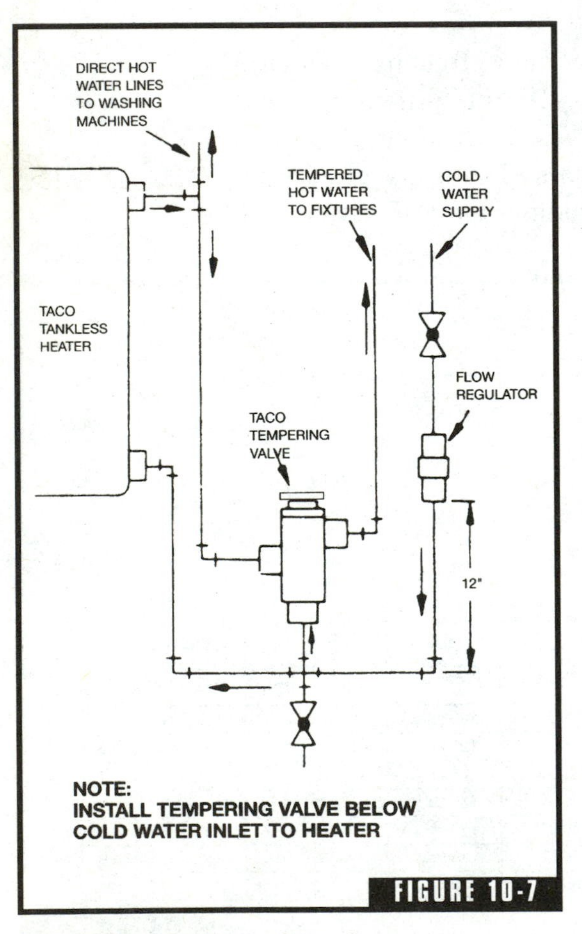

Installation of a tempering valve. (*Courtesy of Taco.*)

and zone valves is a personal matter. Most installers are comfortable using zone valves, and many contractors use them extensively (Fig. 10-8).

There are two types of zone valves. One type uses a small electric motor combined with gears to produce a rotary motion of a valve shaft. The other type uses heat motors to produce a linear push-pull motion. Both types consist of a valve body and an actuator. Either type of valve must be operated either fully open or completely closed (Fig. 10-9).

Zone valves for residential systems are located on the supply pipe of each zone circuit. They are generally positioned near the heat source. In most cases, the zone valves are equipped with transformers that allow them to be wired with regular thermostat wire. There are, however, some zone valves that are designed to work with full, 110-volt power (Fig. 10-10).

Other Types of Valves

Other types of valves are sometimes used in hydronic heating systems. For example, a differential pressure bypass valve might

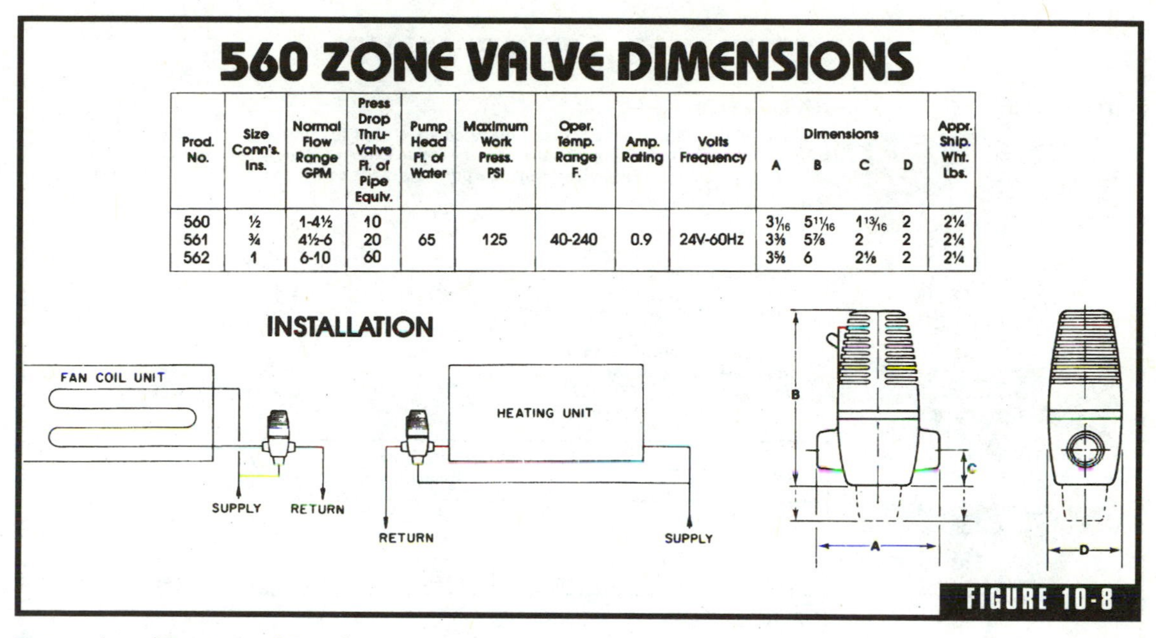

560 ZONE VALVE DIMENSIONS

Prod. No.	Size Conn's. Ins.	Normal Flow Range GPM	Press Drop Thru-Valve Ft. of Pipe Equiv.	Pump Head Ft. of Water	Maximum Work Press. PSI	Oper. Temp. Range F.	Amp. Rating	Volts Frequency	Dimensions A	B	C	D	Appr. Ship. Wht. Lbs.
560	½	1-4½	10						3¹⁄₁₆	5¹¹⁄₁₆	1¹³⁄₁₆	2	2¼
561	¾	4½-6	20	65	125	40-240	0.9	24V-60Hz	3⅜	5⅞	2	2	2¼
562	1	6-10	60						3⅝	6	2⅛	2	2¼

Zone valve. (*Courtesy of Taco.*)

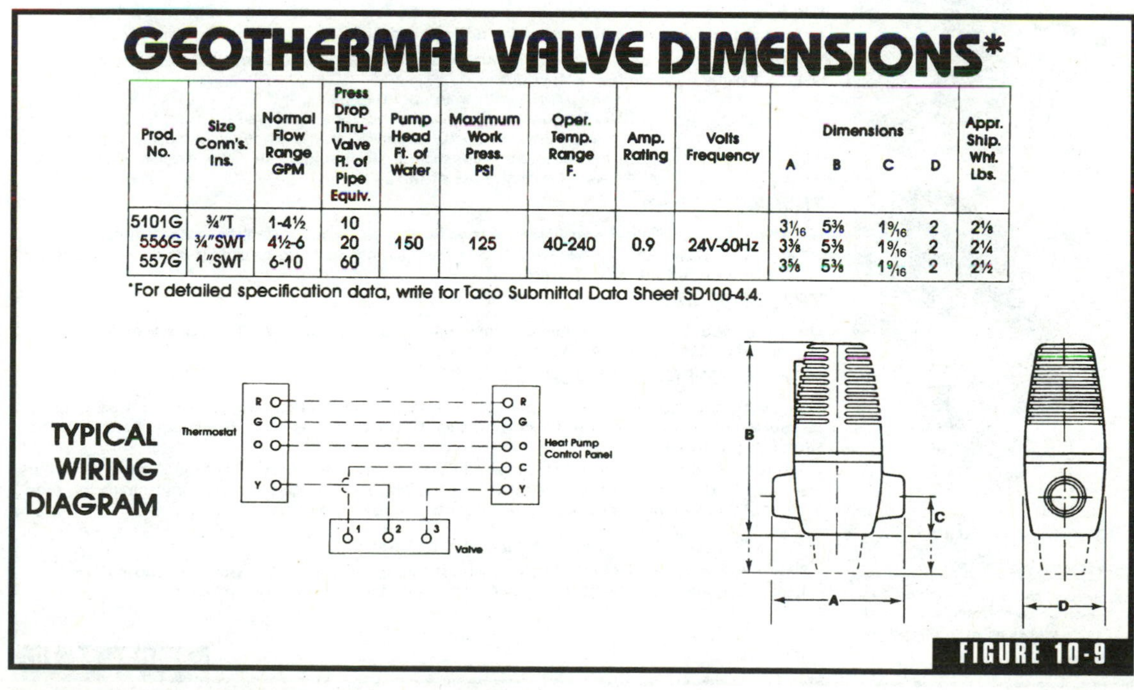

GEOTHERMAL VALVE DIMENSIONS*

Prod. No.	Size Conn's. Ins.	Normal Flow Range GPM	Press Drop Thru-Valve Ft. of Pipe Equiv.	Pump Head Ft. of Water	Maximum Work Press. PSI	Oper. Temp. Range F.	Amp. Rating	Volts Frequency	Dimensions A	B	C	D	Appr. Ship. Wht. Lbs.
5101G	¾"T	1-4½	10						3¹⁄₁₆	5⅜	1⁹⁄₁₆	2	2⅛
556G	¾"SWT	4½-6	20	150	125	40-240	0.9	24V-60Hz	3⅜	5⅜	1⁹⁄₁₆	2	2¼
557G	1"SWT	6-10	60						3⅝	5⅜	1⁹⁄₁₆	2	2½

*For detailed specification data, write for Taco Submittal Data Sheet SD100-4.4.

Geothermal valve. (*Courtesy of Taco.*)

SERVICING A ZONE VALVE

TO SERVICE

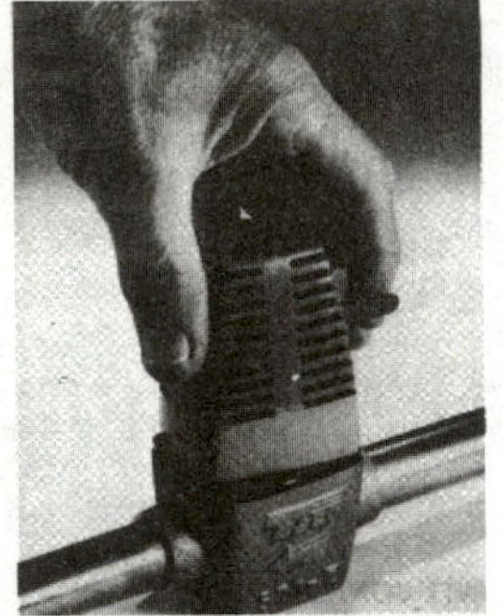

1—Twist off Power Head *

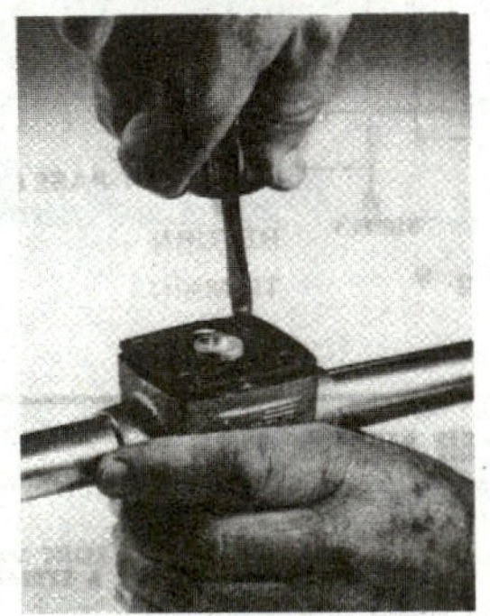

2—Remove the 4 Screws from hold down plate

3—Remove seat assembly

TO RE-ASSEMBLE

Reverse above procedure pushing down on seat assembly and hold down plate against return spring.

APPLICATION

The Taco-Zone Valve is an electricity operated valve used for zone control of Hydronic Heating and/or Cooling Systems. It controls the flow of water in a room or zone in response to the demands of the room or zone thermostat. This valve is a precisely made device and must be installed with care.

RATING

Electrical:	volts	24
	amps	0.9
Temperature range,	max.	240F (115C)
	min.	40F (5C)
Working pressure (at valve, including	pump head)	125PSI (861kPa)
Maximum differential pressure across valve (pump head)		
	555-557	150 Ft. (46m)
	560-562, 571-573	65 Ft. (20m)

Size	Pressure Drop Characteristics			
	C_v	K_v	Equiv. Length of Pipe Feet	Meters
½"	4.2	3.6	8	2.4
¾"	6.1	5.3	20	6.1
1"	7.0	6.1	65	19.8
1¼"	7.2	6.2	150	45.7

INSTALLATION

Valves should be installed vertically, to simplify replacement or cleaning of the seat, if ever required at some future date. The vertical installation permits drawing a vacuum in the system and replacing or cleaning the seat without draining the system.

When installing 560 Series Valves make sure that flow is in at the unit and by-pass connections and out at the main connection as shown in Fig. 8 and Fig. 9.

Valve may be sweat into the line without taking apart, provided, care is taken to prevent overheating. Follow these simple instructions:—

1. Use a torch with sharp, pointed flame.
2. Clean surfaces thoroughly and use a good grade of flux.
3. Use 50-50 or 60-40 solder. If grades of solder requiring higher temperatures are used, such as silver solder, the valve must be dismantled.
4. Avoid excessive use of flux.

THERMOSTAT

Use a No. 568 Taco Thermostat (designed specifically for Taco-Zone Valves) with Heat Anticipator set at " D ". Other suitable two wire (SPST) Thermostats may also be used if Heat Anticipator can be set at 0.9 Amps to match valve rating.

TRANSFORMER

Use a No. 569 Taco Transformer or other make rated at 115/24V-40VA. One transformer can accommodate a maximum of 3 Taco-Zone Valves.

MANUAL OPENING LEVER

For gravity circulation thru valve, push lever in Power Head all the way down. Push back up to restore to automatic operation. Lever moves easily when valve is open. Resistance is encountered when valve is closed.

CAUTION: Addition of certain chemical additives to systems utilizing Taco equipment, voids the warranty. Product can withstand antifreeze additives, ethylene glycol and propylene glycol, provided that there are no hydrocarbon constituents in these antifreezes.

* IMPORTANT NOTE

Never remove Power Head while thermostat is calling for heat. If necessary to remove Power Head, disconnect No. 1 wire from Power Head, wait two minutes, then proceed.

FIGURE 10-10

Servicing a zone valve. (*Courtesy of Taco.*)

be used in a system where there are numerous individual zones. When a large circulating pump is used for the entire system, pressure can build up if several of the zones are shut down. This is not good, due to high flow rates and possible noise in the zones that are running. The bypass valve eliminates the problem.

Metered balancing valves are another type of valve that may be encountered in a hydronic heating system. These valves may be used in multizone systems where more than one parallel piping path exists. The flow rates in the pipes must be balanced to produce desirable heating conditions, and metered balancing valves make this possible.

Another type of valve that is sometimes used is the lockshield-balancing valve. This is a valve that allows a system to be isolated, balanced, or even drained at individual heat emitters. These valves can be purchased in a straight, in-line fashion or in an angle version. The installation of lockshield-balancing valves is not common in typical residential applications, but they may be installed in such systems.

Circulating Pumps

Circulating pumps are to a heating system what the heart is to the human body (Fig. 10- 11). The pumps move fluid through the pipes of a heating system. Closed-loop, fluid-filled, hydronic heating systems may be equipped with a single circulating pump or with many. When zone valves are used, it is common for only a single pump to be installed. Most circulating pumps are centrifugal pumps. The design and types of circulating pumps vary. Many of them are of a three-piece design, while others are in-line designs (Figs. 10-12 and 10-13).

Wet rotor circulators are quite common in small heating systems. This type of pump has a motor, a shaft, and an impeller fitted into a single assembly. The assembly is housed in a chamber that is filled with system fluid, and the motor of the pump is cooled and lubricated by the system fluid. There are no fans or oiling caps. Maintenance of a wet rotor circulator is minimal. Due to their quiet operation and their worry-free maintenance, wet rotor circulators rule the roost when it comes to residential and light commercial heating systems.

Three-piece circulators are also common in residential and light commercial applications. One advantage to this type of pump is that

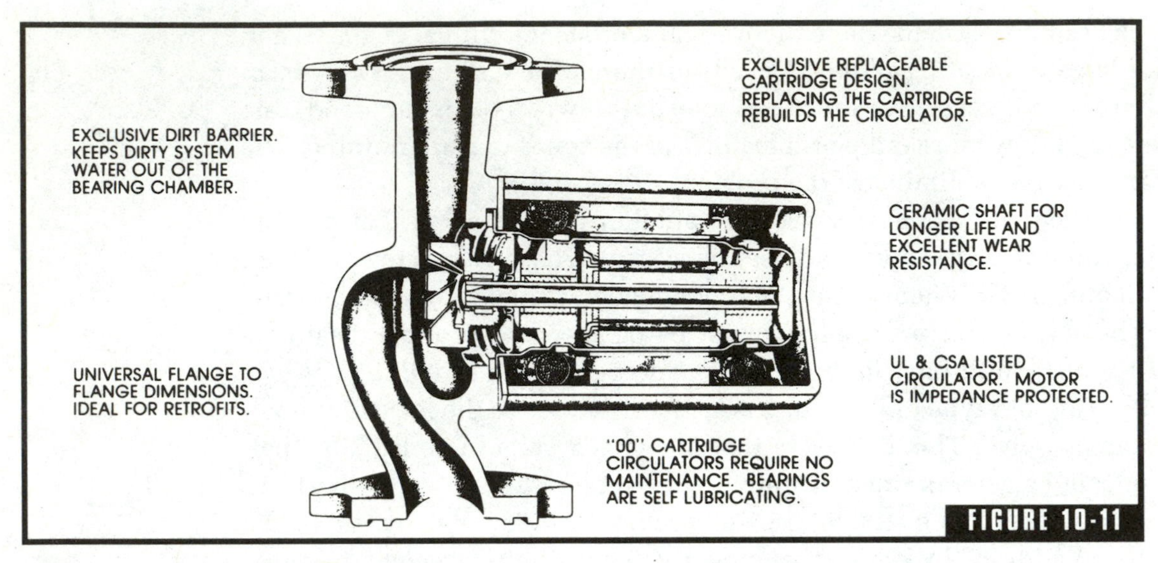

FIGURE 10-11

Cutaway of a circulator. (*Courtesy of Taco.*)

the motor is not housed within the system fluid. If a problem exists, the motor can be worked on without opening the wet system. The disadvantage of a three-piece circulator is that it must be oiled periodically and more noise is present during operation, since the motor is mounted externally.

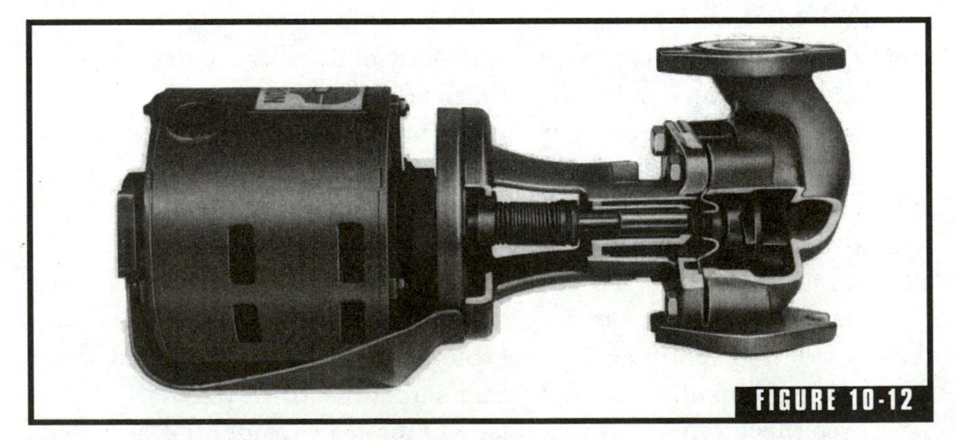

FIGURE 10-12

Typical circulator. (*Courtesy of Taco.*)

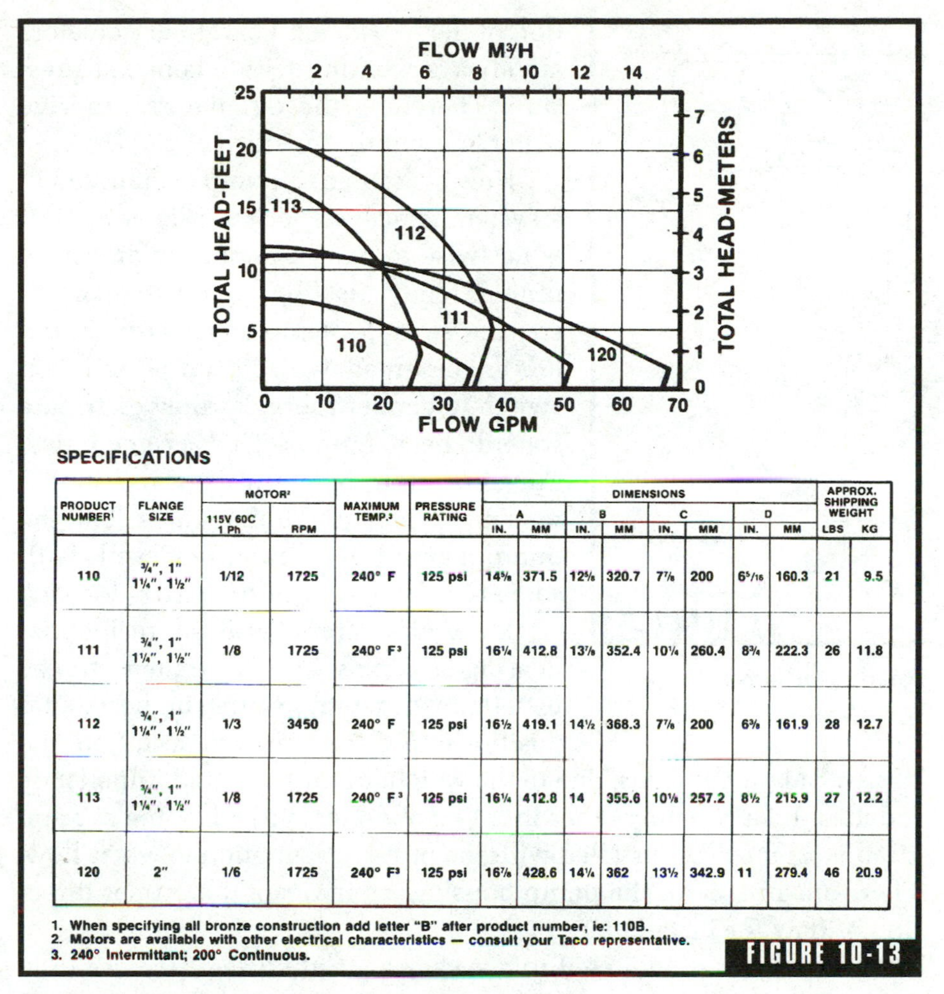

SPECIFICATIONS

PRODUCT NUMBER[1]	FLANGE SIZE	MOTOR[2]		MAXIMUM TEMP.[3]	PRESSURE RATING	DIMENSIONS								APPROX. SHIPPING WEIGHT	
		115V 60C 1 Ph	RPM			A		B		C		D			
						IN.	MM	IN.	MM	IN.	MM	IN.	MM	LBS	KG
110	¾", 1" 1¼", 1½"	1/12	1725	240° F	125 psi	14⅝	371.5	12⅝	320.7	7⅞	200	6 5/16	160.3	21	9.5
111	¾", 1" 1¼", 1½"	1/8	1725	240° F[3]	125 psi	16¼	412.8	13⅞	352.4	10¼	260.4	8¾	222.3	26	11.8
112	¾", 1" 1¼", 1½"	1/3	3450	240° F	125 psi	16½	419.1	14½	368.3	7⅞	200	6⅜	161.9	28	12.7
113	¾", 1" 1¼", 1½"	1/8	1725	240° F[3]	125 psi	16¼	412.8	14	355.6	10⅛	257.2	8½	215.9	27	12.2
120	2"	1/6	1725	240° F[3]	125 psi	16⅞	428.6	14¼	362	13½	342.9	11	279.4	46	20.9

1. When specifying all bronze construction add letter "B" after product number, ie: 110B.
2. Motors are available with other electrical characteristics — consult your Taco representative.
3. 240° Intermittant; 200° Continuous.

Circulator specifications. (*Courtesy of Taco.*)

Pump Placement

Deciding on where to place circulating pumps is not a job that should be taken lightly. Proper placement has much to do with the quality of service derived from a pump. A rule of thumb to follow is to keep all circulators located in a manner so that their inlet is close to the connection point of the system's expansion tank. The reason for this is that the expansion tank is responsible for controlling the pressure of a system's fluid. By keeping circulators near the part of the system that is considered to be the point-of-no-pressure-change, which is the loca-

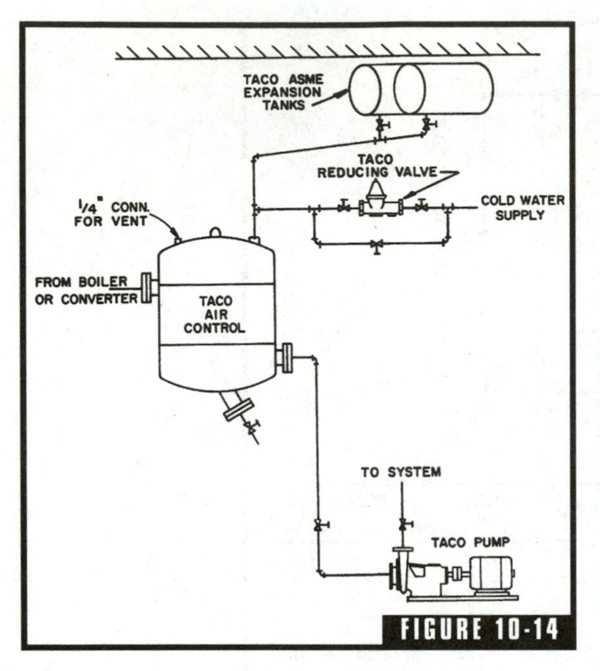

FIGURE 10-14

Air control layout. (*Courtesy of Taco.*)

tion of the expansion tank, the circulators are always working with a constant pressure. Therefore, the circulators can give better, more uniform service.

How a circulating pump is mounted in a system is also important (Fig. 10-14). It is not wise to hang a circulator on flimsy piping. Many installers build their headers, the section of piping where circulators are mounted, with steel pipe and then switch to copper tubing as they distribute water to heat emitters. This is a good idea. The circulators can, however, be mounted in copper pipe or tubing lines, but the pump should be supported well with some type of hanger or bracket.

Most circulators used in residential heating systems are designed to be installed with their shafts in horizontal positions. In doing this, pressure from the thrust load on bushings, due to the weight of the rotor and impeller, is reduced. Like so many other in-line components of a heating system, circulators must be installed with the proper orientation to water flow. There are arrows on the pump housings to indicate the proper direction of flow for installation.

Circulators are mounted to a system with the use of flanges (Fig. 10-15). The flanges are bolted to the circulator so that the pump may be removed for repair or replacement. Smart installers place valves on either side of circulators to make the replacement procedure easier if it is ever required. By having isolation valves both above and below a pump, it is a simple matter to remove the pump without draining the entire system.

Expansion Tanks

Any hydronic heating system must be equipped with an expansion tank (Figs. 10-16 and 10-17). When water is heated in a heating system, the liquid expands. Since thermal expansion of this type is unavoidable, it must be given a means to occur without damage to the heating system. This is done with an expansion tank. The air cushion that is provided in an expansion tank allows water to expand and contract naturally, without fear of damage to the heating system or people in its vicinity. Without an expansion tank, a hydronic heating system could rupture or explode, causing severe damage to both property and people.

The concept behind an expansion tank is easy to understand. The tank has a specified amount of air in it. When water is forced into the tank, the air is compressed. The volume of water in the tank increases, but the air cushion creates a buffer for the expanding water. For example, if the water temperature in the expansion tank is 70° the tank might be half full of water. The same tank holding water at a temperature of 160° might have only one-fourth of its space not filled with water. These are not scientific numbers, just examples. Basically, the cooler the water is, the less water there will be in the tank (Fig. 10-18).

Old-style expansion tanks were basically just metal tanks with an air valve on them where air could be injected. A common problem with this type of tank was related to keeping air in the tank. As air escaped, the tank filled with more water than it should have. This condition is known as waterlogging. It used to be a common problem in both heating and well systems. The problem of waterlogging has been all but eliminated with new technology in the form of diaphragm-type tanks. These tanks have a flexible diaphragm built into them that regulates air charges and greatly reduces, or eliminates, waterlogging (Fig. 10-19).

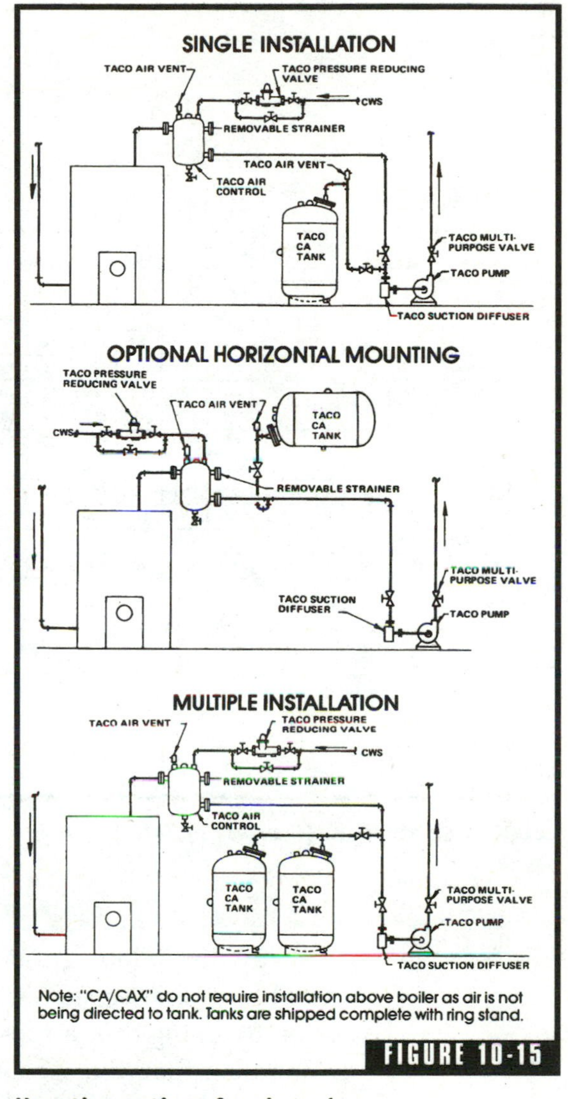

FIGURE 10-15

Mounting options for air tanks. (*Courtesy of Taco.*)

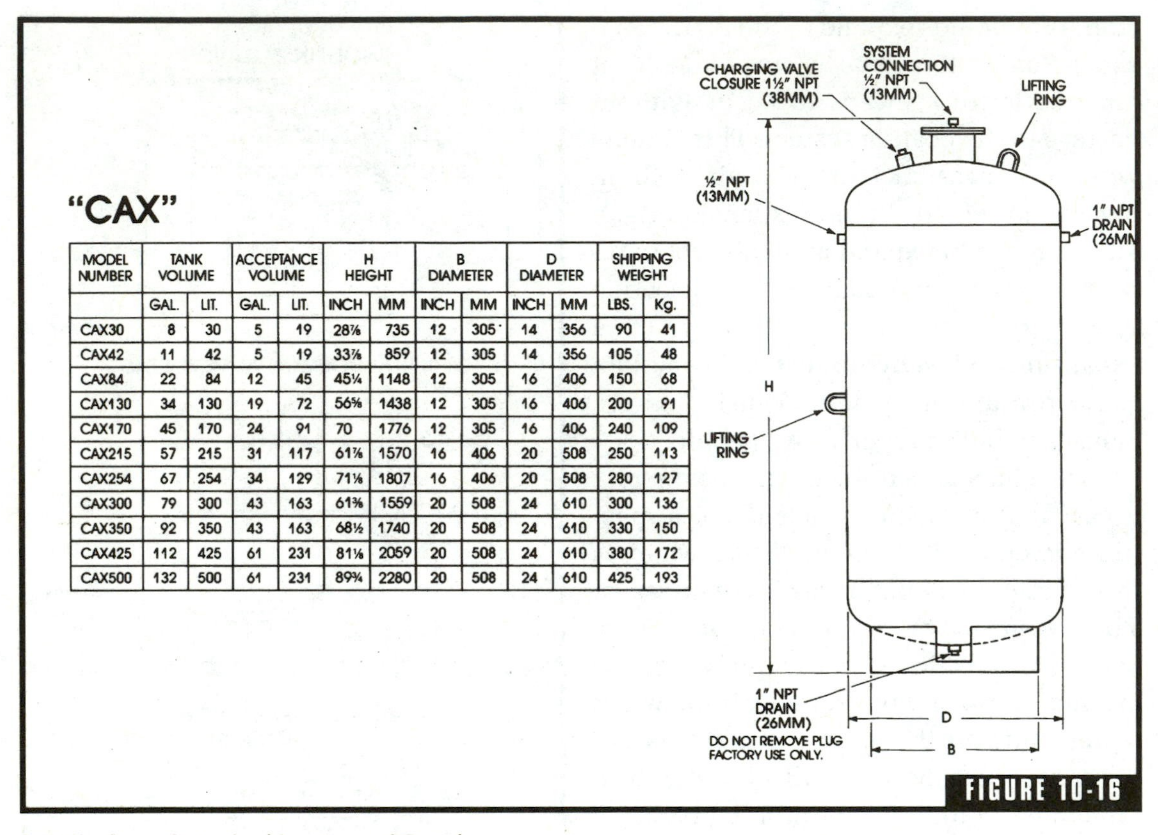

"CAX"

MODEL NUMBER	TANK VOLUME		ACCEPTANCE VOLUME		H HEIGHT		B DIAMETER		D DIAMETER		SHIPPING WEIGHT	
	GAL.	LIT.	GAL.	LIT.	INCH	MM	INCH	MM	INCH	MM	LBS.	Kg.
CAX30	8	30	5	19	28⅞	735	12	305	14	356	90	41
CAX42	11	42	5	19	33⅞	859	12	305	14	356	105	48
CAX84	22	84	12	45	45¼	1148	12	305	16	406	150	68
CAX130	34	130	19	72	56⅝	1438	12	305	16	406	200	91
CAX170	45	170	24	91	70	1776	12	305	16	406	240	109
CAX215	57	215	31	117	61⅞	1570	16	406	20	508	250	113
CAX254	67	254	34	129	71⅛	1807	16	406	20	508	280	127
CAX300	79	300	43	163	61⅜	1559	20	508	24	610	300	136
CAX350	92	350	43	163	68½	1740	20	508	24	610	330	150
CAX425	112	425	61	231	81⅛	2059	20	508	24	610	380	172
CAX500	132	500	61	231	89¾	2280	20	508	24	610	425	193

Detail of an air tank. (*Courtesy of Taco.*)

How do diaphragm tanks work? They are fitted with a synthetic diaphragm that separates the captive air in the tank from water in the tank. By doing this, air loss is greatly diminished. These tanks are the rule rather than the exception in modern heating and well systems. It is important that the diaphragm material used in an expansion tank be compatible with fluids used in the heating system. All diaphragm materials are safe to use with water, but butyl rubber, which is one type of diaphragm material, is not compatible with glycol-based antifreezes, which may be present in some heating systems. If the system you are installing a tank for will contain glycol-based antifreezes, a tank with an EPDM diaphragm material is a better choice. Hydrin diaphragm materials are most commonly used with solar-powered, closed-loop systems (Fig. 10-20).

Check the pressure and temperature ratings assigned to any tank you are planning to use with a heating system. A typical rating for residential use might be 60 psi and 240°. These ratings must be matched to the safety temperature-and-pressure-relief (T&P) valves used with a system. If a typical T&P valve for a residential water heater, rated for 150 psi, were used with an expansion tank system that is rated for 60 psi, the tank could rupture before the T&P valve discharged (Fig. 10-21).

Selection and Installation

The selection and installation of expansion tanks can be confusing. There are tanks available in many shapes and sizes. Most residential heating systems will work properly with tanks having capacities of 10 gallons or less. Sometimes the tanks are mounted vertically. Others are mounted horizontally. Most residential systems have the expansion tank suspended either from piping or ceiling joists. Large expansion tanks are usually floor mounted. In any case, the air-inlet valve for any

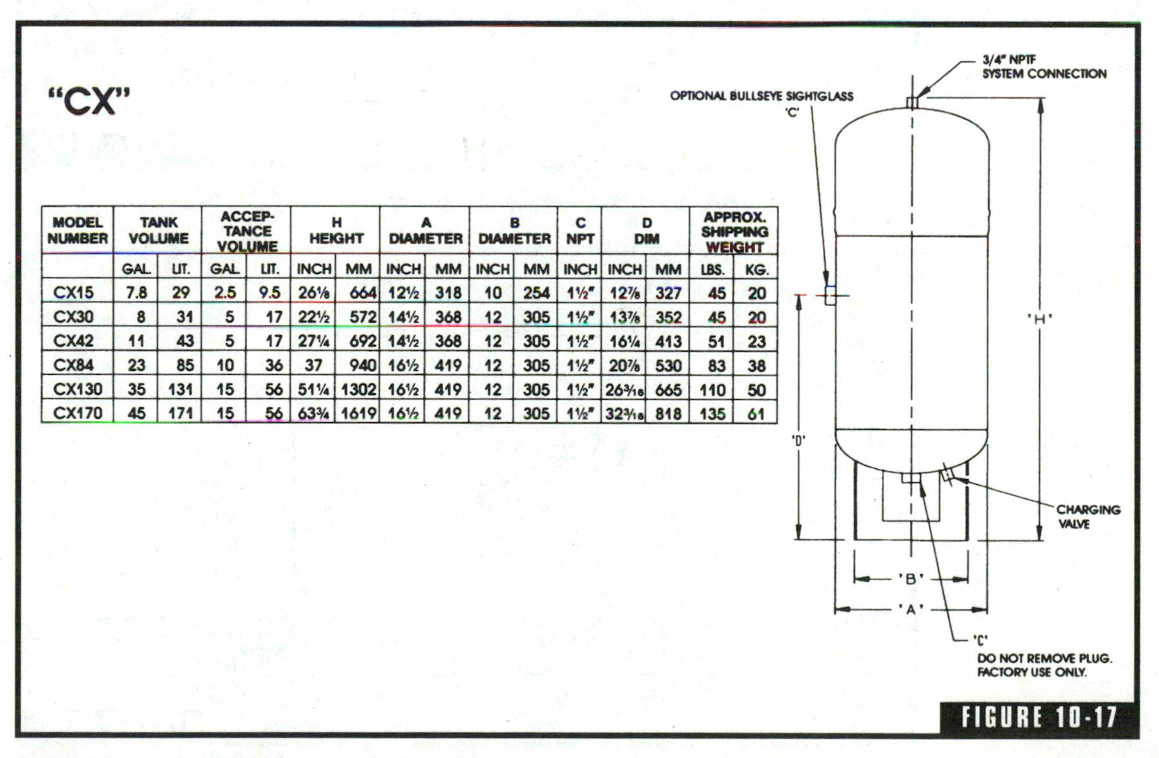

MODEL NUMBER	TANK VOLUME		ACCEPTANCE VOLUME		H HEIGHT		A DIAMETER		B DIAMETER		C NPT	D DIM		APPROX. SHIPPING WEIGHT	
	GAL.	LIT.	GAL.	LIT.	INCH	MM	INCH	MM	INCH	MM	INCH	INCH	MM	LBS.	KG.
CX15	7.8	29	2.5	9.5	26⅛	664	12½	318	10	254	1½"	12⅞	327	45	20
CX30	8	31	5	17	22½	572	14½	368	12	305	1½"	13⅞	352	45	20
CX42	11	43	5	17	27¼	692	14½	368	12	305	1½"	16¼	413	51	23
CX84	23	85	10	36	37	940	16½	419	12	305	1½"	20⅞	530	83	38
CX130	35	131	15	56	51¼	1302	16½	419	12	305	1½"	26³⁄₁₆	665	110	50
CX170	45	171	15	56	63¾	1619	16½	419	12	305	1½"	32³⁄₁₆	818	135	61

FIGURE 10-17

Detail of an air tank. (*Courtesy of Taco.*)

expansion tank should be readily accessible. This is required in case air must be added to the tank. It's a good idea to install a pressure gauge near the inlet of an expansion tank. The gauge makes it possible to monitor the static fluid pressure in the system. In all cases, regardless of tank style, the expansion tank must be supported properly to avoid operation problems (Fig. 10-22).

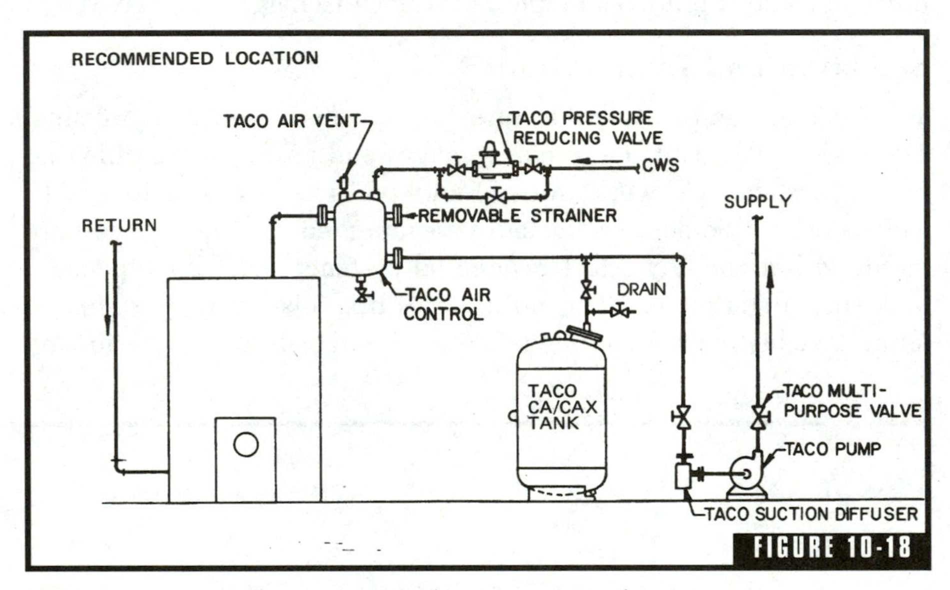

Piping arrangement for an air tank. (*Courtesy of Taco.*)

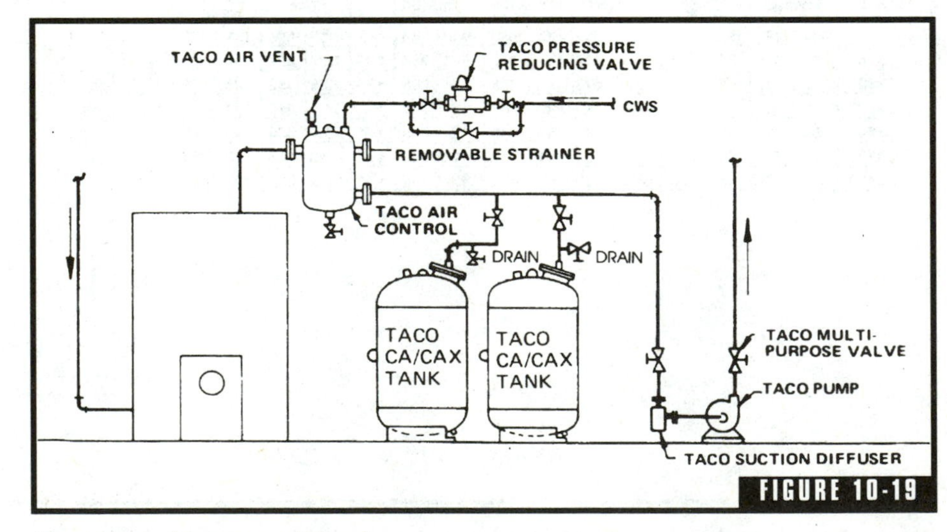

Air control. (*Courtesy of Taco.*)

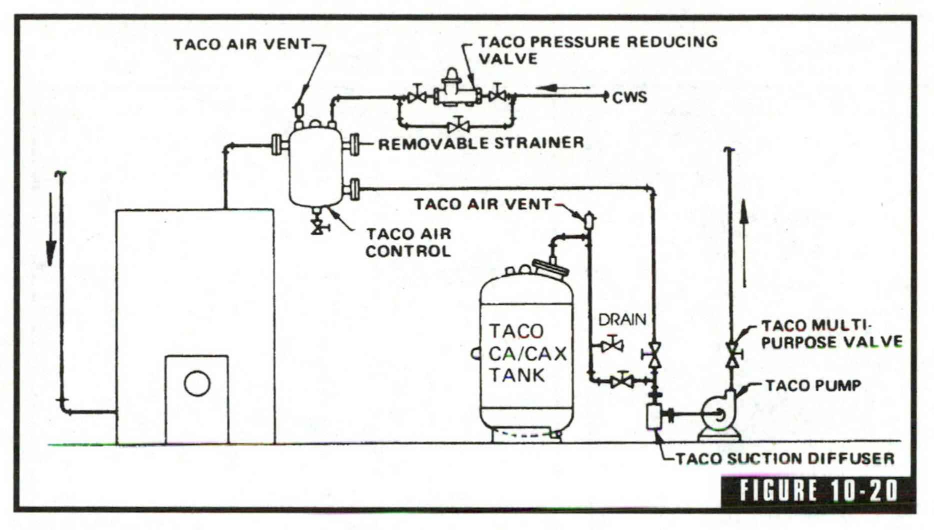

FIGURE 10-20

Common piping arrangement. (*Courtesy of Taco.*)

AIR SCOOP:

Taco Air Scoops are available in 1-inch through 3-inch cast iron threaded and 4-inch flanged cast iron. The 1-inch and 1¼-inch Air Scoop have a vent connection on top, and a diaphram expansion tank connection on the bottom. Air Scoop sizes 1½-inch through 4-inch have an additional tapping on the top for a plain steel expansion tank. The Air Scoop's enlarged design with internal baffles slows the water velocity in order to separate the air from solution.

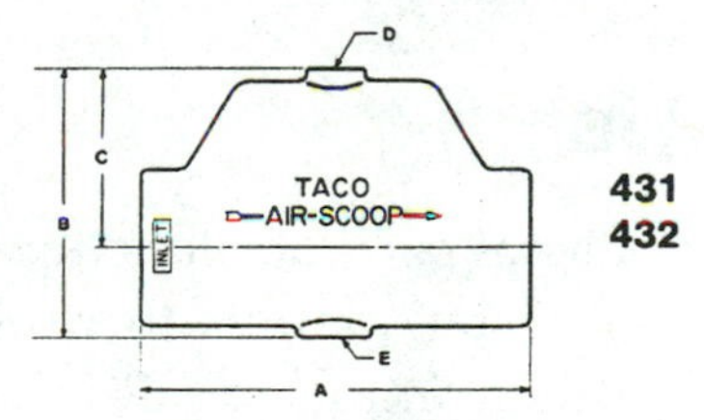

431
432

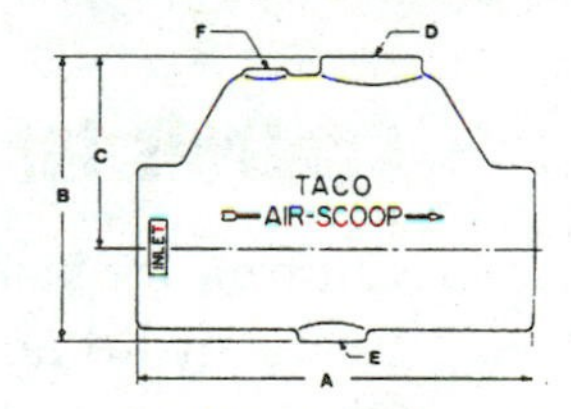

433, 434, 435, 436, 437

SIZE & DIMENSIONS

Prod. No.	Size	A		B		C		D	E	F	Weight	
		in	mm	in	mm	in	mm				lb.	Kg.
431	1"	6	152	4	102	2½	64	N/A	½" NPT	⅛" NPT	4	1.8
432	1¼"											
433	1½"	8	203	6	152	4	102	¾" NPT			7	3.2
434	2"							1" NPT				
435	2½"	10	254	8	203	5½	140				15	6.8
436	3"							1¼ NPT			14	6.4
437	4"	16⁵⁄₁₆	414	11⅝	295	7⅛	181	1½" NPT		¼" NPT	52	23.6

FIGURE 10-21

Air scoop. (*Courtesy of Taco.*)

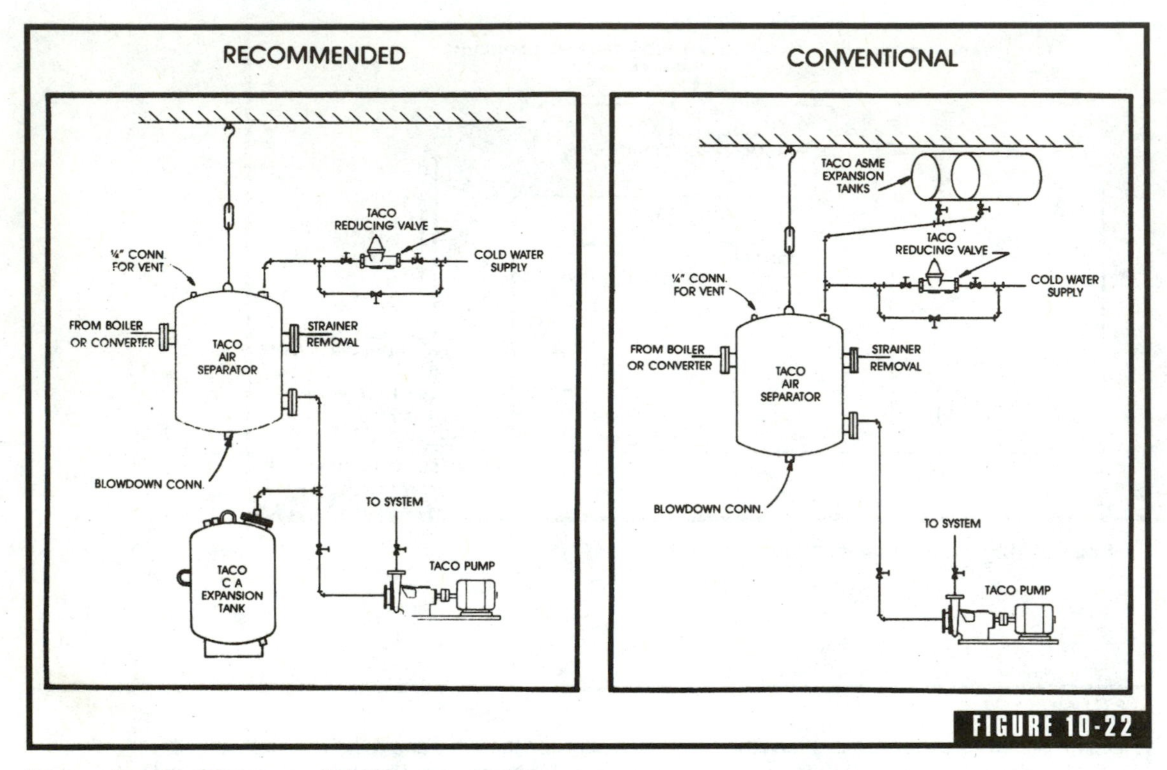

Piping details. (*Courtesy of Taco.*)

Controls for Heating Systems

Controls for heating systems are integral parts of a functioning system. Every hydronic heating system relies on controls to work properly. There are four basic categories for the controls to fall into. There are controls that are merely used to turn a system on or off. Other types of controls are staged, modulating, and outdoor reset. The choice and installation of controls is, to a large extent, a matter of the heating system being fitted with the controls. The best advice is to refer to manufacturers' recommendations and follow them when selecting and installing controls for heating systems. There are far too many facets of controls to cover completely in this chapter. However, we will give an overview of some of the controls in the following sections.

Thermostats

Thermostats (Fig. 10-23), like those on the walls of homes where heating systems are installed, are examples of on-off controls. Devices used to open or close an electrical contact are the most common type of controls used in heating systems. Burner relays and setpoint controls are also examples of on-off controls. These types of devices simply allow a system to cut on or off. The devices do not control or regulate the heat output of a system. This confuses some people. Many people think that a thermostat regulates heat. This is not the case. A thermostat that is set for a high temperature allows a heat source to run longer, but not to burn hotter.

Controlling in Stages

One way to match the output of a heat source with a building's need for heat is to control the heat output in stages. This is known as staged control. It means that the controls used for this type of control cut on and off in stages. The stages can range from zero to maximum heat output. The stages come on and go off in sequence. The use of staged sequences makes it easier to balance heat output with heat need. This type of control is rarely needed in typical residential applications.

Modulating Control

The ultimate in matching output with heat need comes from modulating control. The advantage of modulating control is that heat output is continuously variable over a range from zero to full output. Modulating controls make this possible. Most modulating controls do their job by controlling the temperature of water passing into and through heat emitters. There are several different ways to accomplish modulating control. The most common involves the use of mixing valves or variable-speed pumps. But, electrical elements and even thermostatic radiator valves can produce the effects desired with modulating control.

Outdoor Temperatures

Outdoor temperatures affect the effectiveness of a heating system. When a standard hydronic system is designed, it is set up to provide a certain room temperature based on a certain outdoor temperature

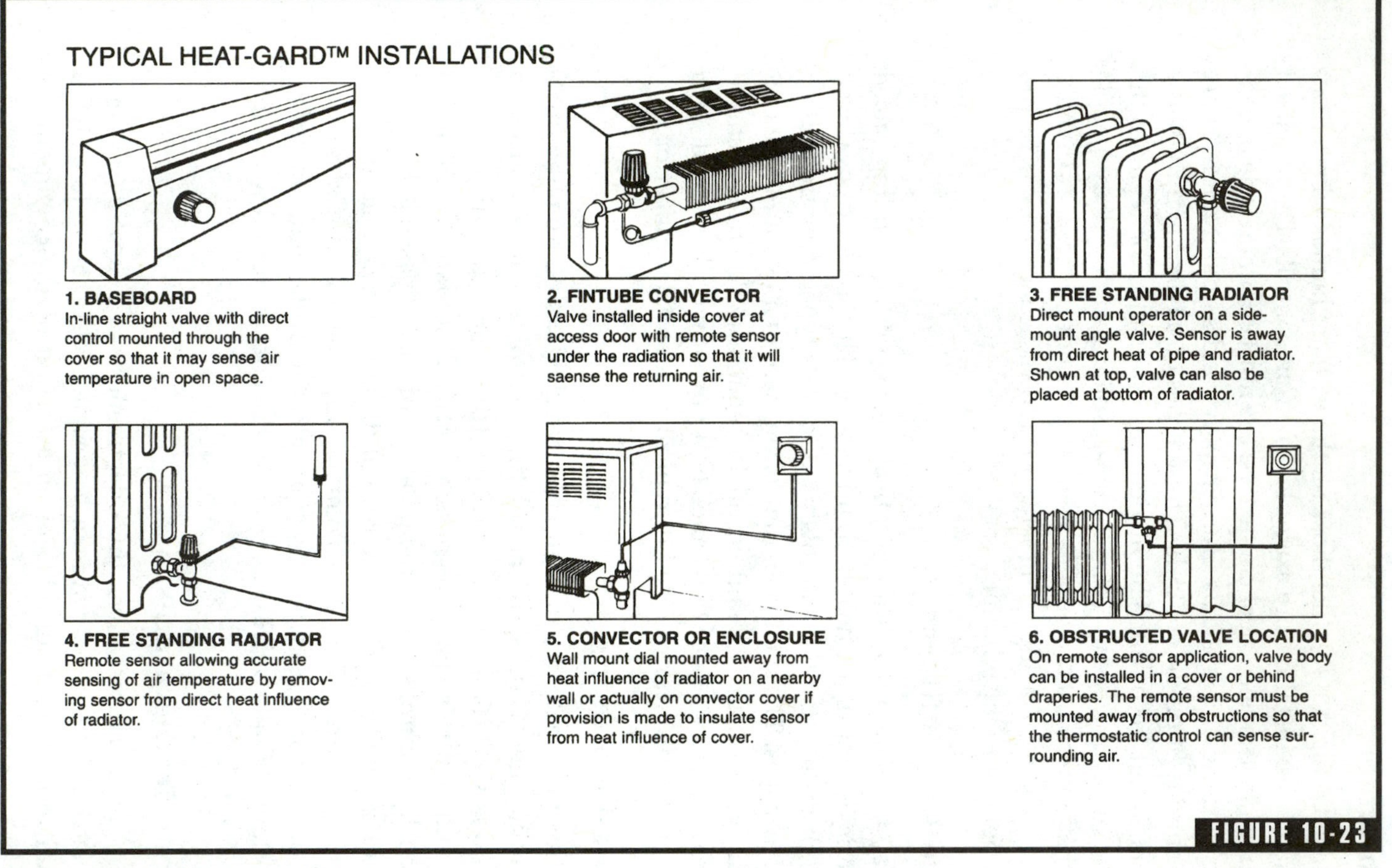

FIGURE 10-23

Heater controls. (*Courtesy of Taco.*)

and the temperature of water in the heating system. An outdoor reset control is ideal for balancing outdoor temperatures with the water temperature in hydronic heating systems. With this type of balancing, a building enjoys near-perfect heating control, even when outside temperatures are uncooperative. When outdoor temperatures plummet, the temperature of water in the heating system is raised through the use of an outdoor reset control. In reverse, if outdoor temperatures warm up considerably, the temperature of water in the heating system is lowered. This balancing act combines to make a more efficient and more comfortable heating system.

There all sorts of controls used in heating systems: switches, relays, time-delay devices, high-limit switches, triple-action controls, multizone relays, resets, and so forth. The safest bet when dealing with controls is to follow the recommendations of manufacturers. The list of potential controls is a long and complicated one. We simply don't have the space here to delve into all of them.

The next chapter is going to show you a lot about radiant floor heating. This is a form of heating that has received mixed reviews. Done properly, radiant floor heating is an excellent way to make the best use of hydronic heating systems. If you or your customers want to know more about keeping floors warm and spreading heat evenly throughout a home, you will enjoy the following chapter.

Hot-Water Radiant Heat

Hot-water radiant heat is the heat of choice in regions where winter temperatures are brutal. This type of heat has proved itself over and over again under such circumstances. I live in Maine, where temperatures often drop well below zero on any given day or night during the winter months. Seeing the outside thermometer dip to 20° below zero, not counting windchill factors, is not uncommon. This type of cold requires a strong heating system if a home or business is to remain at comfortable indoor temperatures. Maine, of course, is not the only region where extremely cold temperatures put pressure on heating systems.

Having been a plumbing-and-heating contractor for over 20 years, I've seen a number of heating systems in a variety of situations. Most heating professionals agree that hot-water baseboard heat is one of the best heating systems available for cold climates. Radiant floor heating is also gaining ground rapidly as a desirable heating system. There are, to be sure, other types of heating systems that will maintain comfortable indoor temperatures. However, the efficiency and operating costs of some systems leave much to be desired when compared to radiant hot-water heat.

The first house I lived in when I moved to Maine had electric baseboard heat as a backup to two wood stoves. As backup heat, the electric baseboard units were fine, but had the electric heat been the pri-

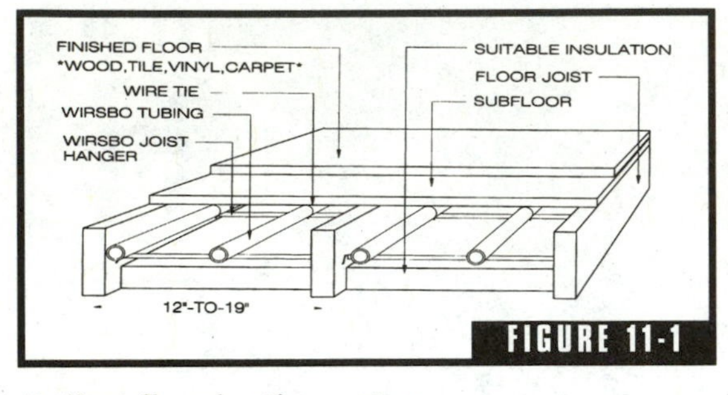

FIGURE 11-1

Radiant floor heating system. (*Courtesy of Wirsbo.*)

mary heating system, the cost of heating during winter months would have been unbearable. I've also lived in Maine homes where forced hot-air was responsible for heating the living space. The forced hot-air systems were able to cope with the extreme temperatures, but the cost of operation was not desirable. Most homes in Maine have either hot-water baseboard heat or radiant floor heat. In some cases, they have both. When I built my home in Maine, I installed hot-water baseboard heat emitters with a cold-start boiler. The combination was very effective, both in terms of comfort and cost-effectiveness.

Radiant floor heating systems (Fig. 11-1) are rapidly growing in popularity in my region, as they are in other areas. This type of heating system heats through flooring, so floors stay warm and heat is evenly distributed when the systems are installed properly. Hot-water baseboard heating units are still the most popular within Maine and probably in many other cold states. Combining the two types of heat emitters makes a lot of sense. In this chapter, we are going to take a look at both types of systems and what the requirements are for each of them. Let's start with baseboard units and work our way into radiant floor heating.

Finned-Tube Baseboard Convectors

Finned-tube baseboard convectors are probably the most commonly used form of heat emitter for hot-water heat. This is true of both residential and commercial applications. This type of heat consists of a housing and an element. The element is made of copper tubing that is surrounded by aluminum fins. The length of an element usually ranges from 2 to 10 feet and is often sold in increments of 1 foot. For example, you can buy a 5-foot section of element or a 6-foot section. The elements can be cut with regular tubing cutters for customized fits. The copper tubing used in elements usually has a diameter of 3/4 inch, but there are 1/2-inch versions available. Commercial elements exist in larger sizes. It is also possible to buy commercial elements in which the fins surround steel pipe. But, due to size and cost, commercial elements are not normally used in residential installations (Fig. 11-2).

Heating elements are housed in sheet-metal enclosures. These enclosures consist of a back, which is screwed to a wall. The back has movable brackets which hold the heating element. There is also a movable damper built into the back section so that the heating housing can be closed to reduce heat output. A front cover snaps into place on special retainer brackets that are a part of the back section. In addition to the front and back covers, there are trim pieces which are used to join sections of baseboard and to terminate runs. The pieces used to join sections of baseboard are called splicers. The termination pieces are called end caps. There are also pieces called inside corners and outside corners which are used when the baseboard sections change direction.

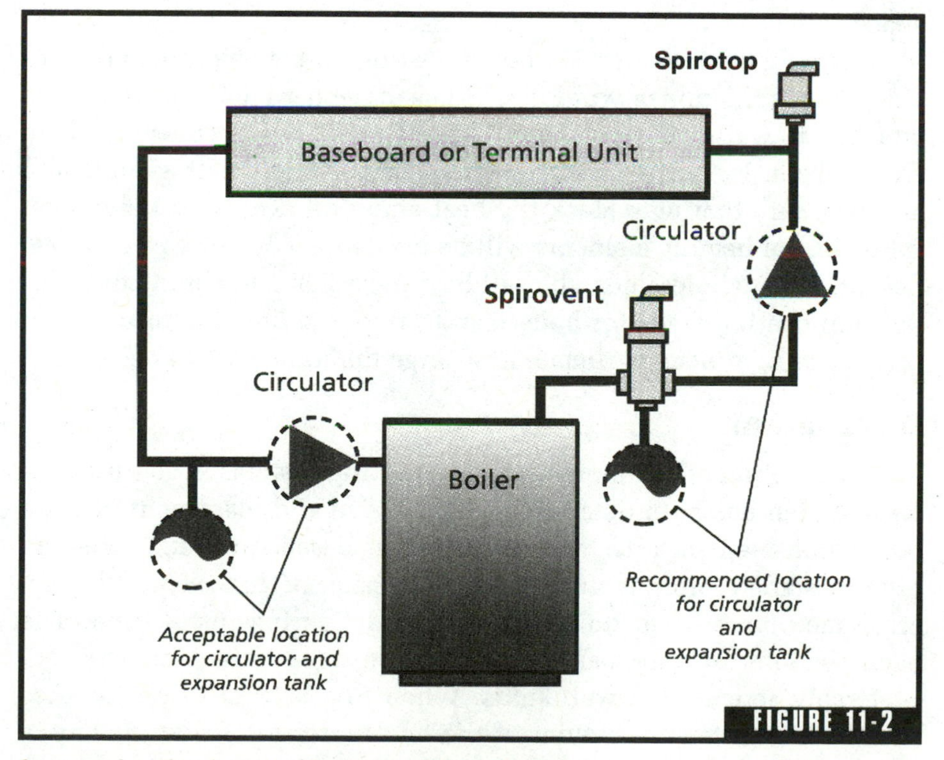

Common heating layout. ***(Courtesy of Spirotherm.)***

Finned-tube heating elements do not rely on a fan for heat distribution. Instead, the convectors operate on the basis that warm air rises. Cool air is pulled in from the bottom of the heating element housing, comes into contact with the finned tube, is warmed, and then rises. It is a simple, yet very effective process that produces good heat. During peak operation, the damper on a heater housing should be open. If the damper is closed, the amount of heat output is reduced by up to 50 percent. Having an upward draft, the warm air from finned convectors negates the effects of downward drafts from exterior walls and windows. This makes the heat feel quite comfortable.

Due to the nature of convection heat, there can be a problem known as stratification. This is basically a situation where warm air rises to a ceiling while cool air lingers lower in a room. When the water temperature in a heating system of this type is very hot, the rising of the air is more extreme. Homes with vaulted ceilings can suffer greatly from stratification. One way to overcome this is the installation of ceiling fans that will push the warm air back down into the living space.

It is most common for baseboard heating units to be placed on outside walls and under windows. Since baseboard heating units are attached to walls, the placement of furniture can be a concern. There should be at least 6 inches of open space between heating units and any furniture that may block the heat emission. Knowing how many linear feet of heating elements will be needed is a major part of a heat design. Most wholesalers who sell heating systems to contractors will develop heating loads for houses and provide a drawing of where to place heating units and dictate how large the units should be.

Installation

The installation of common baseboard heating systems is not difficult. Once a plan has been determined for the size and placement of baseboard units, an installer can set to work. Baseboard enclosures are often installed prior to finish floor coverings. This means that an allowance must be made for the planned flooring installation. The back sections of baseboard housing should be nailed or screwed, preferably screwed, to wall studs. When this is done, an allowance must be made for the bottom of the enclosure. As a rule of thumb, many installers mount the baseboard backs so that the bottom edge is about 1 inch above the subflooring. This allows what is usually adequate room for carpet and pad to be installed. It's best, however, to consult with the builder or flooring contractor to make sure that any baseboard installed will not interfere with forthcoming flooring.

Holes must be drilled to get distribution piping to heating elements. The holes should be oversized, to prevent squeaking in the heating pipes as they expand and contract. Many contractors feel that the holes should be a full 1/2 inch larger than the pipe passing through the hole. Some contractors use flexible sleeves, which do not contribute to noise from expansion, to hold pipes firmly in place.

In most cases, distribution piping will start at a boiler and run in a continuous loop. The piping will enter a heating element, run through it, exit it, and then travel on to the next heating element. In the end, the loop winds up back at the boiler. When multiple zones are used, such as one zone for general living space and one zone for sleeping areas, there will be multiple loops. There are other ways to pipe a system, but the continuous loop is the most popular and it's very effective.

Plumbers who are not used to working with baseboard heating elements often find themselves fixing leaks in their solder joints. Unlike plumbing joints, where both sections of pipe being joined are the same thickness, the thin-wall copper used with heating elements gets hot much faster than the type-M copper tubing usually used to supply water to the element. Since the supply pipe and the heating element will reach soldering temperature at different times, a successful joint requires a little experience. A plumber or heating mechanic who bases the joint on a typical plumbing joint will find leak after leak. The distribution pipe takes longer to bring up to heat for soldering than the heating element does.

Establishing Heat Zones

Establishing heat zones (Fig. 11-3) is as easy as determining what parts of a home or building should be controlled by separate thermostats. In average houses, there may be only two zones. One zone will serve general living space, while a second zone heats sleeping areas. It's not uncommon to find small homes with a single heat zone. Two-story homes may have a heat zone for each level of living area. Larger homes can have many heat zones. In my home, I installed one zone for bedrooms, one zone for general living space on the ground floor, one zone

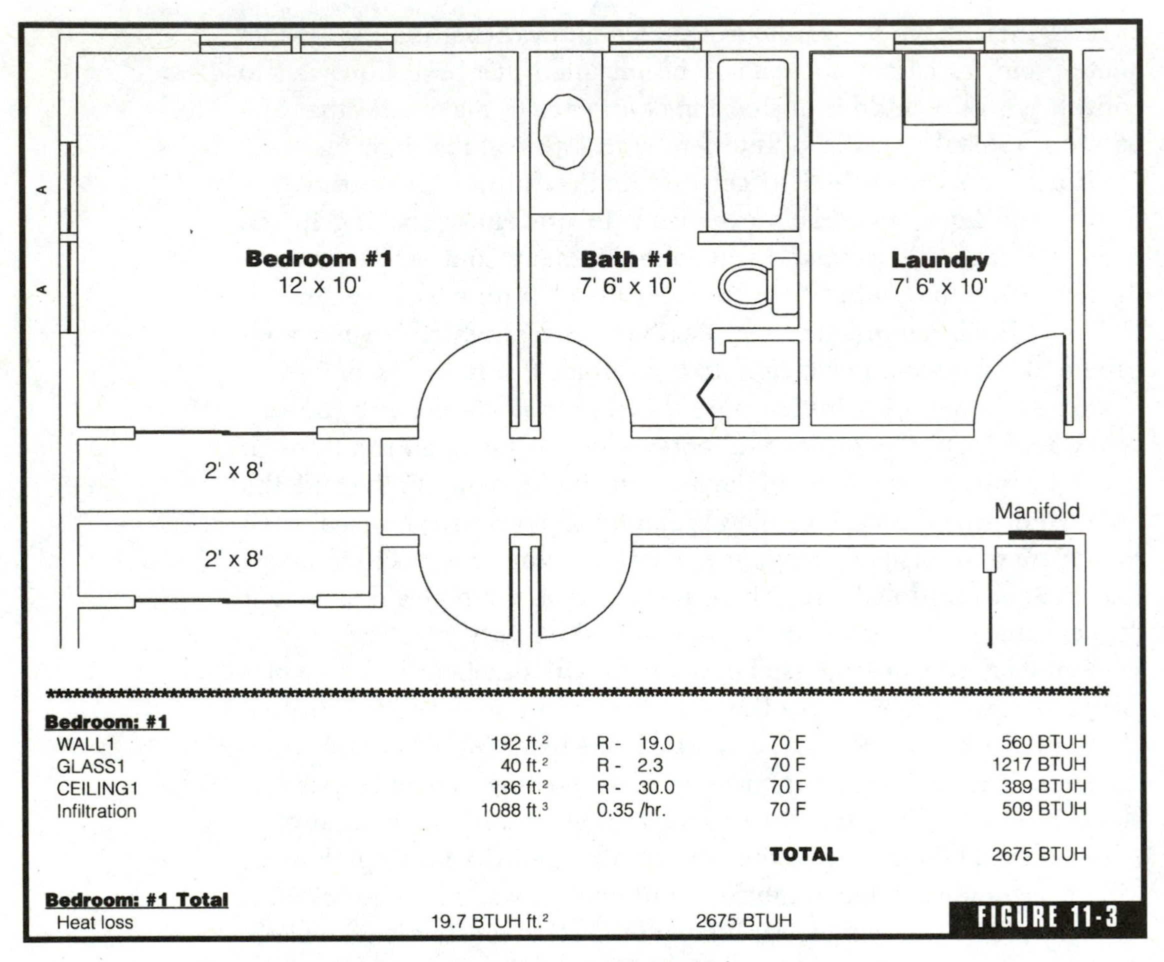

Floor plan used for heating system design. (*Courtesy of Wirsbo.*)

for offices upstairs, and a fourth zone for the heated garage. If I had a basement, it probably would have had a separate heating zone from the rest of the house.

Adding zones to a home make the heating system more expensive. Either additional circulating pumps or zone valves will be needed. More copper tubing is required to feed the various zones. Yet, the money saved in operating cost by heating zones of a home as they are being used can help to recover installation costs. The main thing about zones is to establish them during the planning stages of a job and lay them out accordingly. How a home or building is zoned off for heating is largely up to the choice of the occupant.

Fan-Assisted Heaters

Fan-assisted heaters, or fan-coil heaters as they are also known, see frequent use in kitchens when hot-water heat is installed. Since kitchens have base cabinets in them, it can be difficult to find wall space to mount baseboard units. The solution to this problem is a small, fan-coil heater that is best known as a kick-space heater. This little heater installs under a base cabinet and requires only about 4 inches of height. An average kick-space heater is about 1 foot wide, but it delivers the heat output of up to a 10-foot-long piece of radiant baseboard heating element. Wider kick-space units are available for greater heat output when space allows.

How does such a small package deliver so much heat? The heaters have a built-in coil of 1/2-inch copper tubing. Air is drawn in through the top portion of the heater grill by means of a transverse blower mounted in the back of the unit. This blower reverses air flow and sends it out through the horizontal coil in the lower part of the heater. It is this lower part that contains the heating coil. Since air is moved over the coil via a blower, more air flow is generated, and thus, more heat is created. A low-limit aquastat control prevents the blower from operating until the water temperature in the coil reaches a desired temperature. Normally, the temperature setting is preset by the heater manufacturer and is not adjustable. Most settings range from 110° to 140°. In cases where low-temperature water is used for a system, the low-limit control can be bypassed.

Wall-Mounted Fan-Coil Heaters

Wall-mounted fan-coil heaters are available for use in both residential and commercial applications. There are types that are mounted to walls and types that are recessed into walls. The recessed types take up less floor space and often provide a more attractive appearance. When a recessed unit is used, there must be a cavity built into the wall framing that will accept the size of the unit. Both types usually have copper coils that are connected to 1/2-inch copper tubing. It's possible for a single wall-mounted fan-coil heater to heat an entire room, but this does not come without some potential problems.

Since the wall heaters depend on a blower to push air into a room, two potential problems exist with the design. First, the heating of the room will not be as even as it would be with baseboard convectors. This is obvious, since all of the heat is coming from a single source and location. Secondly, furniture cannot be placed in front of the heating unit.

Wall-mounted heaters are available in different sizes, both in width and height. Even so, finding a location where the heater can operate without blockage may be a problem, especially in a residential home. Since the heaters use blowers, there is some operational noise which must be expected. And, dust circulation may be a problem. Good heaters have dust filters, but dust still escapes in some quantity.

Wall-mounted fan-coil heaters are much more effective than finned-tube convectors when it comes to producing and distributing heat. Due to the design of fan-coil heaters, they can produce more heat with less temperature in the heating coils. This is a big advantage when the heat source is a hydronic heat pump or an active solar system. Using finned-baseboard heat with the heat pump or solar setup would be risky. There is a great chance that either heat source would not produce water hot enough to heat the finned-tube heating elements sufficiently. Finned-tube baseboard units, used with boilers, are much more common than wall-mounted fan-coil units, but the fan-coil units definitely have their place in the heating world.

Heated Garages

Heated garages are a grand convenience for people who live in extremely cold climates. There are many ways to heat a garage. Finned-baseboard can be used, as can a wall-mounted fan-coil unit. However, an overhead fan-coil unit is often the most practical. Just one of these out-of-the-way heaters produces plenty of heat for even a large garage. These same heaters are frequently used in commercial installations. These heaters are sometimes called unit heaters. Louvers on the face of the heaters allow air to be blown in an angled horizontal direction. Overhead unit heaters can be hung from ceiling supports with brackets or threaded rods. The units require very little space, need minimal piping, and produce substantial heat.

Panel Radiators

Panel radiators are a type of hydronic heat that many Americans may never have seen. These heating units have been used for years in Europe, but they are fairly new to the scene in the United States. When horizontal wall space is at a premium, panel radiators are a good choice to consider. Unlike fan-coil units, panel radiators don't depend on blowers for operation. The heaters are both attractive, unobtrusive, and very effective.

Installation of panel radiators is not difficult. In most cases, the radiators are mounted on walls and extend only about 2 inches into living space. The units are generally mounted so that the bottom of the heater is between 6 and 8 inches above finished-floor level. Wall brackets, which must be secured to wood members or securely into solid masonry, hold the panels in place. When the installation of panel heaters is planned in conjunction with new construction, wood backing blocks can be installed horizontally between wall studs to provide a strong surface for attachment.

Water connections to the panels may be made with either 1/2-inch or 3/4-inch piping. It is common to install a manual or thermostatic radiator valve in the upgoing riser directly to the inlet connection of the radiator to allow individual control of the heating unit. Valves and unions can be installed on risers to allow the removal of radiator panels when the time comes to paint the room where the heater is located.

Radiator panels can be purchased as either vertical units or in a horizontal design. This allows maximum flexibility in mounting locations. Since there are numerous size combinations available, it's possible to install panel radiators in almost any location. Panel radiators are subject to rust and should be used only in closed-loop systems.

Radiant Baseboard Heat

Radiant baseboard heat is another relatively new addition to the U.S. market. The baseboard is only about 1 inch wide and about 5 inches high. PEX tubing is most often used in the baseboard, but copper tubing is sometimes used. There are no fin-tube elements in radiant baseboard. The heat output from radiant baseboard is about 85 percent thermal radiation and about 15 percent convection. Radiant baseboard heats in such a way that stratification is hardly a problem. And, the baseboard units are sold in lengths from 1 to 10 feet, in 6-inch increments.

Due to the thin-line design, radiant baseboard heat is intended to act as a replacement for traditional baseboard. Essentially, the heating units are installed in a full perimeter around a room. When baseboard units are connected to each other, the connections are made with brass compression fittings. This is an advantage because there is no soldering required. It's common to run the supply and return for each room from a central manifold. Radiant baseboard heat does not produce as much heat on a per-foot basis as fin-tube baseboard, but because radiant baseboard is installed completely around the room, this problem turns out not to be a problem at all.

There are plenty of choices when it comes to using hot water for a heating system. While fin-tube baseboard is currently the most popular way to heat homes with hot water, there are certainly other options available, and some of them are gaining popularity quickly. People are looking into the use of heat-pump and solar-powered systems. Radiant baseboard and radiator panels are on the move. Another major mover in the market is radiant floor heating. Since this is a large subject, let's move to the next chapter to discuss it.

CHAPTER

12

Radiant Floor Heating Systems

Radiant floor heating systems have been around for quite awhile. If you want to go back to the origins of heating buildings by having heated floors, you can trace the process all the way back to the Roman Empire. The Romans heated their floors by directing exhaust gases from wood fires to open space under raised floors. Today's radiant floor systems depend on hot water, rather than exhaust gases, to heat floors. For many years, the concept of heating homes and buildings by putting heat in the floors of the structures was largely ignored. This fact is changing. With modern technology, the cost of installing radiant floor heat is lower. Additionally, the problems that were often associated with threaded pipe connections and steel pipe, as well as the pin-hole leaks sometimes experienced with copper tubing, have been greatly reduced with the introduction of PEX tubing.

The benefits of radiant floor heat are numerous (Figs. 12-1 to 12-6). Such systems give unsurpassed thermal comfort, produce no noise, can operate with low-temperature water, and often use less energy to function. Another advantage of radiant floor heating is the fact that the heating pipes are out of sight, out of mind, and don't interfere with furniture placement. Dust can be a problem with fin-tube baseboard heat, and it can be a big problem with some fan-assisted heating units. This is not a problem with radiant floor heating. Stratification is another problem that may occur with baseboard heating systems, but it is not

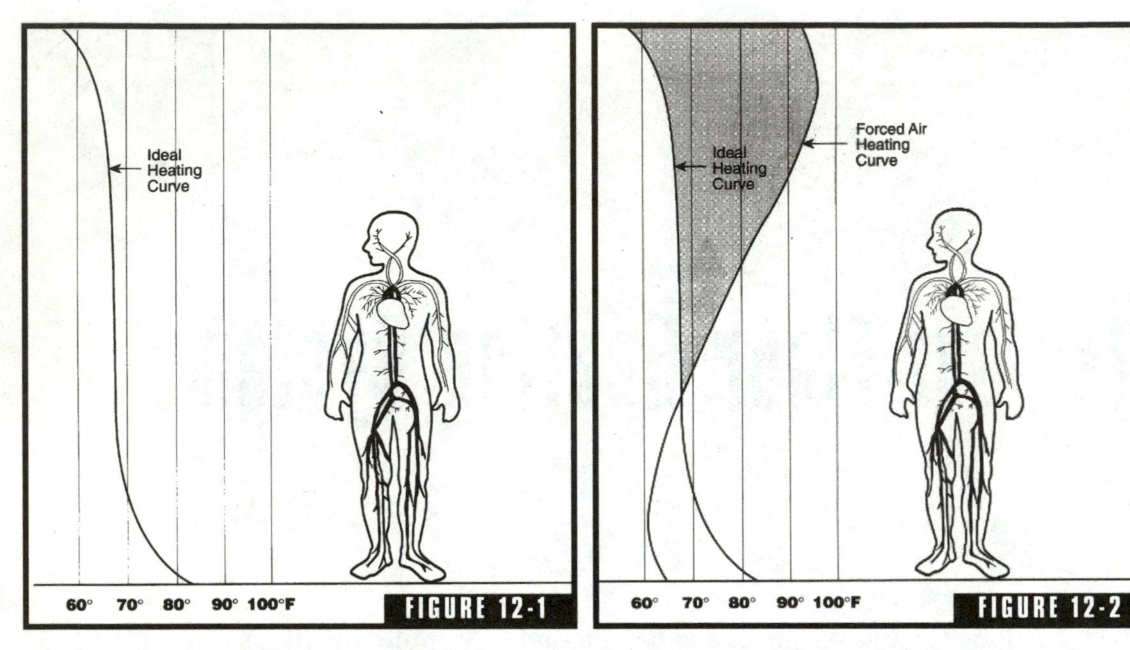

Ideal heating curve. (*Courtesy of Wirsbo.*)

Heating curve with forced air heat. (*Courtesy of Wirsbo.*)

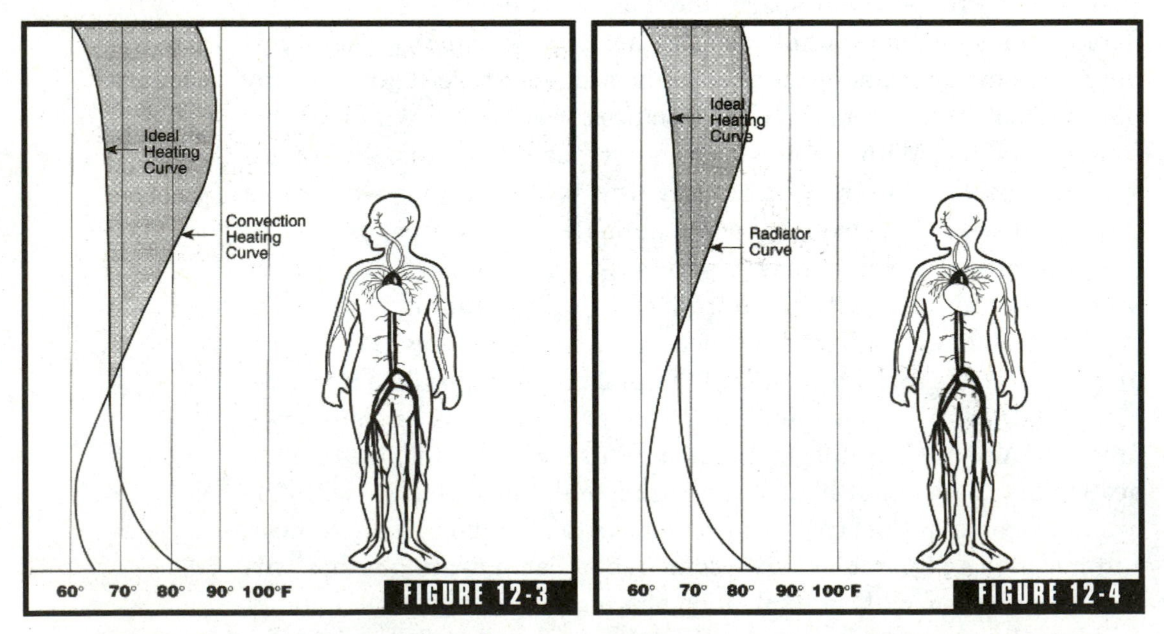

Convection heating curve. (*Courtesy of Wirsbo.*)

Radiator heating curve. (*Courtesy of Wirsbo.*)

a problem with radiant floor heating. The key to having an enjoyable experience with radiant floor systems is to have the system designed and installed properly.

Radiant floor heating is not new, but the methods and materials used to install the systems are new. When this type of heating system first gained acceptance in the United States, it was most often installed in concrete slabs. Concrete is a good place to embed radiant heat pipes. The concrete gathers the heat and stores it. As the desire for in-floor heating increased, so did new installation methods. The three types of installations used today are slab-on-grade systems, thin-slab systems, and dry systems. For a long time, slab-on-grade systems were the most often used type of radiant floor heating. This is still a popular type of heating installation, but other ways are growing in popularity, as people want more of the advantages possible with radiant floor heating (Figs. 12-7 and 12-8).

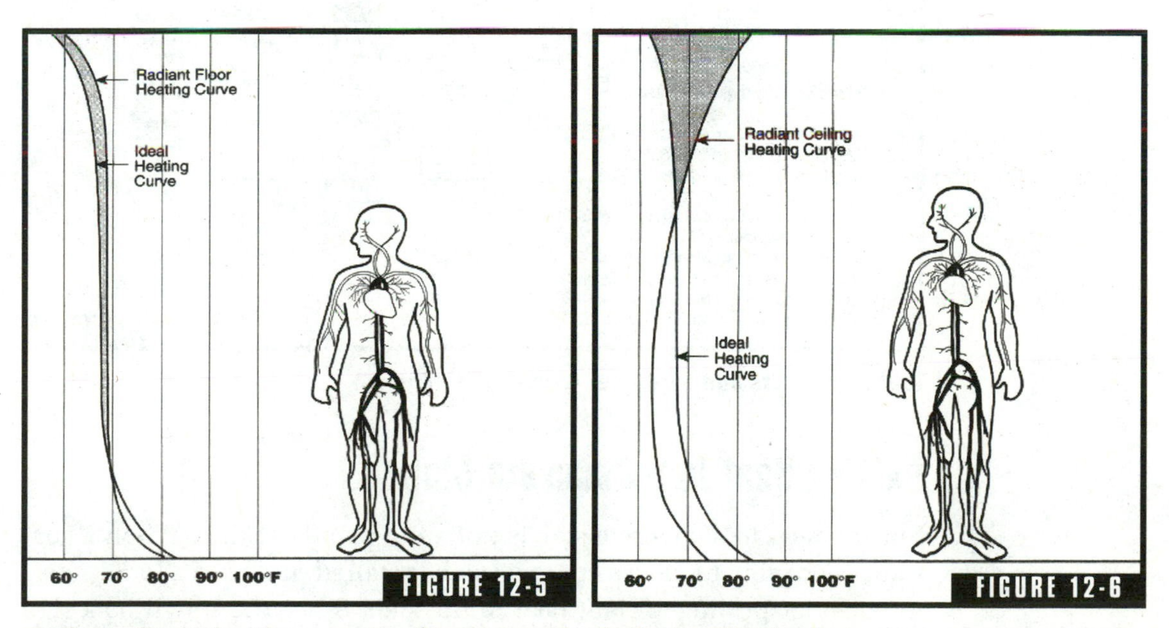

Radiant floor heating curve. (*Courtesy of Wirsbo.*)

Radiant ceiling heating curve. (*Courtesy of Wirsbo.*)

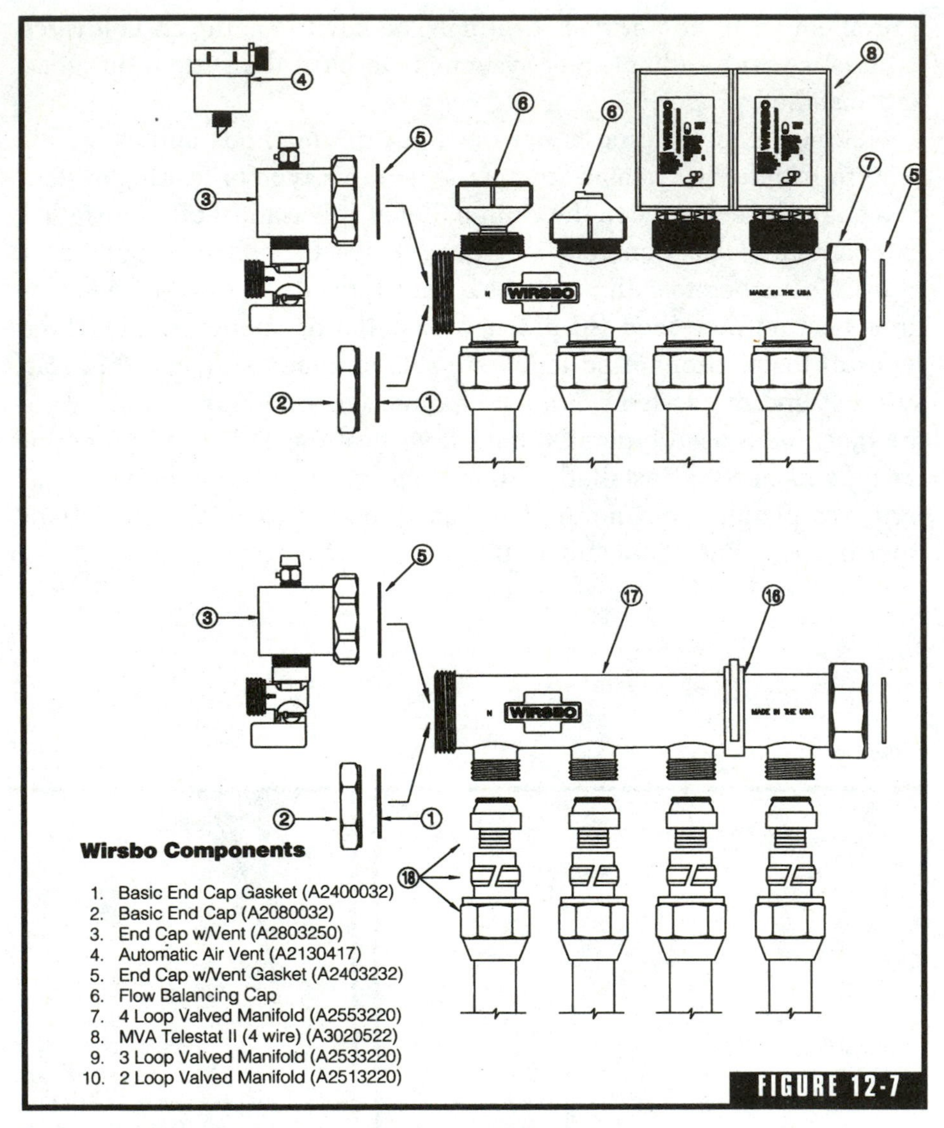

Tubing products and hardware. (*Courtesy of Wirsbo.*)

Putting Heat in a Concrete Slab

Putting heat in a concrete slab is not a huge undertaking. When a customer is paying to have a concrete slab installed, it makes a lot of sense to consider putting radiant heat in the slab. Since there will be a slab regardless of whether there is heat or not, the only real cost of adding

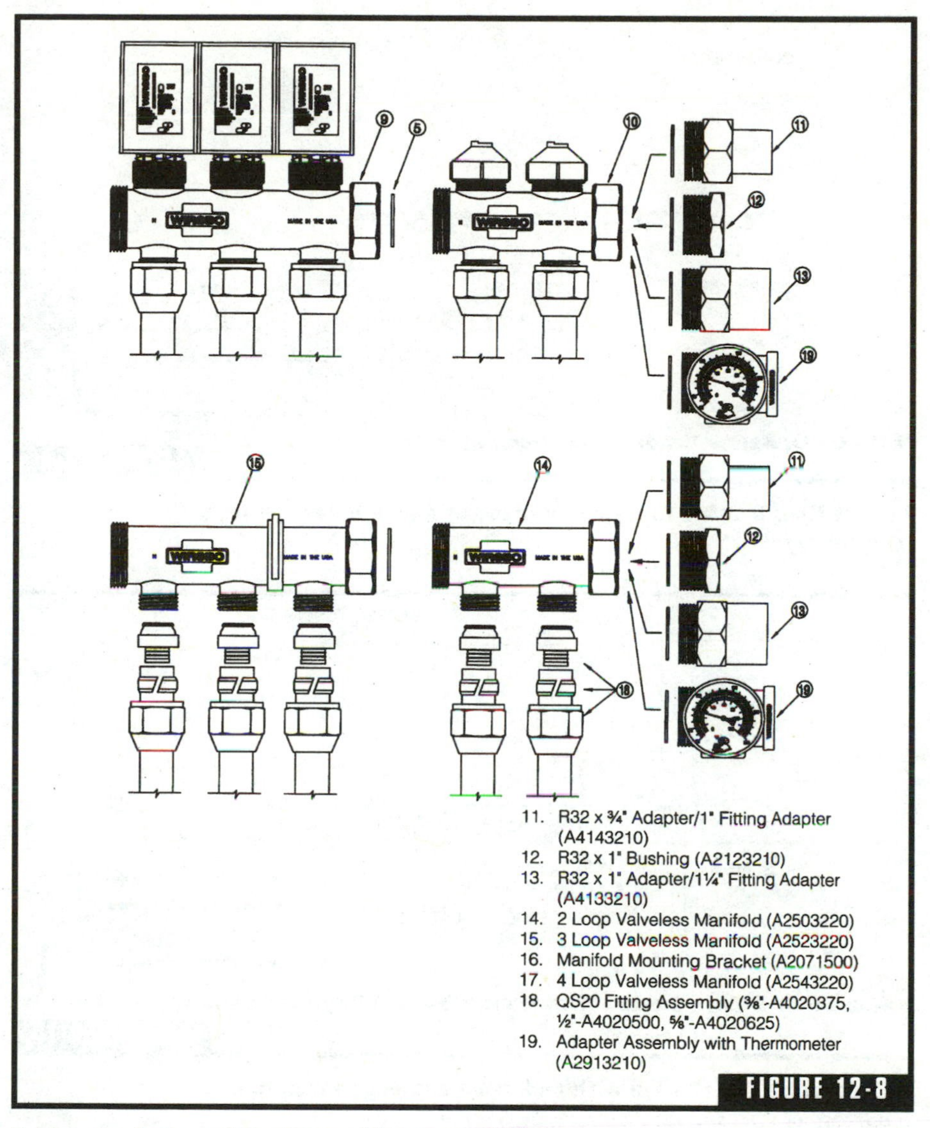

FIGURE 12-8

Tubing products and hardware. (*Courtesy of Wirsbo.*)

the heat is some inexpensive PEX tubing and some labor cost in having it installed. This, of course, is assuming that some form of hot-water heat will be used for the remainder of the heating system. If the slab will hold the entire living space of a home, the situation is even better. This means that there will be no need for any other type of above-floor heating units (Figs. 12-9 and 12-10).

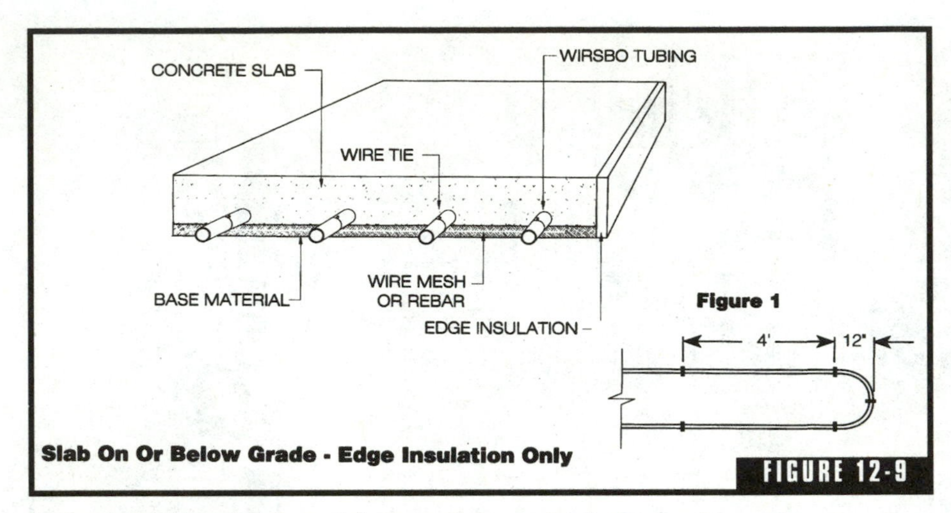

Radiant floor heating in a slab on grade or a slab below grade. (*Courtesy of Wirsbo.*)

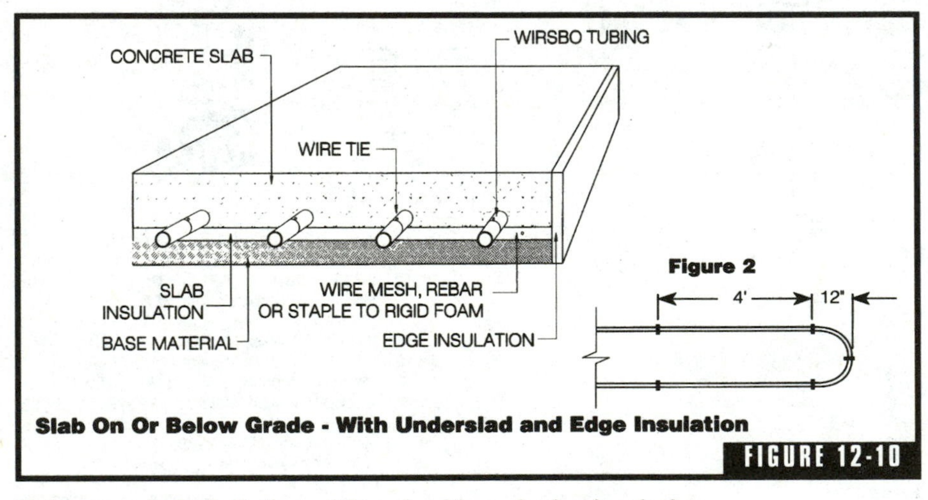

Slab-on-grade installation with underside and edge insulation. (*Courtesy of Wirsbo.*)

A very important part of installing radiant heat in a concrete slab is the protection of the tubing used to transport hot water through the concrete. It is normal to install PEX tubing on the reinforcement wire that is generally placed in the slab area prior to pouring concrete. The tubing is looped throughout the slab area and can be attached to reinforcing wire with wire ties that are specifically designed for the purpose (Fig. 12-11). Another method, and one that many contractors feel

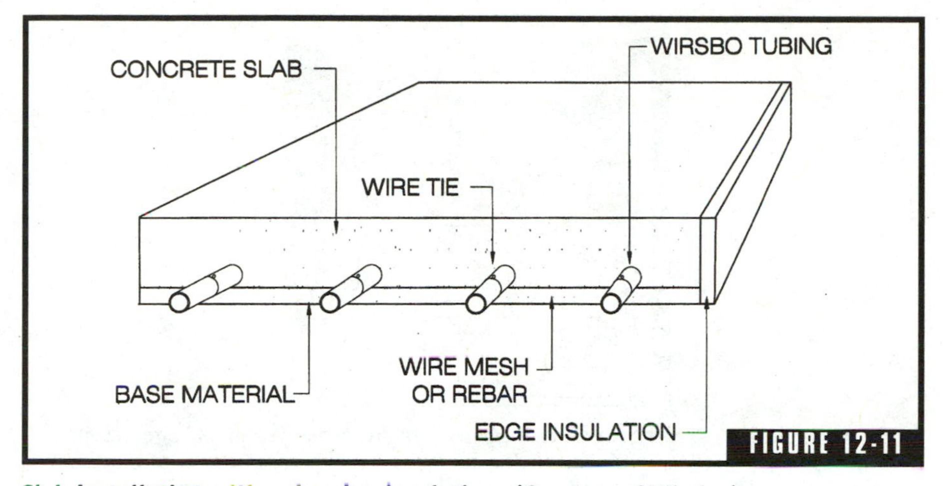

Slab installation with only edge insulation. (*Courtesy of Wirsbo.*)

is better, is to use special clips to support the PEX tubing. The clips are available as individual clips or as bars that are notched to accept the tubing. Regardless of your preference for securing the tubing, refer to the instructions provided by the tubing manufacturer. Failure to secure the pipe in compliance with manufacturer's recommendations can result in a voided warranty.

The spacing between loops of radiant piping can vary a great deal (Fig. 12-12). Loops that are placed close together will boost heat output. Loops may be placed as far as 2 feet apart in some jobs and much closer in different jobs. Most installers attempt to keep the loops about 1 foot apart. This makes bending PB or PEX tubing much easier, without as much fear of kinking the tubing. The distance between loops is determined when a heat load is computed and a system is designed (Fig. 12-13).

Just as the distance between loops varies, so does the depth at which radiant tubing is installed. It's generally considered best to position tubing so that it is concealed approximately midway in a slab. However, tubing is often set lower in concrete. This can be due to many factors, such as reinforcing wire sinking during the pouring of concrete. When tubing is placed deep within a concrete slab, the heating efficiency is reduced. This means that hotter water is needed to produce the same amount of heat that cooler water would produce if the tubing was closer to the surface of the slab. If a system has been

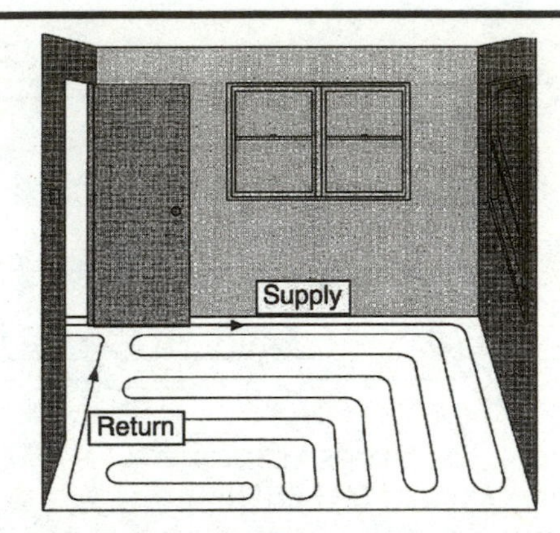

Double Wall Serpentine—Used when there are two adjacent walls representing the major heat loss of the room. The supply is fed directly to either of the heat loss walls and then serpentined toward the lower heat loss area in an alternating pattern against the two heat loss walls. Start tubing runs 6" from walls or nailing surfaces. A 6" on-center tubing run is often installed along outside walls to improve response time.

FIGURE 12-12

Serpentine tubing installation. (*Courtesy of Wirsbo.*)

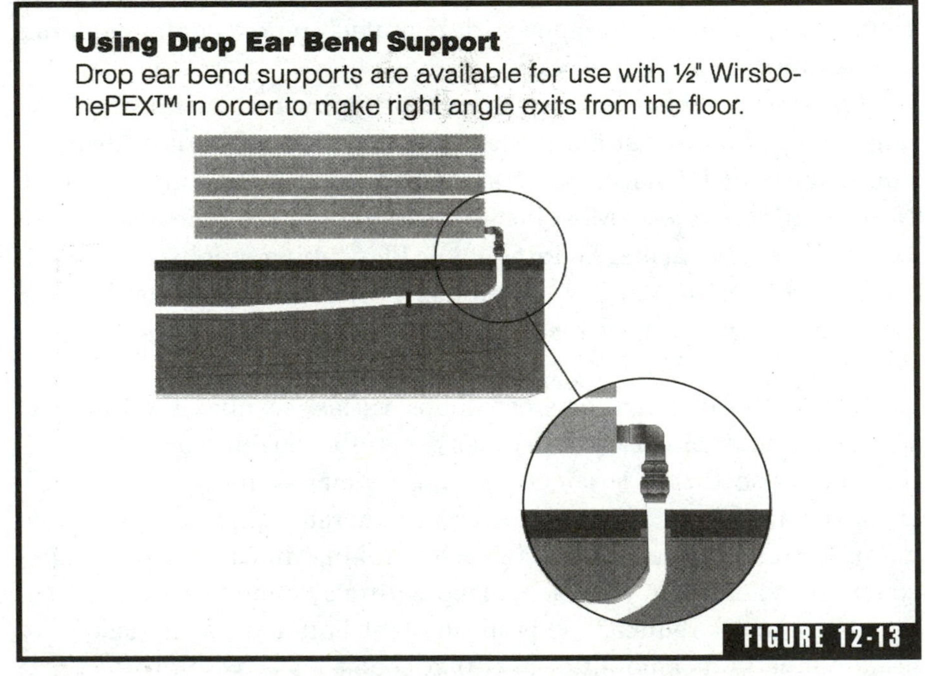

FIGURE 12-13

Drop ear bend support. (*Courtesy of Wirsbo.*)

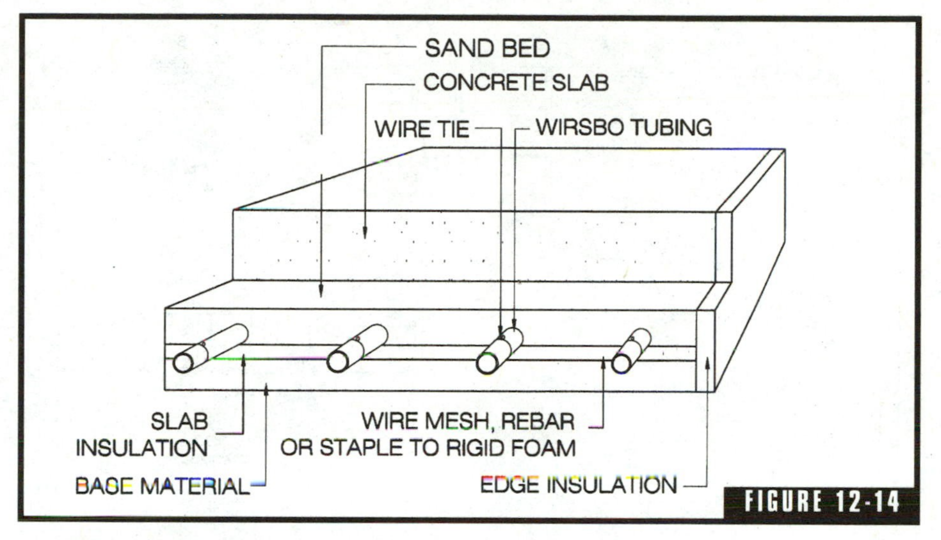

Slab installation over a sand bed. (*Courtesy of Wirsbo.*)

shut down for awhile, it will take longer for tubing that is deep within a slab to warm the surface of the concrete (Fig. 12-14).

In order to maximize heating efficiency, rigid foam insulation should be installed between the earth under the slab and the tubing installed for heating (Fig. 12-15). Older homes where radiant floor heating was installed often did not have the benefit of such insulation. As a result, the heating systems were forced to work much harder to

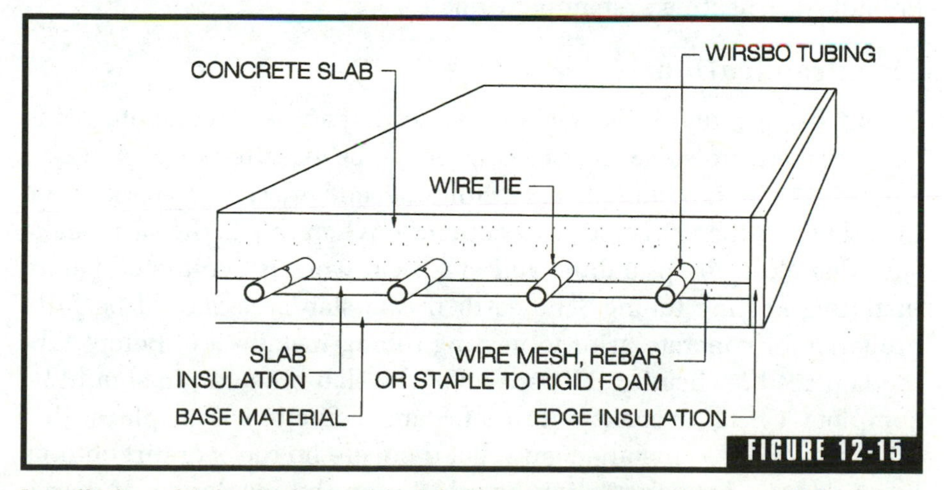

Installation with insulation under slab and around edges. (*Courtesy of Wirsbo.*)

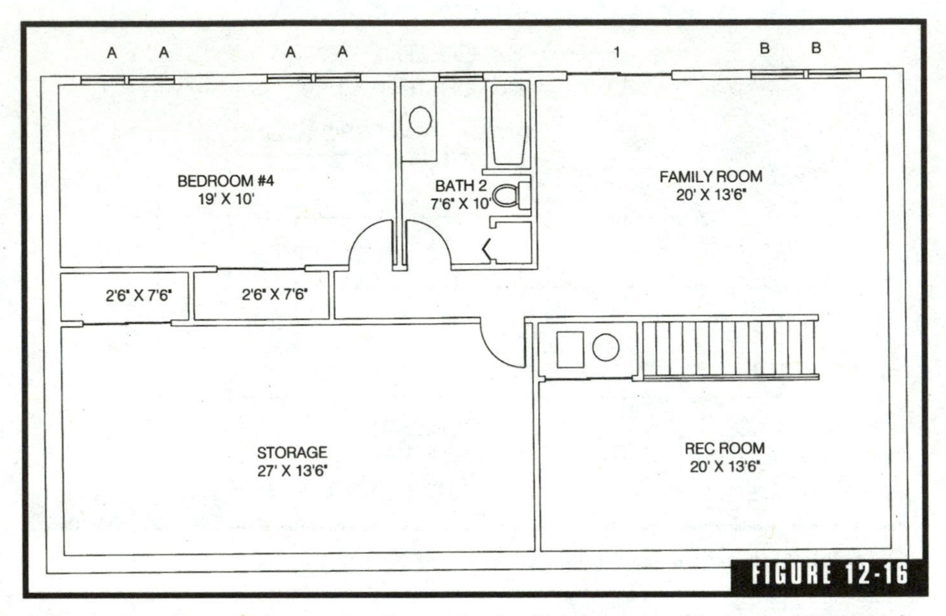

FIGURE 12-16

A floor plan that can be used to lay out a heating system. (*Courtesy of Wirsbo.*)

maintain a comfortable heating temperature. Without the insulation, much of the heat produced by the tubing is lost to the earth. In time, the ground reaches a level temperature that allows the tubing to heat the slab fairly well, but this lost heat can be greatly reduced when insulation is installed below the tubing. A foam insulation board with a thickness of just 1 inch will make a huge difference in how well a radiant floor heating system performs.

Site Preparation

Before heating tubing is installed in anticipation of concrete being poured, there are some site preparations to be attended to (Figs. 12-16 and 12-17). It is common for plumbing and electrical work to be installed and then covered with concrete. When this is the case, make sure that all of the plumbing and electrical work is completed before installing heating tubing. The earth in the slab area should be fully prepared for concrete prior to heating tubing installation. Before tubing is placed for heating, all aspects of the slab preparation should be complete. Check to make sure that foam insulation is in place, that reinforcing wire is installed, and that there are no rocks or dirt clumps lying on top of the insulation board. Sweep the insulation, if neces-

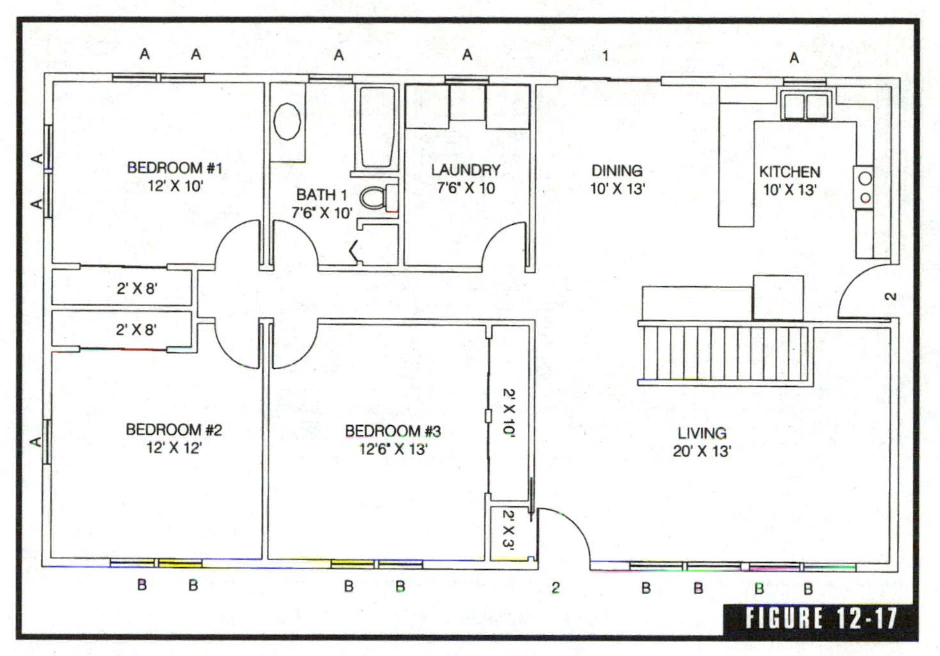

FIGURE 12-17

A floor plan that can be used to lay out a heating system. (*Courtesy of Wirsbo.*)

sary, to provide a clean surface for your tubing. Rocks or other sharp objects could cut or crimp tubing when concrete is poured (Figs. 12-18 and 12-19).

Manifold Risers

Manifold risers are the sections of tubing that are turned up to be exposed above a slab when concrete is poured. These are the feed and return pipes for the heating system. Assuming that you are working with a detailed heating design, and you should be, the locations for manifold risers will be shown on the layout drawing (Fig. 12-20). Precise placement of the manifold risers is usually critical. These pipes are typically designed to turn up in wall cavities. Since installers are working prior to full construction, the walls that the risers are to turn up in are not yet in place (Fig. 12-21). This requires the installer to read plans perfectly and to create mock walls. The mock wall locations can be created with string and stakes. Some installers use what are called block guides. These guides are usually the same width as the walls to be installed. Not everyone likes the idea of using

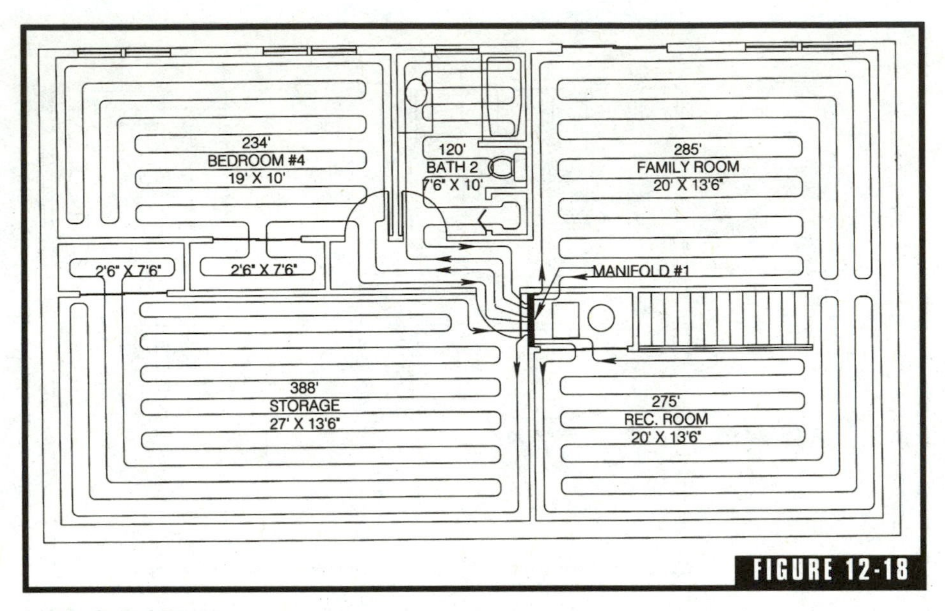

An example of heating plan for installation. (*Courtesy of Wirsbo.*)

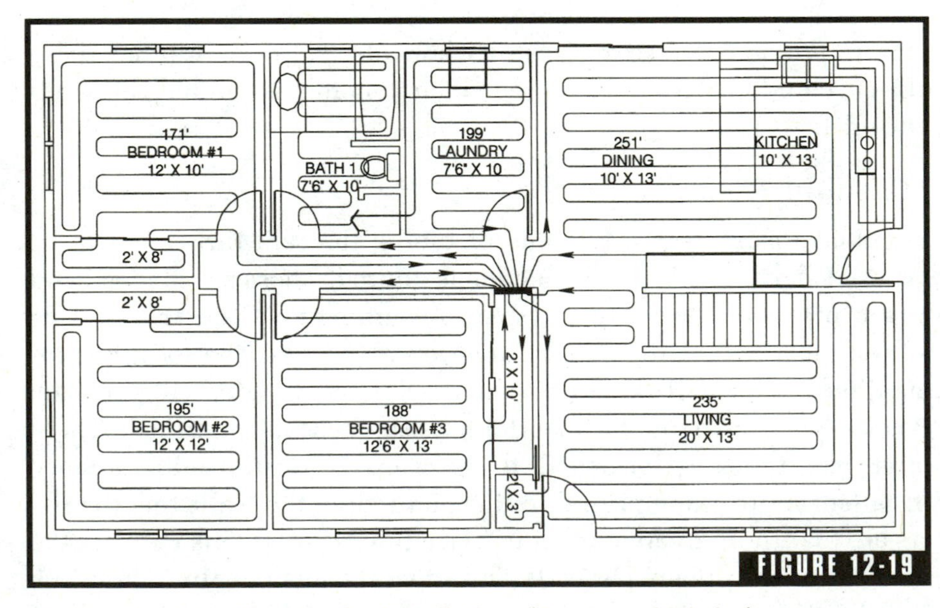

An example of heating plan for installation. (*Courtesy of Wirsbo.*)

wood below grade. This is due to the potential risk of termite infestation. For this reason, a lot of installers pull strings to indicate wall locations and then stake the risers with nonwood stakes.

Plastic bend supports are used to turn risers up, through the concrete. The supports act as sleeves to protect tubing that is to penetrate concrete. Another purpose of the supports is to guide the tubing through a tight bend to an upright position without kinking the tubing. If stakes are used to hold risers in place, the supports are generally attached to the stakes with several wraps of duct tape. It is essential that the risers not move during the pouring and finishing of the concrete. If the tubing is moved accidently, it may wind up sticking up through the finished floor in a place that is not acceptable. The result of this could be using a jackhammer and losing a good deal of money to correct the problem. To avoid this, make sure that all risers are secured in a way that is sure to produce desired results.

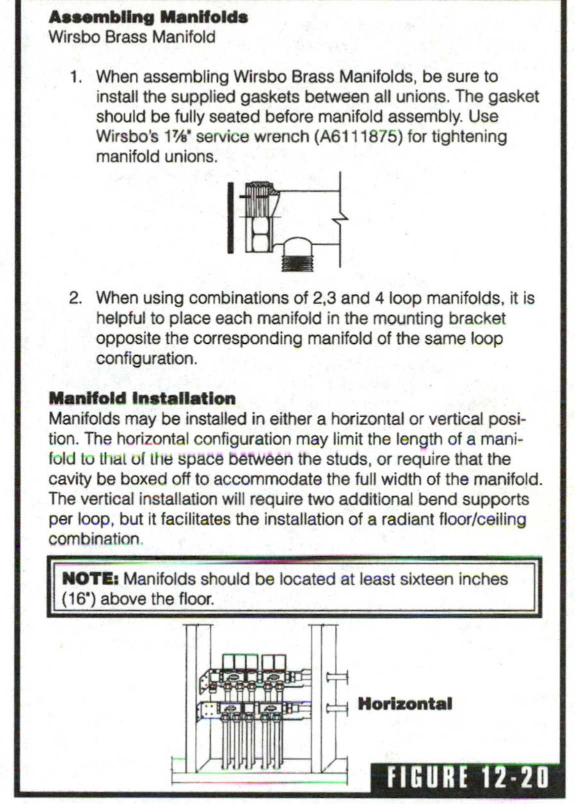

Assembling Manifolds
Wirsbo Brass Manifold

1. When assembling Wirsbo Brass Manifolds, be sure to install the supplied gaskets between all unions. The gasket should be fully seated before manifold assembly. Use Wirsbo's 1⅞" service wrench (A6111875) for tightening manifold unions.
2. When using combinations of 2,3 and 4 loop manifolds, it is helpful to place each manifold in the mounting bracket opposite the corresponding manifold of the same loop configuration.

Manifold Installation
Manifolds may be installed in either a horizontal or vertical position. The horizontal configuration may limit the length of a manifold to that of the space between the studs, or require that the cavity be boxed off to accommodate the full width of the manifold. The vertical installation will require two additional bend supports per loop, but it facilitates the installation of a radiant floor/ceiling combination.

NOTE: Manifolds should be located at least sixteen inches (16") above the floor.

FIGURE 12-20

Manifold installation. (*Courtesy of Wirsbo.*)

Follow the Design

When you begin to lay out tubing in a groundworks (Figs. 12-22 and 12-23), the prepour slab area, you should always follow the heating diagram provided to you by your design engineer. After finding locations for manifold risers, it's time to run the tubing system in the groundworks. All tubing should be installed without joints or couplings. When it is absolutely necessary to install a joint below a slab, there are approved fittings for the job, but they should only be used in extreme circumstances. Use full lengths of tubing whenever you can. When joints don't exist, they can't leak, and this is a big advantage, especially in underfloor systems (Figs. 12-24 and 12-25).

A typical slab will contain a lot of heating tubing installed in it. PEX is the tubing most often used for heating, but PB tubing is also

Wood Frame Construction

The manifold bracket is equipped with bending guides to allow the bracket to be bent around the outsides of 16" center stud walls (Method A). The installer can off-set the manifolds to allow tubing destined for the return manifold to run behind the supply manifold.

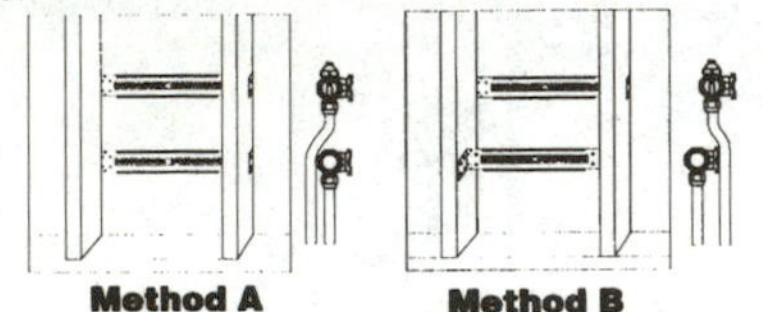

Method A

1. Using suitable hardware, secure both brackets around the outside of the wall stud. Be sure there are at least 8 inches between manifold brackets to allow the tubing to reach the upper manifold and connect squarely.

Method B

1. Install the return manifold bracket (upper) by bending at the two outside bending guides and attaching to the outside of the 16" O.C. stud wall using suitable hardware.
2. Install the supply manifold bracket (lower) by bending at the two middle bending guides and attaching to the inside of the 16" O.C. stud wall using suitable hardware. Allow enough space behind the bracket for the tubing to fit (at least the O.D. of the tubing). Make sure there are at least 10" in between the two brackets.
3. The tubing for the return manifold will run behind the supply manifold.

NOTE: Do not tighten compression fitting on the manifold solely against manifold support bracket. Hold the manifold while tightening fittings.

FIGURE 12-21

Methods of securing manifold tubing. (*Courtesy of Wirsbo.*)

used and was used more than PEX for a long time in the United States. Both types of tubing are easy to work with and neither of them crimp quickly. But, crimping or kinking is a possibility that must be avoided. Since there is a large volume of tubing needed for most jobs, you need some way to manage the tubing. A good way of doing this is using an uncoiler. The uncoiler is basically a rotating spool that allows tubing to come off a large roll with little likelihood of kinking. It's similar to the uncoilers so often used by electricians for their large rolls of electrical wire (Figs. 12-26 and 12-27).

All tubing should be installed in compliance with the manufacturer's recommendations. In typical, straight runs, the tubing should be secured in increments of 24 to 30 inches. It is common practice to install three fasteners on return bends. But, it is important to read and follow the fastening instructions provided by manufacturers. If you are securing tubing to reinforcing wire, make sure that the wire, where it comes into contact with tubing, is not sharp

All tubing should be protected from any anticipated damage (Fig. 12-28). If you have tubing in an area where damage may occur, such as if it is passing through an area that will become a foundation wall or is in an area where expansion joints or sawn control joints might exist, sleeve the tubing. Any sleeve used should be at least two pipe sizes larger than the tubing it is protecting. The pipe used by plumbers for drains and vents works well for sleeving PEX or PB tubing. This could be polyvinyl chloride (PVC) or acrylonitrile butadiene styrene (ABS) plastic pipe. If tubing is already installed before you realize the need for a sleeve, you can cut a section out of the sleeve material and place it over the tubing. The main thing is to make sure that all tubing is protected from potential damage.

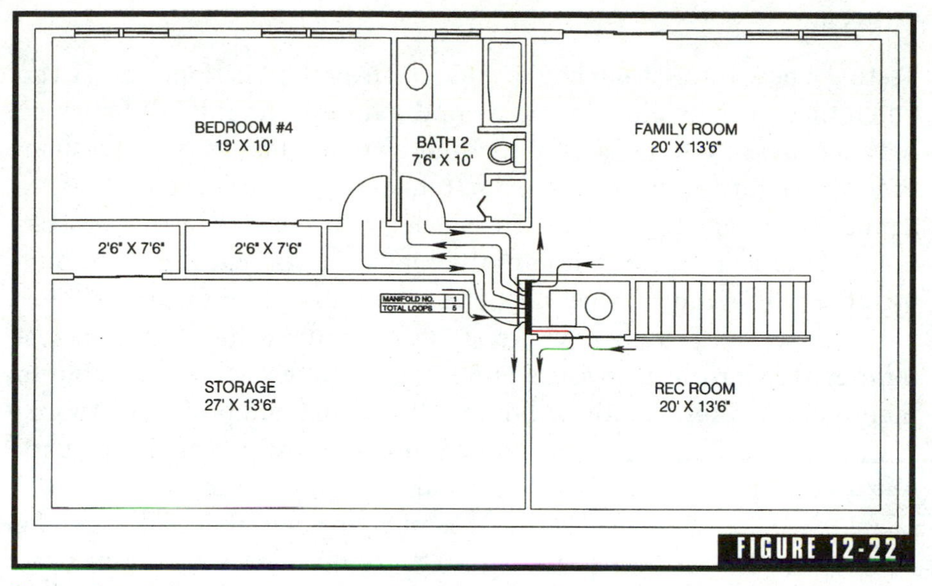

FIGURE 12-22

Floor plan showing basic heating directions required. (*Courtesy of Wirsbo.*)

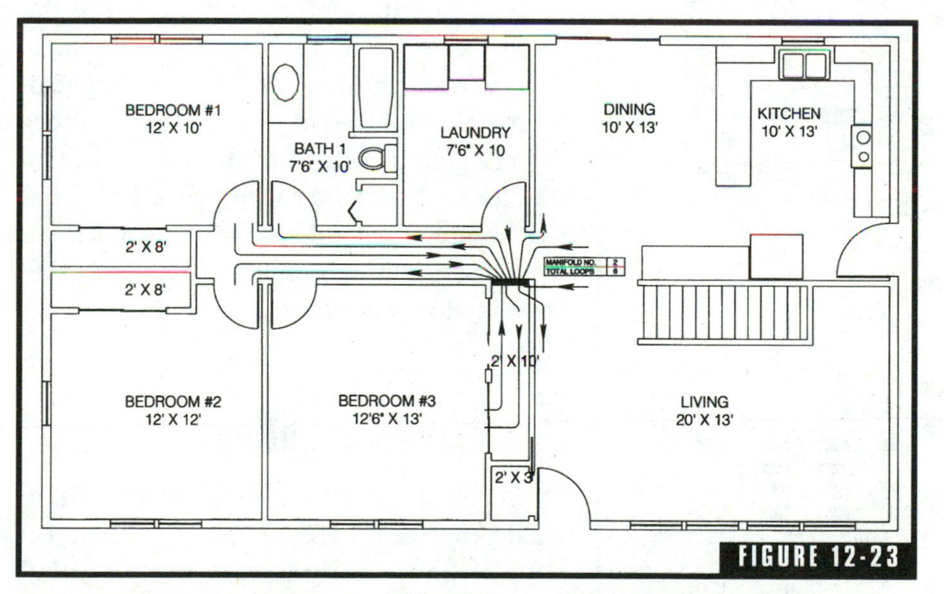

FIGURE 12-23

Floor plan showing basic heating directions required. (*Courtesy of Wirsbo.*)

Testing

Testing a new installation before concrete is poured is important. Even when there are no joints in a system, the tubing should still be tested with air pressure to make sure that there are no defects in the system. Procedures for testing are easy, don't take long to perform, and assure a much better job. Once you create a test rig, or rigs, the testing process is fast. You can test each line individually, or you can tie the tubing together with temporary connections to test all of it at once.

What amount of air pressure should you use when testing a system? A standard test pressure is 50 psi. A system should be able to maintain this pressure for 24 hours. It's not uncommon for the tubing to expand during a test period. If this happens, the air pressure on an air gauge will drop. It is also possible that leaks may be present in the temporary connections used to join a system for testing. If you suspect a leak at a test joint, use some soapy water to paint the connection points. When a leak is present, bubbles will appear in the soapy water. Don't omit the test stage. Even without underfloor joints, tubing can have holes in it. This could be a factory defect, a hole that was incurred due to storage and transportation, or a hole that was made during installation. Check all the pipes. It's much easier to make replacements before concrete is poured.

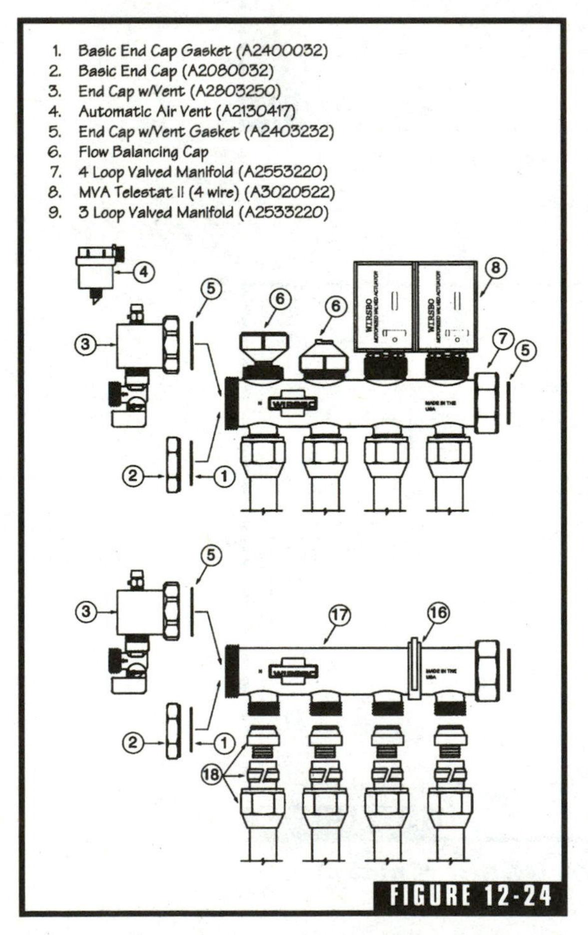

Heating system components. (*Courtesy of Wirsbo.*)

Thin-Slab Installations

Thin-slab installations are pretty much what they sound like (Figs. 12-29 and 12-30). Radiant heating systems installed above a large slab can be done with a thin-slab approach. This means installing the heating tubing on a wooden floor and then pouring a thin slab of concrete over the heating tubing. This procedure is common and effective. A thin-slab system is not the only way to use radiant floor heating above a slab, but it is a fine choice. The thickness of concrete used in thin-slab systems doesn't usually exceed 1 1/2 inches.

Piping procedures for a thin-slab system are about the same as those used for larger concrete slabs. One difference is how the tubing is attached. In a slab-on-grade system, the tubing is usually attached to reinforcing wire or with special clamps. When installing a thin-slab system, the tubing is secured to the wood subflooring. The floor joist cavities below the subflooring should be filled with insulation. Batt glass fiber insulation is the typical choice for insulation (Figs. 12-31 and 12-32).

Manifold risers are needed with a thin-slab system, just as they are with a large slab. The risers usually turn up either in walls or inside closets. When the risers are in a wall, there should be an access door provided for service and repair. All tubing must be attached securely to the subflooring. Since thin slabs don't have much concrete cover for tubing to be protected and covered, the tubing must be held tightly to the floor. Any loose-fitting tubing can rise up and protrude above the finished concrete covering. General principles call for tubing to be secured at intervals of no more than 30 inches on straight runs. It is a good idea to keep the fasteners closer together.

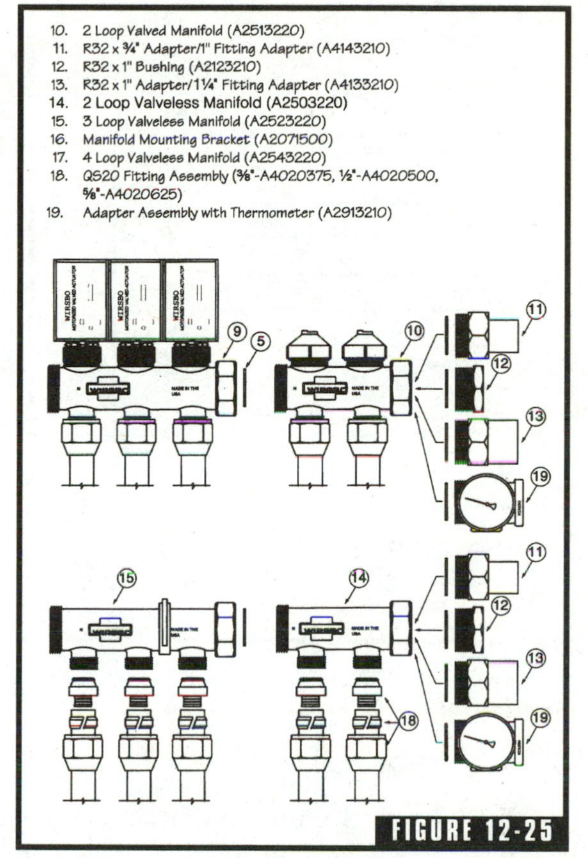

FIGURE 12-25

Heating system components. (*Courtesy of Wirsbo.*)

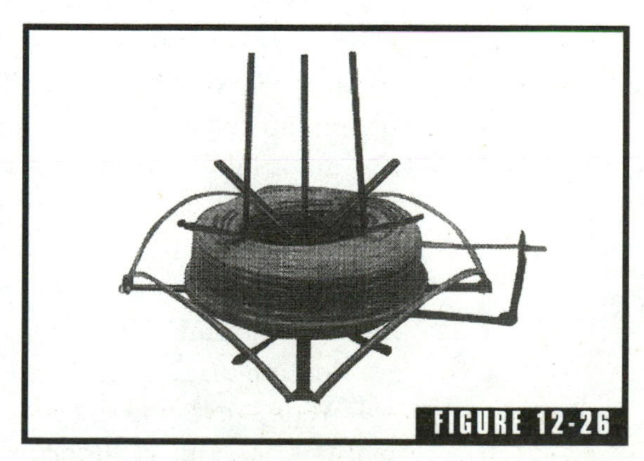

FIGURE 12-26

Uncoiler. (*Courtesy of Wirsbo.*)

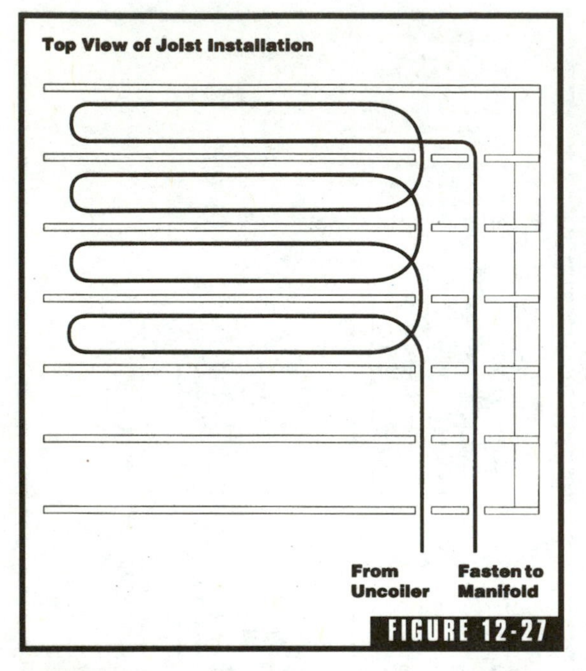

FIGURE 12-27

Example of routing tubing from an uncoiler. (*Courtesy of Wirsbo.*)

Construction Considerations

Special construction considerations have to be addressed when a thin-slab system is used. Weight is one major concern. The weight added when concrete is poured over subflooring can be quite substantial. This has to be considered during the planning stage of construction. Larger floor joists or other means of additional support are needed to accept the added weight of concrete. While weight is a major consideration with thin-slab systems, it is not the only thing to think about.

Adding about 1 1/2 inches of concrete to a subfloor raises the level of a finished floor. This is not a big problem when it is planned for in advance, but it can mean trouble if allowances are not factored in for the increased height. For example, door openings and thresholds will be off if the added height is not allowed for. Plumbing fixtures and base cabinets will be affected by the increased height. As long as adjustments are made during rough construction, the finished product should be fine.

Concrete or Gypsum?

There are two types of cover to place over heating tubing in a thin-slab system. In my region, lightweight concrete is the leading cover for thin-slab systems. But, gypsum-based underlayment is another popular material for covering heating tubing. Lightweight concrete in a typical mix will add up to about 14 pounds per square foot to the dead-load weight rating of a floor. Believe it or not, this is actually a little less weight than what gypsum-based material will add. Either type of cover material will work; it's a matter of regional preference as to which material is most likely to be used.

When gypsum-based materials are used, a sealant is generally sprayed on the subflooring. By spraying a sealant and bonding agent on the subflooring, the floor surface is strengthened and made more

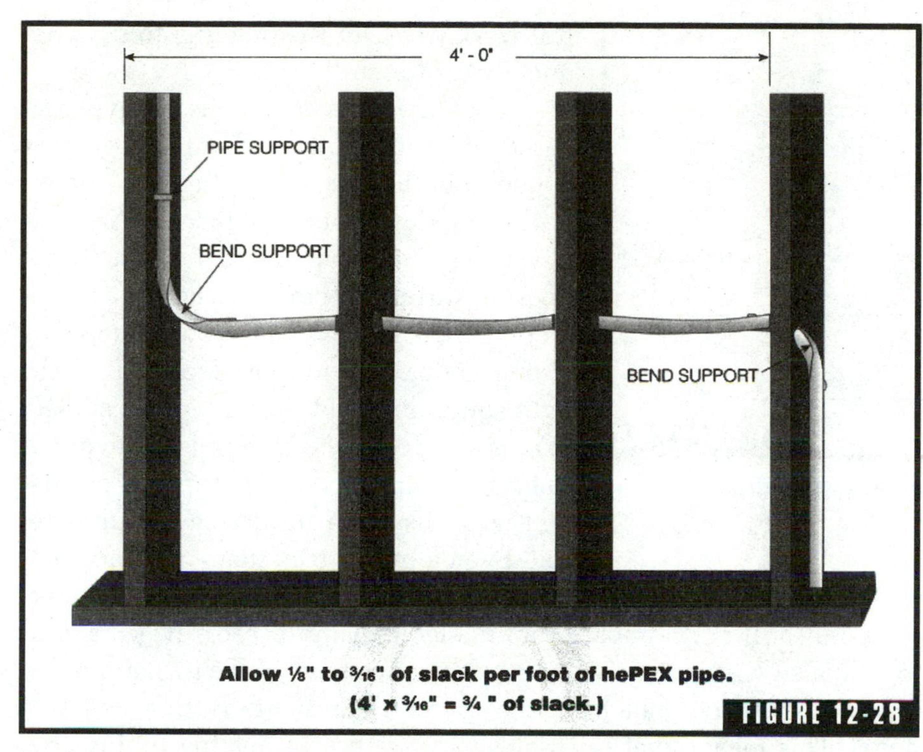

Bend supports used to prevent crimping of tubing. (*Courtesy of Wirsbo.*)

resistant to moisture problems. The spraying process is done after all tubing is installed. Once the floor is prepared, the gypsum material is mixed in a concrete-type of mixer and pumped to the location of the heating system through a hose. The material is loose enough in consistency to flow under and around the tubing. Material is pumped in until the level of it is equal to the tubing's diameter. This is called the first coat or the first lift. More material will be needed after the first coat is dry.

The first layer of cover material should dry, depending upon site conditions, within a few hours. After the material can be walked on (it could be ready in as little as 2 hours), you can apply the second lift, or coat. When the second level is distributed, it should cover the tubing by at least 3/4 inch. After a few hours, the second coat should be dry enough to walk on, but this does not mean that it is dry enough to install a finished floor covering. Depending upon site conditions, meaning air temperature, humidity, and so forth, it can take up to a

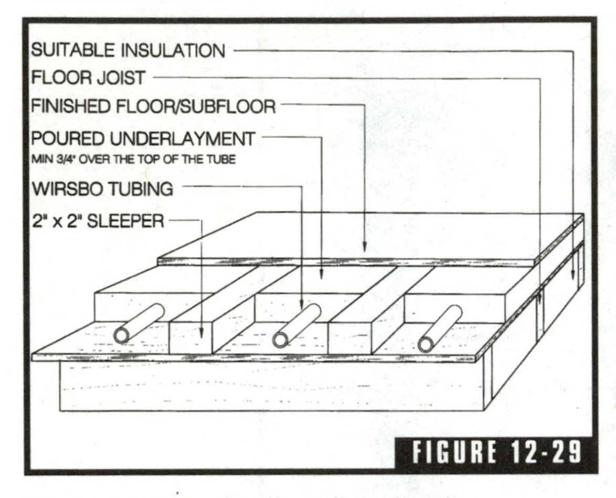

Sleeper system of radiant floor heating. (*Courtesy of Wirsbo.*)

full work week for the material to dry adequately for finish flooring.

There are pros and cons to gypsum-based cover material. A big advantage over concrete is the ease of application offered with gypsum-based material. Another advantage of the gypsum material is that it doesn't shrink or crack as badly as concrete might. The gypsum floor material is strong enough to support foot traffic and light equipment, but the material is subject to cuts and gouges, which must be avoided during construction. Any type of consistent water leak can ruin the gypsum material. Even a minor leak that goes undetected for a period of time can turn the material to mush. It is common for a finished gypsum-based cover material to be sealed with a sealant. Another disadvantage of the gypsum material is that it does not have the thermal conductivity that concrete does. This means that the water temperature in heating tubing covered with gypsum material is likely to be higher than it would need to be in a concrete floor.

Lightweight Concrete

Lightweight concrete is a common cover material for thin-slab systems. The concrete can be delivered to the point of installation in buckets or wheelbarrows. It can also be pumped to the distribution point with a hose and grout pump. Since concrete is what it is, the material is not affected greatly by moisture, as a gypsum-based produce would be. But, concrete does have a tendency to crack, and this can be a problem for finished floor coverings. Most contractors install control joints in the material to reduce, or to at least control, the effects of cracking. Control joints are often placed under door openings.

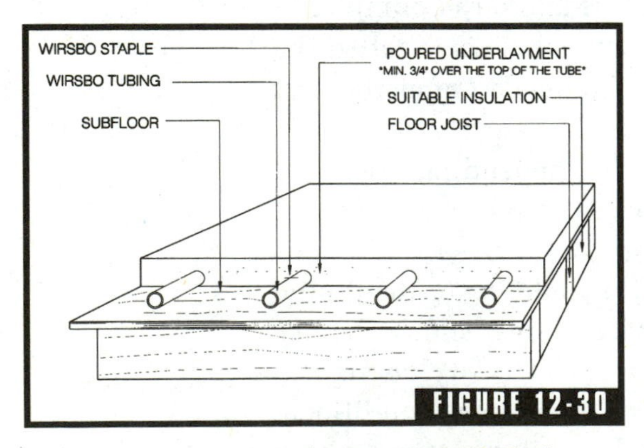

Thin-slab installation. (*Courtesy of Wirsbo.*)

The finished floor covering on a floor containing radiant heat must be taken into consideration. Some types of floor coverings allow heat to rise better than others. For example, a tile floor will allow much more heat to escape from a radiant system than a padded, carpeted floor covering. To control this, insulation can be placed under the subflooring of a radiant system. When the R-value of the insulation under the floor is greater than the R-value of the finished floor covering, heat will rise through the floor, rather than escaping below it.

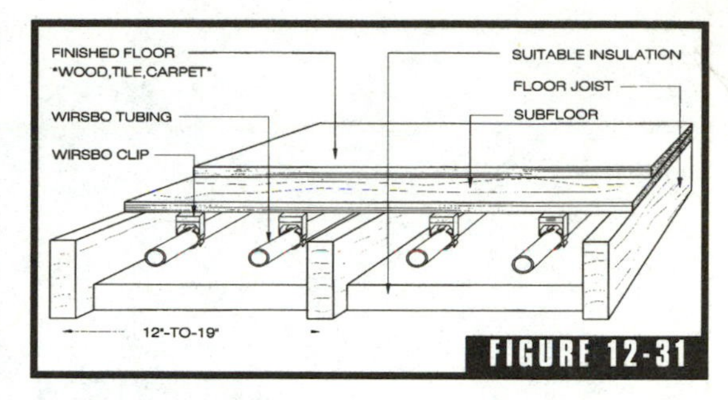

Joist heating. (*Courtesy of Wirsbo.*)

Dry Systems

When radiant heat is installed beneath a floor (Fig. 12-33) and is not covered with concrete or gypsum-based material, the installation is known as a dry system. The reason for this is simple. The fact that no material is poured over the tubing makes the system dry. These systems are great in that they don't add any appreciable weight to a flooring system. Since weight is not added in substantial amounts, a dry system can be used in remodeling without the need for additional floor support. However, a dry system does need some help in producing heat in desirable quantities.

Since dry systems do not have concrete or gypsum to conduct heat, it's common to install heat transfer plates for lateral heat conduction. The plates may be installed above or below subflooring, depending upon the design. Heat plates are made of aluminum and work very well. The installation of below-floor systems is less expensive than above-floor systems. There is much less labor involved in below-floor systems, and this is one reason why the cost is less (Fig. 12-34).

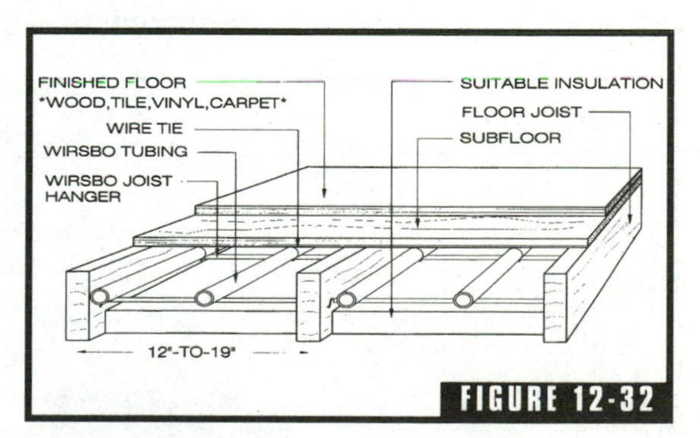

Bracket hangers used for joist heating. (*Courtesy of Wirsbo.*)

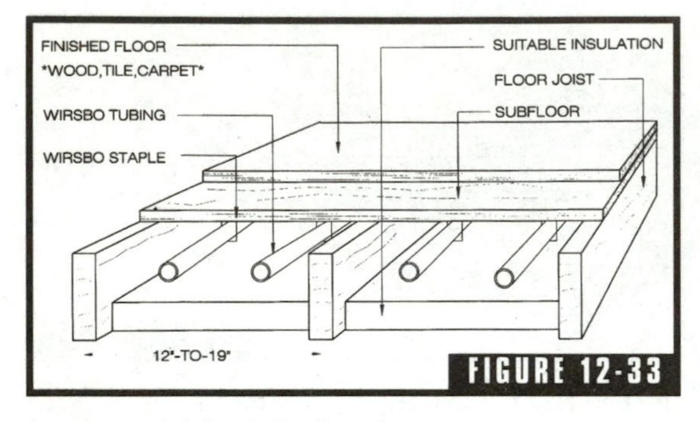

Staples used for joist heating. (*Courtesy of Wirsbo.*)

Above-Floor Systems

Above-floor systems are installed above subflooring and below finished flooring (Fig. 12-35). Since space is needed for tubing, a sleeper system is required for above-floor systems. The sleeper system is a series of wood strips that provide cavities for the tubing to run through and give flooring installers a nailing surface for an additional layer of subflooring. Heat transfer plates are installed for straight runs of tubing. The plates are placed between the sleeper members and stapled down on one side. Once the plates are in place, the tubing can be installed in them. Since sleepers and a second subflooring are installed in an above-floor system, the floor height is raised by an inch or more. This can be a problem for doors, plumbing fixtures, and base cabinets. Just as with a thin-slab system, these problems can be avoided with proper planning and adjustments.

Below-Floor Systems

Below-floor systems require less time and labor to install. They also require less material, since sleepers and a second subflooring are not needed. Whether remodeling or building, below-floor systems make a lot of sense. Heat tubing is usually placed in heat transfer plates that are stapled to the bottom of subflooring. When tubing penetrates floor joists, the holes should be drilled near the middle of the joist, rather than the top or bottom edge. This helps to maintain structural integrity. Also, the holes should be somewhat larger than the tubing diameter, to avoid squeaking as the tubing expands.

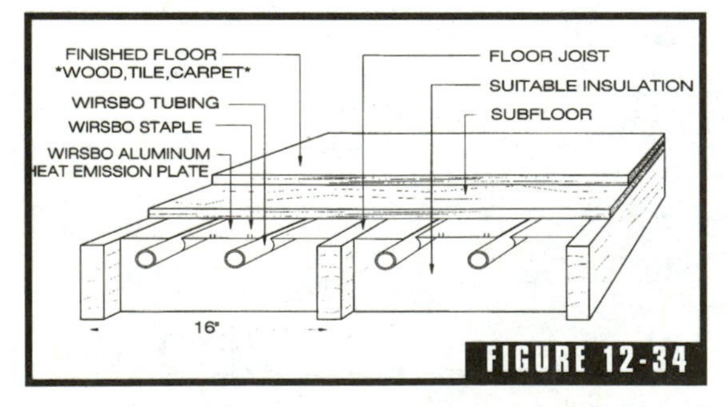

Heat emission plates. (*Courtesy of Wirsbo.*)

When heat tubing is attached to the underside of subflooring, it is at risk of damage from the work of others. A plumber or electrician who is not aware of the heat tubing may drill through it accidently. It's not practical to protect all tubing with nail plates, but it is wise to make

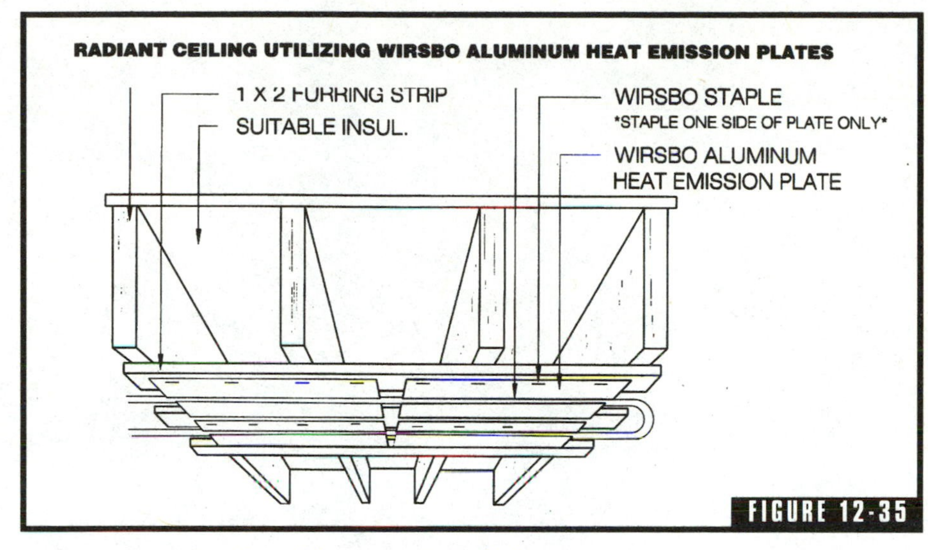

Radiant ceiling heating. (*Courtesy of Wirsbo.*)

all workers aware of the tubing under the subfloor. If a section of tubing is damaged, repair couplings can be used to correct the problem. However, whenever possible, avoid joints and connections in heating systems. It's best to maintain full lengths of tubing, rather than coupled sections.

Radiant floor heating systems are gaining popularity quickly. They can be quite cost-effective to install and operate. The comfort offered from a radiant system is, in many ways, unmatched. If you are not familiar with this type of heating system, you owe it to yourself and your customers to learn more about it. Radiant heating systems in floors are well worth consideration.

CHAPTER 13

Troubleshooting Oil-Fired Boilers

Oil-fired boilers are used in many parts of the country. The boilers are usually very dependable. But, like any type of equipment, boilers can suffer from mechanical problems. If a boiler fails to run during the cold of winter, plumbing and heating systems in a home can freeze. Much damage can occur if a boiler fails and no one is aware of the problem. Granted, this doesn't happen often, but it does happen. When the heat goes off in a house or building, it is usually noticed by the occupants of the building. However, if a house is being marketed for sale and no one's living in it, a boiler failure can be disastrous. Even when there is someone to notice a boiler failure, the risk of damage to plumbing and heating systems exists. Whether property damage is imminent or not, the comfort level of residents in buildings with a failed heating system is at stake. Knowing what to look for when a boiler fails and how to correct the problem quickly is a big asset for anyone in the heating business.

There are many items associated with a boiler that can fail. However, most of the problems can be traced with relative ease. Still, pinpointing a problem quickly can be tricky. The job is easier if you use a logical approach in troubleshooting. Anyone with a basic understanding of boilers can eventually find the cause of failures by using a trial-and-error method. This type of troubleshooting takes time. A systematic approach is faster. Knowing what to look for and where to look for it is half the battle. Once a problem is identified, correcting it is usually not very difficult.

Some problems encountered with oil burners can be potentially dangerous. Smoke or fumes can fill a home quickly when a boiler become defective. The smell and residue left behind can be hard to live with. In some cases, the risk of fire may exist. If a safety control malfunctions, big trouble could be right around the corner. The types of complaints common with oil burners range from noisy operation to no heat. And, there's a lot of possible trouble in between. Getting to the heart of a problem quickly is something of an art. In the case of boilers, the service technician is the artist (Fig. 13-1).

Noise

Noise is not an uncommon complaint with oil-fired boilers. Since many boilers are located under or near living space, excess noise can be a real problem. A boiler that is properly tuned should not produce a lot of noise. When a noise is noticed, it is a sign that something is wrong with some aspect of the boiler. The root cause of the problem can be one of any number of things. The type of noise heard can be an indicator of what to look for. There are three typical types of noise associated with oil burners. Boilers sometimes give off pulsing sounds. Thumping noises are also common. Rumbling is another type of noise that you might experience with an oil burner. Once you have a noise to deal with, what do you do? The nozzle in the burner is always a good place to start your investigation.

A systematic approach (Fig. 13-2) is best when troubleshooting, but there are some times when trial and error is the only way to determine the exact cause of a problem. The trick is to use a systematic method of trial and error. In the case of noise, you should systematically turn to the nozzle in the burner. Then, you have to gamble with some trial-and-error work. Start by replacing the existing nozzle with one that has a wider spray angle. Try the unit and see if it is still producing excess noise. With a little luck, you will have solved the problem. If not, you have to take a few more steps.

When a nozzle with a wider spray angle will not quiet the noise made by a burner, try a nozzle that has a smaller opening. Then instal-

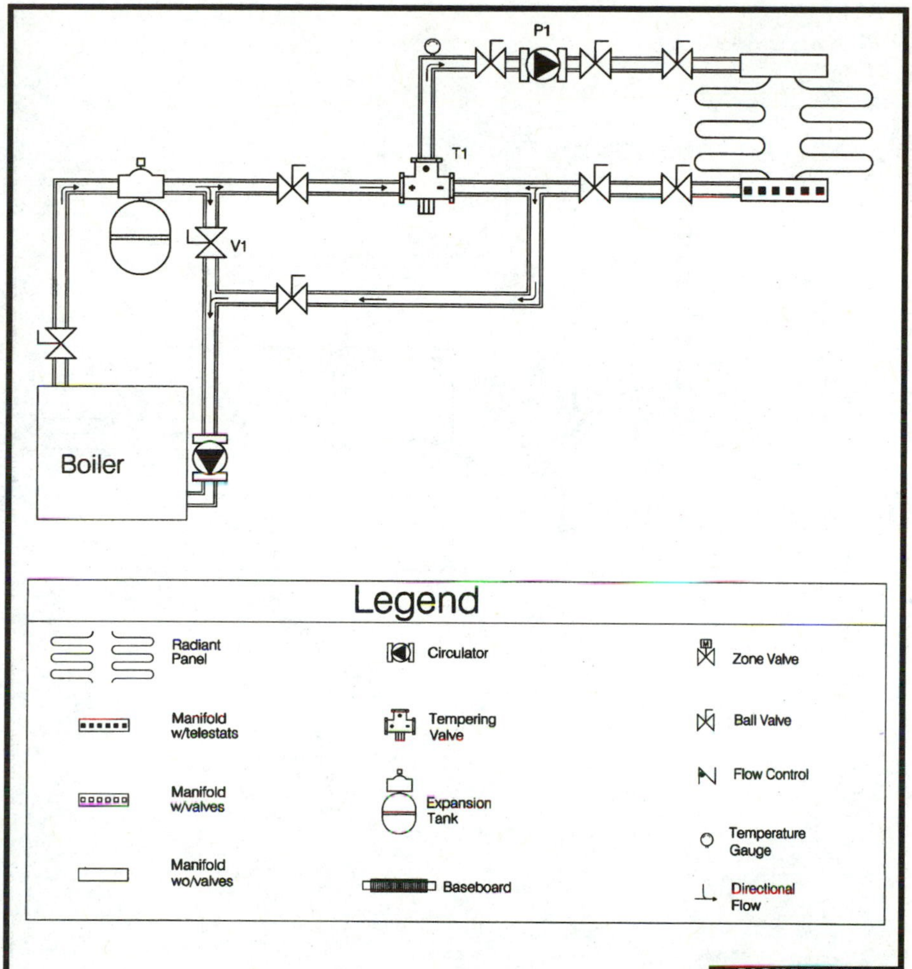

FIGURE 13-1

Boiler piping schematic. (*Courtesy of Wirsbo.*)

lation of a new nozzle that is one size smaller than the one previously installed could be all it will take to quiet the boiler. It is necessary to make adjustments and test after each one to see if the problem is corrected. Another course of action, especially if the oil burner is producing pulsating noises, is to install a delayed-opening solenoid on the nozzle line.

There is another consideration. If you are getting a noisy fire and the nozzle replacements don't fix the trouble, check to see where the

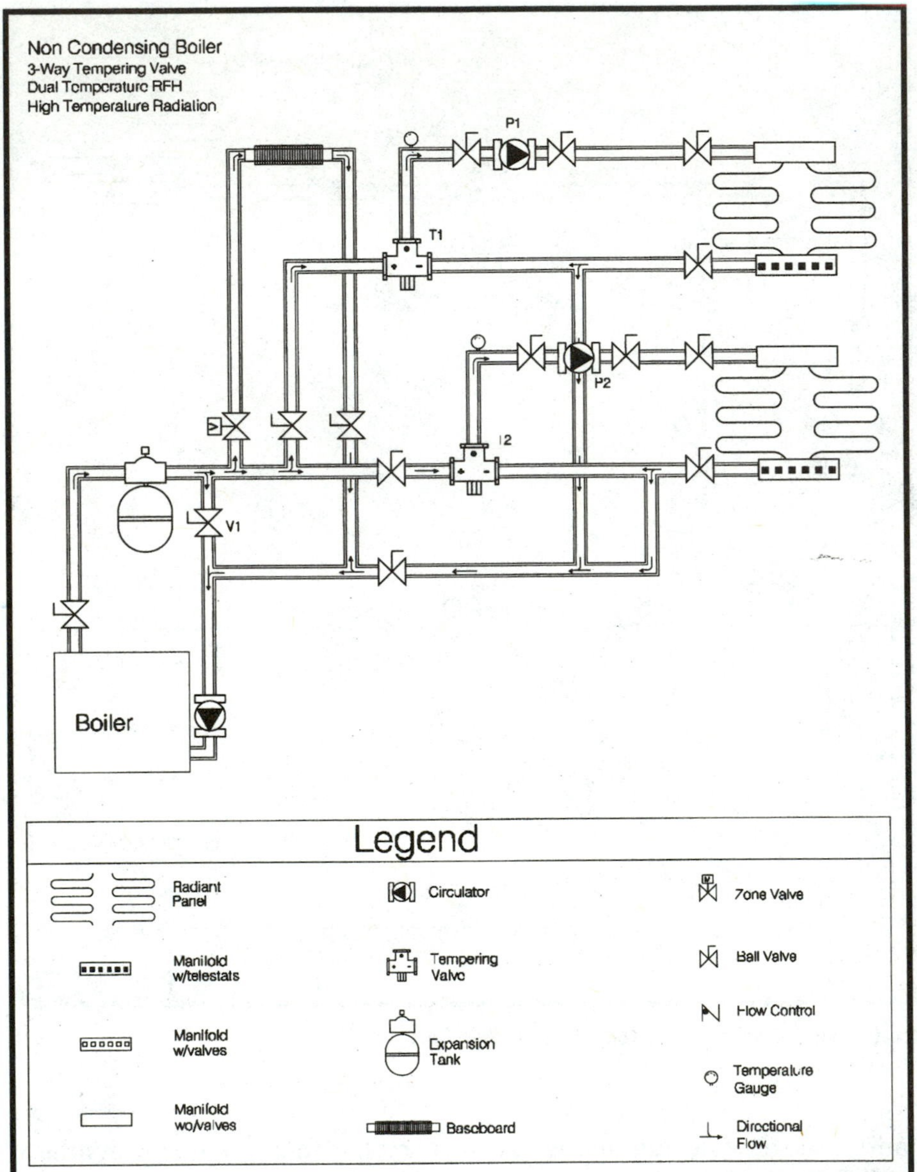

Non Condensing Boiler
3-Way Tempering Valve
Dual Temperature RFH
High Temperature Radiation

P1
T1
P2
T2
V1
Boiler

Legend

Radiant Panel
Manifold w/telestats
Manifold w/valves
Manifold wo/valves
Circulator
Tempering Valve
Expansion Tank
Baseboard
Zone Valve
Ball Valve
Flow Control
Temperature Gauge
Directional Flow

FIGURE 13-2

Boiler piping schematic. (*Courtesy of Wirsbo.*)

oil supply tank is located. Assuming that nozzle replacement and a delayed-opening solenoid have not satisfied you, the problem might be cold oil. When an oil tank is installed outside, the oil can get cold enough to create a noisy burner. To remedy this, pump the fuel oil through a nozzle that is the next size smaller at a pressure of about 125 psi. One of these four courses of action should put the burner into good operating condition.

A Smokey Joe

If you run across a boiler that smokes like a barbeque grill, there could be serious problems at hand. The problem could be as simple as a defective nozzle to something as major as a damaged combustion chamber. If the problem turns out to be the combustion chamber, your customer is going to be very unhappy, due to the high cost required to get the heating system back up and running. Don't just assume that the problem is in the combustion chamber or in the burner tube. Start with the simple fixes.

When you are called to troubleshooting a smoking boiler, check first for dirt. If the air-handling parts of the oil burner are dirty, smoke output can be present. The dirt prevents the burner from functioning properly. A good cleaning may be all that it will take to solve the problem. Assuming that dirt is not the problem, turn your attention to the nozzle in the burner. If it's the wrong size, the nozzle could be the cause of the smoke. The worst scenario for your customer is a cracked combustion chamber.

The air-handling parts to check for dirt include the fan blades, air intake, and air vanes in the combustion head. When these components are clean, they allow the system to work much more efficiently. Nozzle problems are not difficult to fix. In the case of a smoking boiler, the replacement of an existing nozzle with a smaller one can be all it takes to stop the smoking. Another tactic with the nozzle is to install one that has a narrower spray angle. If the cause of the smoke is a cracked combustion chamber, it's likely that the boiler will need to be replaced. It may be possible to repair the combustion chamber, but if repairs are made, you must make sure that there is no smoke or gas escaping the combustion chamber.

It Stinks

There are times when boilers just simply stink. Fuel odors can occur when oil burners are used. Many people are very sensitive to the smell of oil, and the odor can be difficult to get rid of once it is present. There are three basic reasons why oil burners produce odors. Chimney obstructions are one of the first things to consider when you are faced with a boiler that is stinking up a mechanical room. If the chimney checks out, the next logical step is to check the ignition time on the burner. If it is delaying, that can account for the buildup of excess fumes. The third possibility is that there is too much air going through the boiler.

Since chimney problems are the first consideration when odors are present, let's start with troubleshooting them. The draft in a chimney or flue over a fire from a boiler should not be less than .02 to .04. If you do a draft test and find the draft to be below suitable levels, you will have to do a physical inspection of the chimney or flue. Maybe a bird has started building a nest in the flue. It's possible that leaves or other obstructions are blocking the draft. There is a good chance that odors being produced from oil burners will be associated with a poor draft.

Once you know that the chimney or flue is clear, look for delayed ignition. This is a sure cause of odor buildup. There are many factors which may contribute to delayed ignition. A simple test may prove that the electrode setting is not correct. If there are insulator cracks, delayed ignition can result. Dirt, such as soot or oil, can foul an electrode and cause it to fire improperly. When a pump is set with an incorrect pressure setting, you can experience delayed ignition. A bad spray pattern for a nozzle can also be responsible for delayed ignition. If a nozzle becomes clogged, it can cause slow ignition. Another potential cause of delayed ignition is an air shutter that is opened too far.

Electrode settings and nozzles are the main issues to consider. Sometimes replacing an existing nozzle with one that has a solid spray pattern will correct the problem. More likely than not, a simple nozzle replacement will solve the problem. If this is not the case, resetting the electrode settings can solve the problem. Check manufacturer suggestions for nozzle angles, flow rates, and electrode settings. If you have a burner that is using a hollow spray pattern and firing 2.00 gallons per hour (gph), try replacing it with a nozzle that produces a solid spray pattern.

In addition to the problems that we have just discussed, there are many other types of problems that arise from time to time with oil burners. To make this information as accessible as possible and as easy to use as we can, let's group the remainder of the troubleshooting steps together under the headings which are most appropriate.

Noisy Operation

We talked earlier about some of the most common causes of noisy operation in an oil burner. In addition to what you've already learned, you should consider the possibility that there may be a bad coupling alignment in the system. To correct this, you must loosen the fuel-unit mounting screws slightly and shift the unit in different positions until the noise is eliminated. If the noise stops, tighten the mounting screws and pat yourself on the back.

If air gets into an inlet line, noise can occur during operation. To eliminate this possibility, check all connections for leaks. All connections should be made with good flare fittings. It may be worth checking the flaring of all tubing. If you find a leak, either tighten the connection or replace it with a new one.

Have you ever run into a tank hum on a two-pipe system with an inside tank? If you haven't, you may. This type of problem is fixed by installing a return-line hum eliminator in the return line. As you can see, there are not a lot of causes for noisy boilers, so this is one aspect of service and repair that shouldn't take long to troubleshoot.

No Oil

It's common knowledge that oil burners need oil to function. As basic as this is, sometimes the possibility of a lack of oil is ignored. When you've got a burner that will not fire, check the supply tank to see that it contains an adequate supply of oil. Don't trust the oil gauge; it may stick. Use a dip stick to probe the tank and to determine the fuel level. If the tank has oil in it, check the fuel filter. It may be clogged. Remove and inspect all fuel filters and strainers. Clean them if they appear to be blocked.

Nozzles frequently become clogged. Obviously, an obstructed nozzle is going to produce less oil for the burner to fire on. This is easy enough to fix. Just replace the nozzle with a new one. If you've followed all of these steps and still have a problem, look for a leak in the intake line. If you find one, try tightening all fittings where you suspect the leak. Check the unused intake port plug to make sure that it is tight and not leaking. Air can also invade a system around a faulty filter cover or gasket, so check these components to make sure that they are in good shape and are not leaking. When you are checking the intake line for leaks, make sure that there are no kinks in the line which may be obstructing it.

Occasionally, a two-pipe system will become airbound. If this happens, you must check the bypass plug. Insert one if necessary and make sure that the return line is below the oil level in the supply tank. When you have an airbound single-pipe system, you should loosen the port plug or easy-flow valve and bleed oil for about 15 seconds after all foam is gone from the bleed hose. Then check all connections and plugs to make sure that they are tight. Getting the air out is not hard, and it can be all it takes to get a system back up and running.

There are a couple of other things to check if you are having trouble getting oil to the burner. For example, check to see that there are no slipping or broken couplings. If there are, they must be tightened or replaced. It may sound weird, but you may find a situation where the rotation of a motor and fuel unit is not the same as indicated by the arrow on the pad at the top of the unit. Should this happen, install the fuel unit so that the rotation is correct. The last thing to consider is a frozen pump shaft. If you find this to be the problem, the unit must be sent out for approved servicing or factory repair. Before you install a replacement unit, check for water and dirt in the oil tank.

Leaks

Leaks happen in oil systems. The vibration of equipment can create small leaks. There are many reasons why a system may suffer from leakage, but the leaks should not be left unattended. Sometimes removing a plug, applying a new coat of thread sealant, and reinstalling the plug will take care of a leak. When leaks show up at nozzle plugs or near pressure adjustment screws, the washer or o-ring at the connection may need to be replaced. Time can take its toll on both washers and o-rings. The covers on housings sometimes leak. When this happens, you may get by with just tightening the cover screws, but be prepared to replace a defective gasket. If a seal is leaking oil, the fuel unit must be replaced.

Blown seals can occur in both single-pipe and two-pipe systems. A blown seal in a single-pipe system may mean that a bypass plug has been left out. Check it and see. You may have to install one. You may find that you will have to replace the whole fuel unit. Two-pipe systems with blown seals could be affected by kinked tubing or obstructions in the return tubing. If this is not the case, plan on replacing the fuel unit.

Bad Oil Pressure

Bad or low oil pressure might be as simple as a defective gauge. It's worth checking. If the gauge proves to be defective, replace it and you are home free. Assuming that the gauge is reading true, look for a nozzle where the capacity is greater than the fuel unit capacity. Should you find this to be the case, replace the fuel unit with one that will offer an adequate capacity.

Pressure Pulsation

Pressure pulsation may be caused by a partially clogged strainer or filter. When this is the problem, all you have to do is clean the strainer and the filter element. Air leaks are another possible cause for pulsating pressure. If you are experiencing an air leak in an intake line, you should tighten all fittings where leaks may be present. Another place to look for air leaks is the cover of the unit. If you suspect leaks on a cover, tighten the cover screws first and then check for leaks. If leaks persist, remove the cover and install a new cover gasket. Then, of course, replace the cover and tighten the cover screws.

Unwanted Cutoffs

Unwanted cutoffs can be caused by several factors. If you have a burner that is cutting off when it shouldn't be, start by testing the nozzle port with a pressure gauge. Put the gauge in the nozzle port of the fuel unit and run the system for about 1 minute. Then, shut the burner down. If the pressure drops from normal operating pressure and stabilizes, the fuel unit is fine. This means that air is the culprit of your troubles. But, if the pressure drops to zero, the fuel unit has given up the ghost and needs to be replaced.

Filter leaks can cause unwanted cutoffs. Check the face of the cover and the cover gasket to see if there are any leaks. If there are, replace the gasket and secure the cover tightly. Another thing to check is the cover strainer. This can be fixed by tightening the cover screws of the strainer. Sometimes an air pocket builds up between the cutoff valve and the nozzle. If this happens, run the oil burner while stopping and starting it until all smoke and afterfire disappear.

An unwanted cutoff could be the result of an air leak in the intake line. Check all fittings for possible leaks. Tighten any suspicious connections. Check the unused intake port and return plug to make sure that the plug is tight. If you are still having trouble, look for a partially clogged nozzle strainer. Clean obstructed strainers or replace the nozzle. If there is a leak at the nozzle adapter, change the nozzle and the adapter.

Well, there you have it. You are now ready to set forth and troubleshoot oil burners. Oh, well, maybe not quite; it helps to have on-the-job-experience, but the approaches given here will get you going in the right direction on a fast track. If you follow the systematic procedures discussed in this chapter, you should have less lost time and a more productive work day, not to mention fewer headaches in the field. Our next topic of conversation is the general maintenance of gas-fired boilers. So, if you're ready, let's turn to the next chapter and dive into that subject.

Troubleshooting Gas-Fired Boilers

Troubleshooting gas-fired boilers is a bit different from the same job done with oil-fired boilers. There are significant differences in the ways in which the two types of boilers function. Gas-fired boilers are very common in many parts of the country. The boilers may be fueled by either natural gas or manufactured gas. Fuel oil burns but will not explode under normal conditions. Gas, on the other hand, can explode. This in itself makes the troubleshooting process somewhat different. A leak in a gas line can prove fatal, either from inhaling the fumes or from being blown up. This doesn't happen with oil-fired systems.

The type of fuels used is not the only difference between oil- and gas-fired boilers. Since the fuels are different, the firing mechanisms are also different. This requires different troubleshooting techniques for the two types of boilers. One of the first considerations when troubleshooting a gas-fired system is safety. If excessive gas fumes have collected in the boiler area, the least little spark could ignite disaster. Gas is not a fuel to be taken lightly (Fig. 14-1).

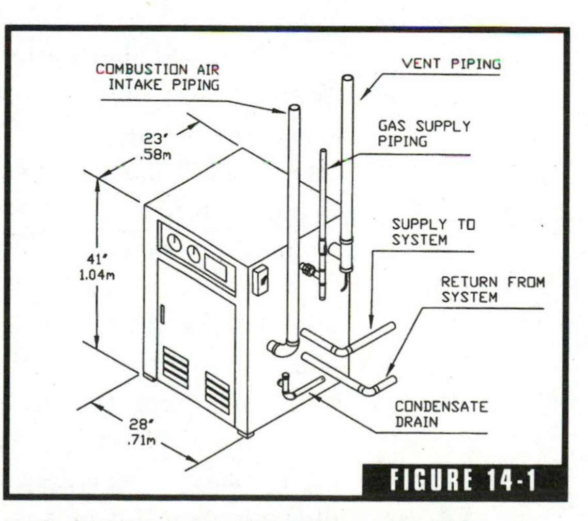

Typical boiler setup. (*Courtesy of Dunkirk.*)

Standard Safety Measures

There are certain standard safety measures to take into consideration whenever you are working with a gas-fired heating system. The first rule is to pay attention to any odors. If you smell any appreciable amount of gas or other strange odors, ventilate the area before working in it. Avoid electrical devices, if possible. Open windows when you can. If gas accumulates heavily, it can explode. The quicker you can rid the area of fumes, the safer you will be.

Always shut off the gas supply to any device you are planning to work on. Cut the gas valve off and wait several minutes before proceeding with your work. If there are multiple cutoff valves, close all of them. Wait at least 5 minutes to give residual gas time to vacate the system. You may already know, but if you don't, be advised that LP gas is heavier than air and will not dissipate upwardly in a natural fashion. LP gas lingers below air, gathering in low areas and creating a possibility for an explosion risk. In minute quantities, this is not a problem, but it is a fact that any service technician should be aware of (Figs. 14-2 and 14-3).

If you will be working with or near electrical wiring, you should turn off the power supply. This may mean removing a fuse, but in most homes it is as simple as flipping a circuit breaker. There may also be an individual disconnect box near the heating unit. If this is the case, you can turn off the power by throwing the switch on the disconnect box.

If you will be working with electrical matters, don't attempt to jump or short the valve coil terminals on a 24-volt control. Doing this can short out the valve coil or burn out the heat anticipator in the thermostat. Also, never connect millivoltage controls to line voltage or to a transformer. If you do, you may be burning out the valve operator or the thermostat anticipator.

If you have any reason to suspect a gas leak, check all connections where leaks may be possible. Don't use any type of flame to test for leaks. I've known plumbers who tested joints on natural-gas piping with their torch flames, but don't do this, and, certainly, never do it with LP gas. Natural gas usually just burns in a leak situation, but LP gas is likely to explode. I say again, don't search for leaks with any type of flame. Soapy water is the right substance to use when looking for

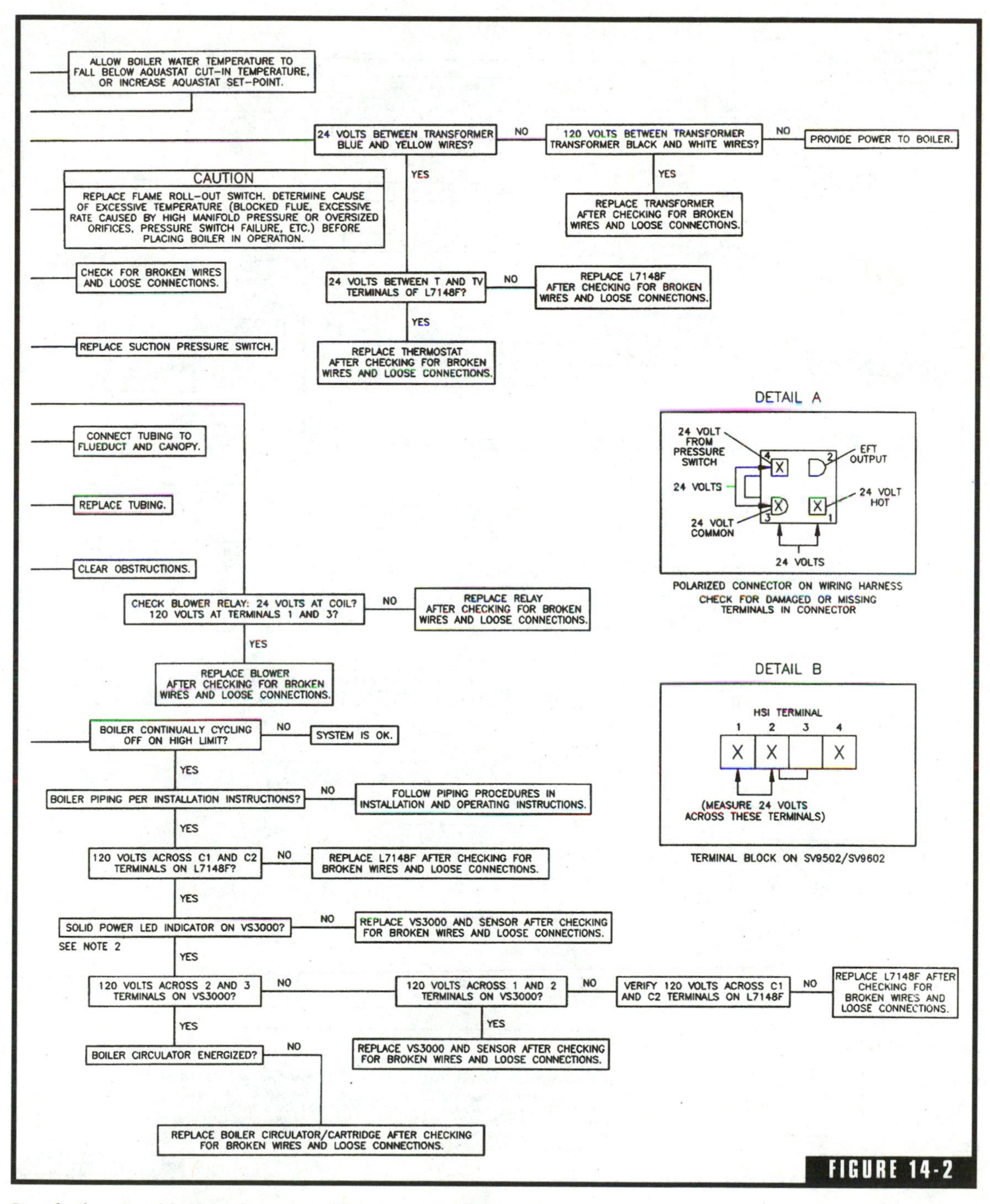

Revolution troubleshooting tree. (***Courtesy of Burnham.***)

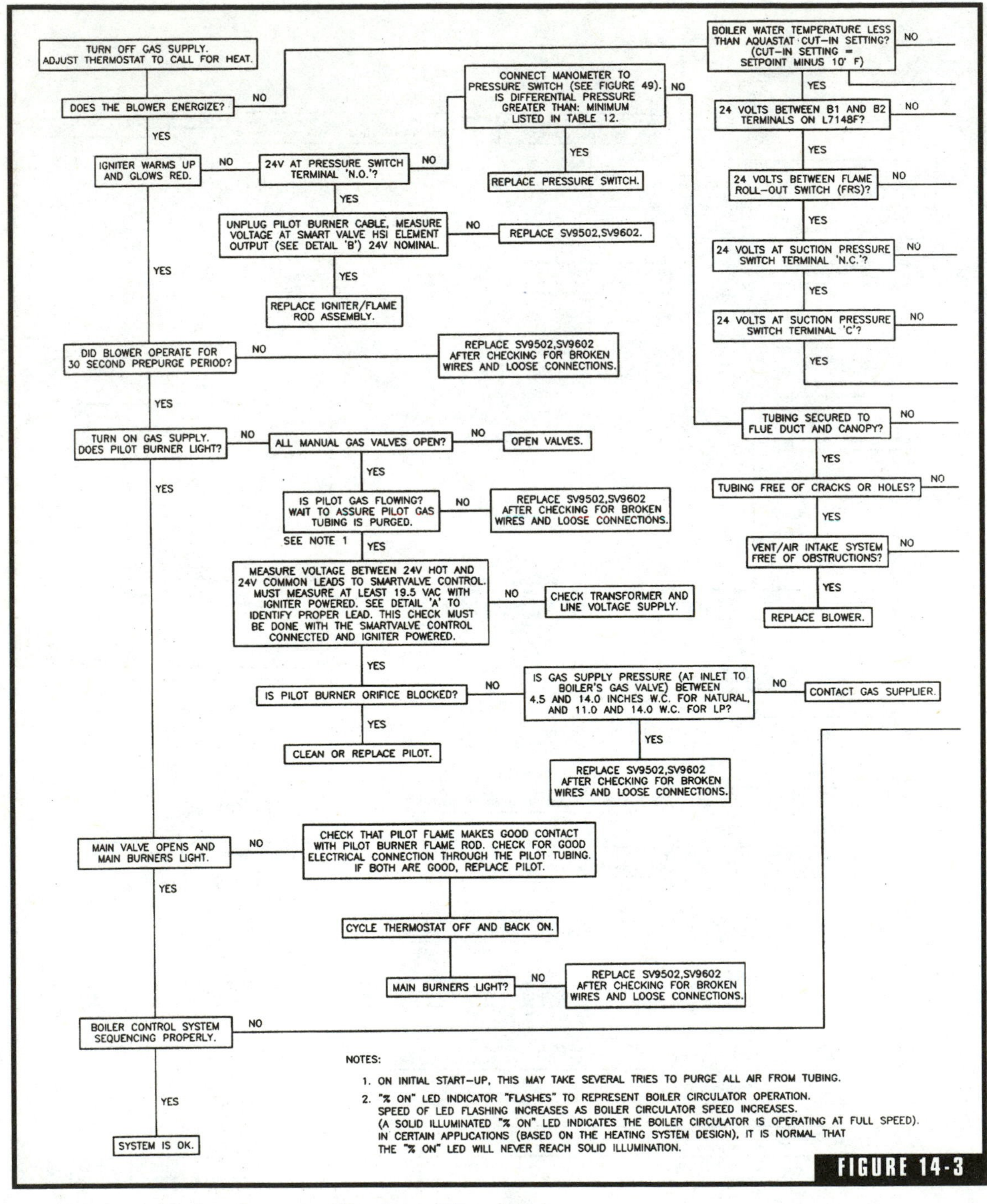

Revolution troubleshooting tree. (*Courtesy of Burnham.*)

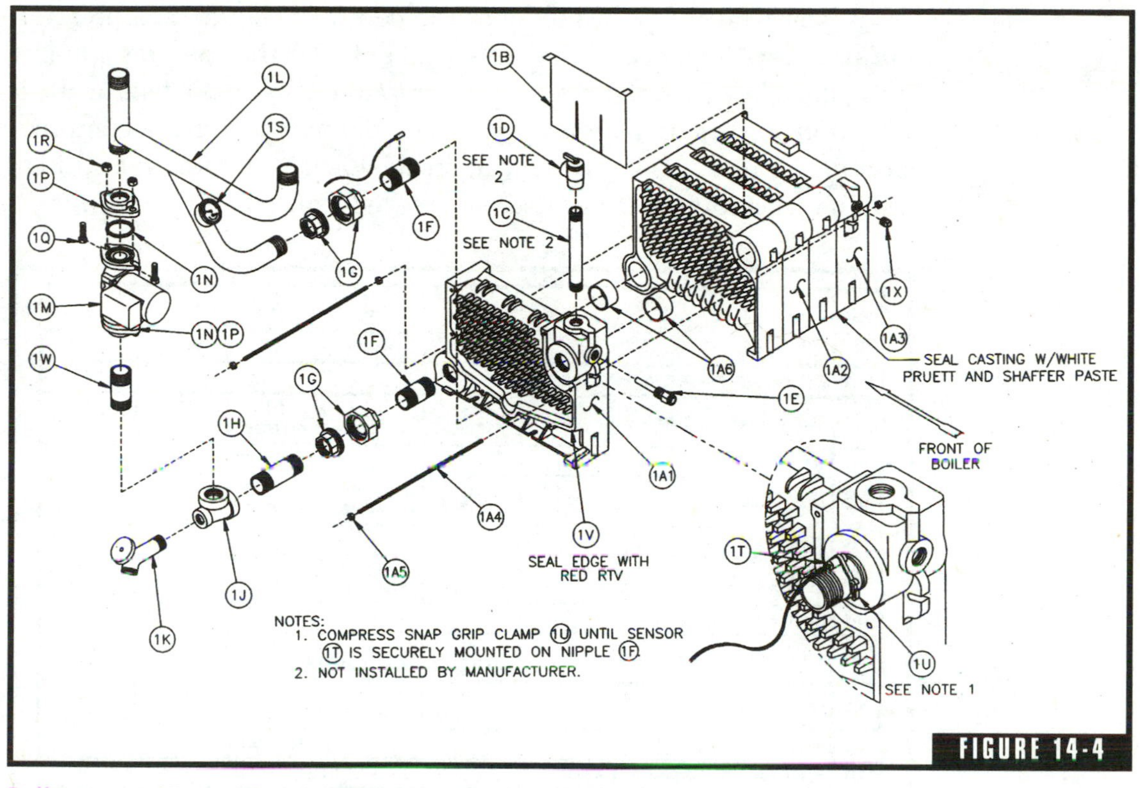

Boiler parts. (*Courtesy of Burnham.*)

leaks in gas piping. Obviously, to find a gas leak the gas supply must be turned on (Figs. 14-4 and 14-5).

When in search of a leak, ventilate the area by opening windows and other ventilating openings. With the gas on, apply a generous quantity of soapy water to all gas connections. When the water is applied to a connection where a leak exists, the escaping gas will create bubbles in the soapy solution. Take note of the leak and continue testing the rest of the gas connections. Once you have checked all connections, you can turn off the gas supply and fix the leaks. When fixing leaks, don't take shortcuts. Treat each leak with respect. Loosen all leaking joints, apply thread sealant as required, and tighten the joints. Sometimes just tightening a fitting is enough. Regardless of what you have to do to correct a leak, always test the connection carefully after it is repaired. Test the joint just the way you did to find the leak in the first place.

Never bend the pilot tubing at the control once the compression nut has been tightened. If the tubing is bent after the compression nut is tightened, a leak can result. Compression nuts don't take well to vibration and manipulation. Any substantial movement of a compression fitting after it is tightened can create a leak. Compression fittings are fine when they are installed and treated properly (Fig. 14-6).

Key No.	Description	[Quantity] Part Number				
		RV-3	RV-4	RV-5	RV-6	RV-7
1. SECTION AND WATER PIPING ASSEMBLY						
1A	Section Assembly, Complete	[1] 6170603	[1] 6170604	[1] 6170605	[1] 6170606	[1] 6170607
1A1	Left End Section	[1] 7170601				
1A2	Center Section	[1] 7171703	[2] 7171703	[3] 7171703	[4] 7171703	[5] 7171703
1A3	Right End Section	[1] 71717021				
1A4	Shipping Rod, ¼" - 20	[2] 80861040	[2] 80861041	[2] 80861042	[2] 80861043	[2] 80861044
1A5	Hex Nut, ¼" - 20, Heavy	[4] 80860407				
1A6	Push Nipple, #3	[4] 806600009	[6] 806600009	[8] 806600009	[10] 806600009	[12] 806600009
1B	Flue Gas Baffle	[2] 71106001	[3] 71106001	[4] 71106001	[5] 71106001	[6] 71106001
1C	Nipple, ¾" x 8" Lg.	[1] 806600221				
1D	Relief Valve, ¾", ConBraCo 10-408-05	[1] 81660319				
1E	Immersion Well, ½", Honeywell 123869A	[1] 80160456				
1F	Nipple, 1¼" x 3" Lg. w/Teflon Tape	[2] 6066016				
1G	Union, 1¼"	[2] 806604003				
1H	Nipple, 1¼" x 4½" Lg. w/Teflon Tape	[1] 6066041				
1J	Tee, 1¼" x 1¼" x ¾"	[1] 806601002				
1K	Drain Valve, ¾", ConBraCo 31-612-02 w/Teflon Tape	[1] 6066025				
1L	Water Manifold	[1] 80606001				
1M	Circulator, Taco 007F, Position 4	[1] 8056107				
1N	Gasket, Circulator Flange,Taco '00' Series	[2] 806602006				
1P	Circulator Flange, 1¼"	[2] 806602013				
1Q	Cap Screw, Hex Head, 7/16" - 14 x 1½"	[4] 80861301				
1R	Hex Nut, 7/16" - 14	[4] 80860406				
1S	Temp/Pressure Gauge, Short Shank	[1] 8056169				
1T	BVS #071 Univ. Fluid Temp. Sensor	[1] 80160218				
1U	Snap Grip Clamp	[1] 80861691				
1V	Silastic Sealant, RTV 6500	[A/R] 9056060				
1W	Nipple, 1¼" x 4" Lg. w/Teflon Tape	[1] 6066014				
1X	¼" NPT Black Pipe Plug, Sq. Head	[1] 806603516				

FIGURE 14-5

Description of boiler parts. (*Courtesy of Burnham.*)

The Troubleshooting Process

The troubleshooting process for gas-fired boilers doesn't involve a lot of steps. There are only about nine types of failures that are typically encountered with gas-fired boilers. While there are not a lot of poten-

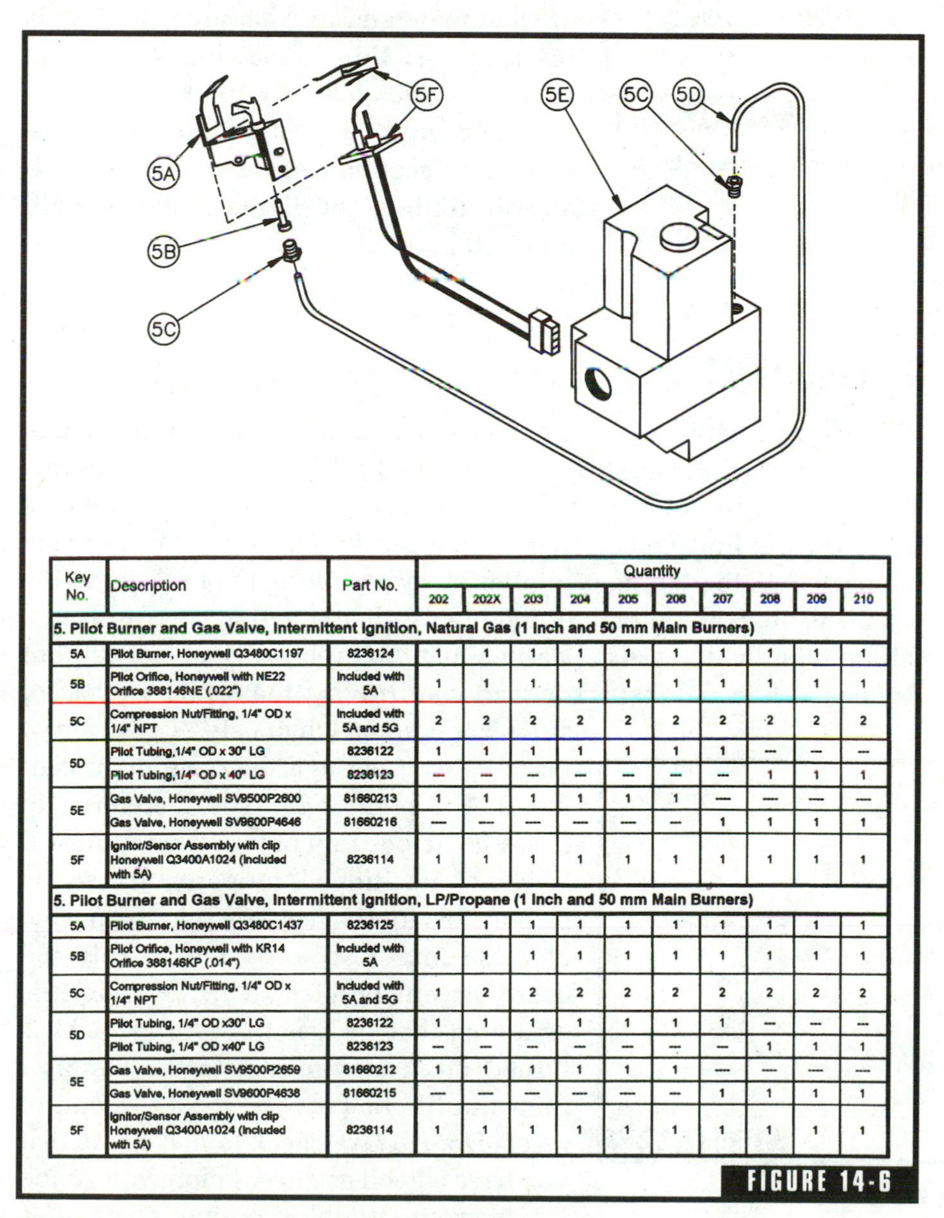

Key No.	Description	Part No.	Quantity									
			202	202X	203	204	205	206	207	208	209	210
5. Pilot Burner and Gas Valve, Intermittent Ignition, Natural Gas (1 Inch and 50 mm Main Burners)												
5A	Pilot Burner, Honeywell Q3480C1197	8236124	1	1	1	1	1	1	1	1	1	1
5B	Pilot Orifice, Honeywell with NE22 Orifice 388146NE (.022")	Included with 5A	1	1	1	1	1	1	1	1	1	1
5C	Compression Nut/Fitting, 1/4" OD x 1/4" NPT	Included with 5A and 5G	2	2	2	2	2	2	2	2	2	2
5D	Pilot Tubing,1/4" OD x 30" LG	8236122	1	1	1	1	1	1	1	—	—	—
	Pilot Tubing,1/4" OD x 40" LG	8236123	—	—	—	—	—	—	—	1	1	1
5E	Gas Valve, Honeywell SV9500P2600	81660213	1	1	1	1	1	1	—	—	—	—
	Gas Valve, Honeywell SV9600P4646	81660216	—	—	—	—	—	—	1	1	1	1
5F	Ignitor/Sensor Assembly with clip Honeywell Q3400A1024 (Included with 5A)	8236114	1	1	1	1	1	1	1	1	1	1
5. Pilot Burner and Gas Valve, Intermittent Ignition, LP/Propane (1 Inch and 50 mm Main Burners)												
5A	Pilot Burner, Honeywell Q3480C1437	8236125	1	1	1	1	1	1	1	1	1	1
5B	Pilot Orifice, Honeywell with KR14 Orifice 388146KP (.014")	Included with 5A	1	1	1	1	1	1	1	1	1	1
5C	Compression Nut/Fitting, 1/4" OD x 1/4" NPT	Included with 5A and 5G	1	2	2	2	2	2	2	2	2	2
5D	Pilot Tubing, 1/4" OD x30" LG	8236122	1	1	1	1	1	1	1	—	—	—
	Pilot Tubing, 1/4" OD x40" LG	8236123	—	—	—	—	—	—	—	1	1	1
5E	Gas Valve, Honeywell SV9500P2659	81660212	1	1	1	1	1	1	—	—	—	—
	Gas Valve, Honeywell SV9600P4638	81660215	—	—	—	—	—	—	1	1	1	1
5F	Ignitor/Sensor Assembly with clip Honeywell Q3400A1024 (Included with 5A)	8236114	1	1	1	1	1	1	1	1	1	1

FIGURE 14-6

Pilot burner and gas valve. (*Courtesy of Burnham.*)

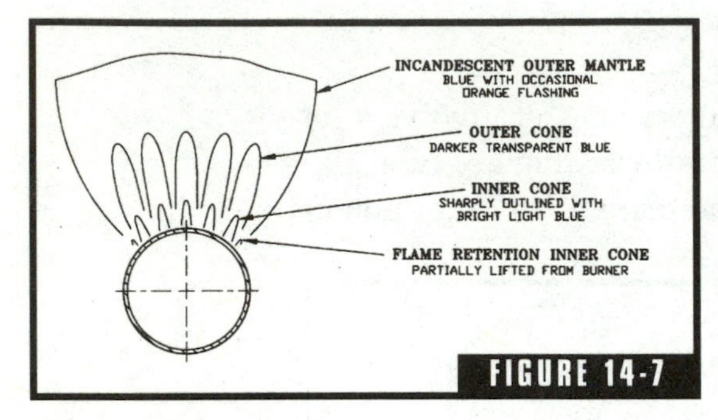

Burner flame for a 50-millimeter burner. (*Courtesy of Burnham.*)

tial problems, there are several possible solutions. But, using a methodical plan during the troubleshooting process makes the job easier. Experience is the best teacher in many ways, but a good checklist of what to do and in what order to do it is the next best thing. And, that's what I'm about to give you. So, with the safety issues covered, let's move into the cause-and-repair section to give you some guidance in finding and fixing problems with gas-fired boilers.

No Pilot Light

When a gas-fired boiler fails to operate, it could be due to the lack of a functioning pilot light (Figs. 14-7 to 14-12). There are four basics reasons why this problem might exist. First, check to see if there might be air in the gas line. Open the connections and purge any air that may be trapped in the tubing or piping. A second thing to check for is the gas pressure. Is it too high or too low? Either situation can result in a failing pilot light. It's also possible that the problem is a blocked pilot orifice. Do a visual inspection and confirm that the orifice is clear of any obstructions. Obviously, if gas cannot make its way through the opening, it cannot fuel a pilot light. The last thing to check is the position of the flame runner. If it is not positioned properly, it can be responsible for the lack of a pilot light. One of these four reasons is most likely the cause of your problem. This, of course, is assuming that the gas to the pilot light is turned on. Sometimes it is the simplest things in life that are the most difficult to recognize. Always check to make sure that you have a good gas flow before you go too far in your troubleshooting steps. Gas

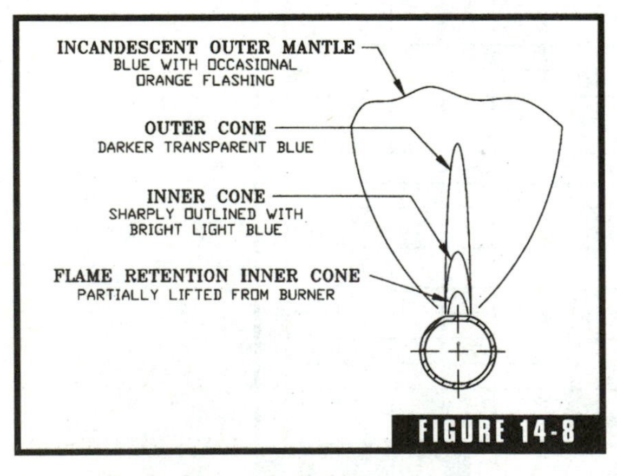

Burner flame for a 1-inch burner. (*Courtesy of Burnham.*)

valves can be turned off for many reasons, and someone may forget to turn the valve back on. It's also possible that children will somehow turn off a valve and not realize what they are doing. Just remember to make sure that there is a gas flow before you begin the more complex aspects of troubleshooting a problem with a pilot light (Fig. 14-13).

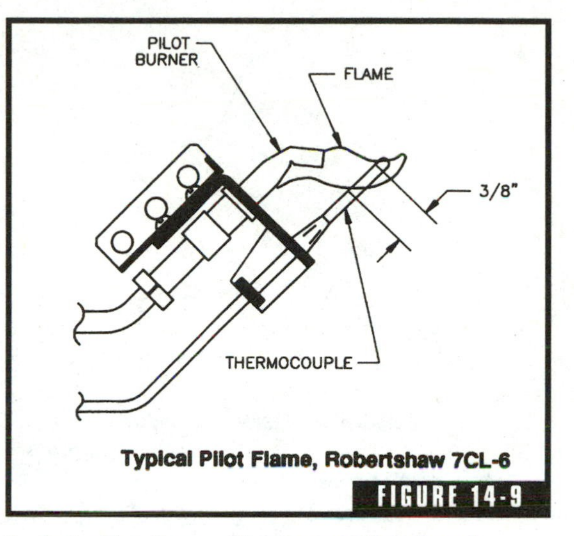

Typical pilot flame. (*Courtesy of Burnham.*)

When a Pilot Light Goes Off during a Standby Period

When a pilot light goes off during a standby period, it might be that there is some type of obstruction in the tubing leading to the pilot light. Look to see if there are any kinks in the tubing. If not, remove the section of tubing and confirm that it is not blocked. This can be done just by blowing air through it. If there is a kink or obstruction, either replace the tubing or remove the obstruction. Low gas pressure can also cause a pilot light to go off during standby. Check the gas pressure and confirm that it is within acceptable levels for the equipment that you are working on.

You already know that a blocked pilot orifice can prevent a pilot light from burning. But, did you know that a blocked orifice can also be responsible for a pilot light that goes off during standby? It can, so check the orifice if your problem persists. If there is a loose thermocouple on 100 percent shutoff, a pilot light can go off during a standby period. It's also possible that the thermocouple is defective. When a pilot safety is not working properly, it can cause a pilot light to go off during standby. Another consideration to check is the draft condition. If there is a poor draft, a pilot

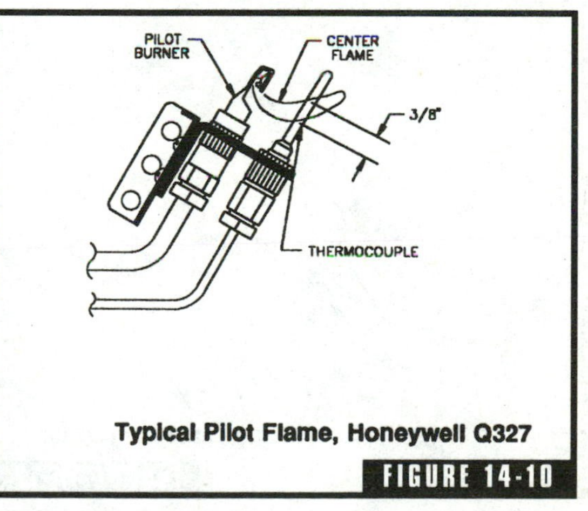

Typical pilot flame. (*Courtesy of Burnham.*)

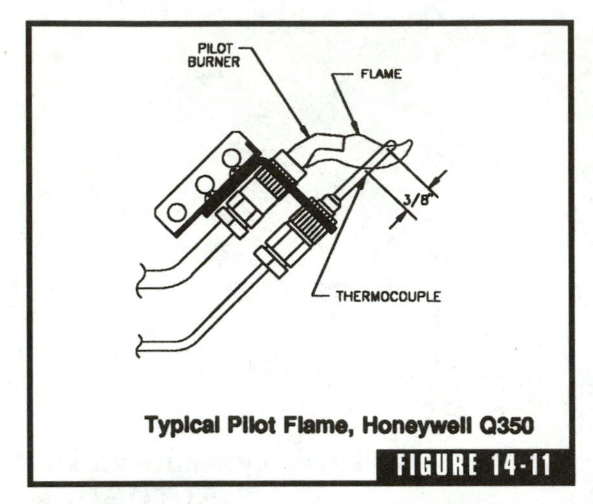

Typical pilot flame. (*Courtesy of Burnham.*)

light may not stay lit. A final possibility is a draft tube that is set into or flush with the inner wall of the combustion chamber. These same seven reasons for failure can apply to a safety switch that needs frequent resetting.

If a Pilot Light Goes Off When the Motor Starts

If a pilot light goes off when the motor of a unit starts, there are only three probable causes. The first one to look into is the possibility of some type of restriction or obstruction in the tubing to the pilot light. You need to loosen the connection on the tubing and confirm gas flow. If the tubing is blocked or kinked and cannot be repaired, replace the tubing. Once again, gas pressure could be at fault. If the pressure is too high or too low, the pilot light function can be affected. Confirm that the gas pressure is within suitable limits for the equipment that you are working with. A final consideration is the possibility that there may be a substantial pressure drop in the gas piping when the main gas valve opens. Putting the equipment through cycles until you can observe what happens as the main gas valve opens might be necessary.

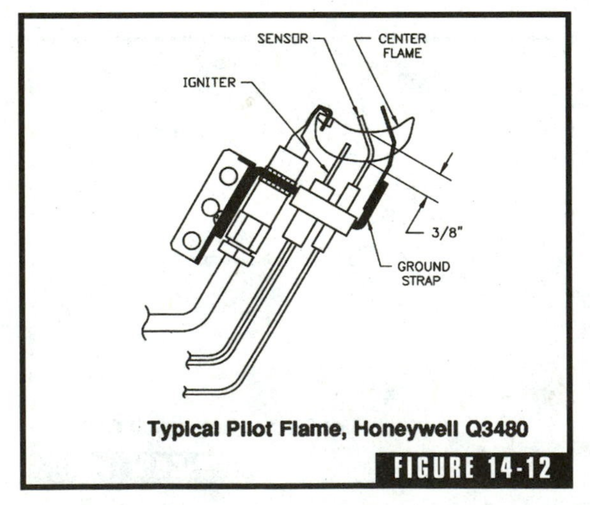

Typical pilot flame. (*Courtesy of Burnham.*)

Motor Won't Run

The first thing to check when you encounter a boiler with a motor that will not run is the electrical circuit (Figs. 14-14 and 14-15). If there is a local disconnect box, check it first to see that it has not been turned off. Next, go to the electrical panel for the home or building and check all

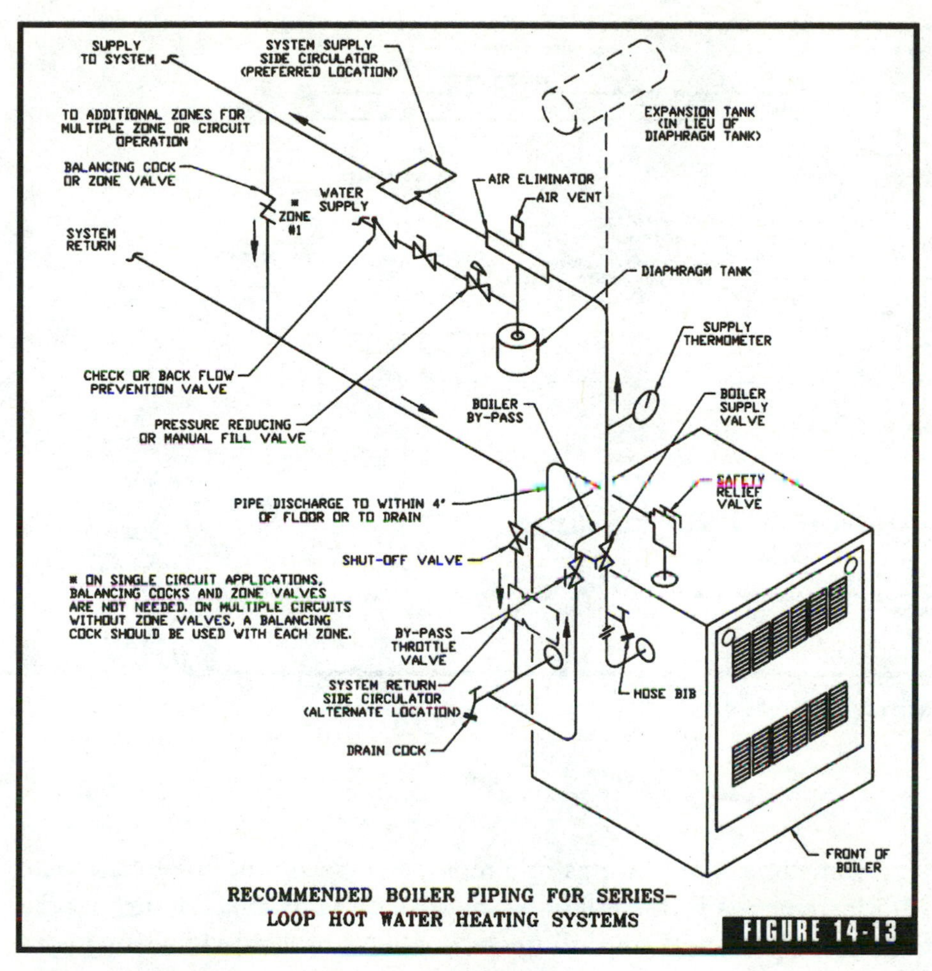

RECOMMENDED BOILER PIPING FOR SERIES-LOOP HOT WATER HEATING SYSTEMS

FIGURE 14-13

Boiler piping diagram. (*Courtesy of Burnham.*)

fuses or circuit breakers. If all appears to be in order, use your electrical meter to confirm that power is reaching the motor. Also make sure that the wiring is connected to the controls properly. Before you get too involved, move around the building and check the thermostats. Make sure that they are turned up to a sufficient temperature to be calling for heat. A mechanic can feel quite foolish after spending considerable time performing major troubleshooting steps and then finding out that the thermostats were turned down for some reason. Assuming that the thermostats are calling for heat, you must check to see that the thermostats and limit controls are not defective or calibrated improperly (Figs. 14-16 and 14-17).

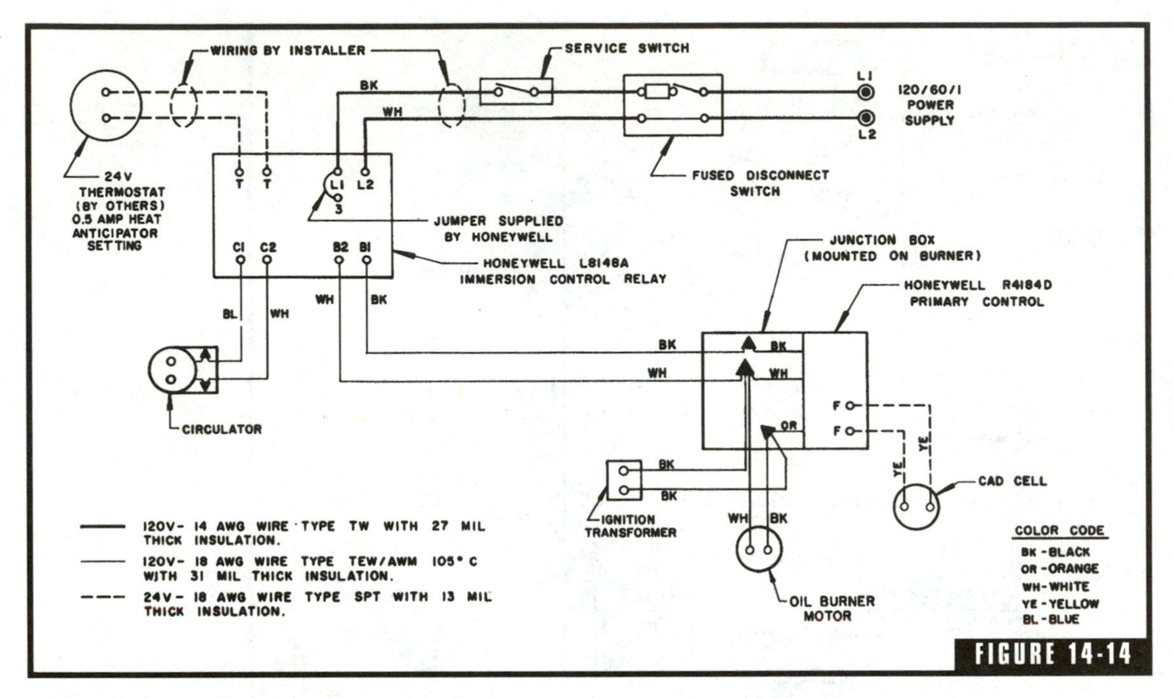

Wiring diagram. (*Courtesy of Burnham.*)

Sometimes the bearings in a motor will seize up, due to a lack of lubrication. See if this might be the case on your job. Try turning the equipment. If it will not turn freely, you have probably identified your problem. In a worst-case scenario, you are dealing with a motor that has burned out, which will, of course, require replacement.

It Runs But Doesn't Heat

If you have a boiler that runs but doesn't heat, check to see if the pilot light is burning. If the pilot is out, relight it. You may discover that the root of the problem lies within a defective pilot safety control. This could be as simple as resetting the safety or it could require replacement of the control. Another possible cause for the problem of a motor that runs with no flame is that the thermocouple might not be generating enough voltage for the system. The causes are limited, so the troubleshooting should go quickly. Gas pressure has a large effect on a gas-fired system. Check the pressure to make sure that it is within recommended working parameters. If it is, and you still have the problem, check to see if maybe the motor is running slower than it should be. This could be the cause of the motor running without a flame.

A Short Flame

A short flame in a gas-fired boiler is not good. It might be caused by any one of about five possible problems. The place to start is with the pressure regulator. If it is set too low, the flame will be short. Also, check the air shutter. An air shutter that is open too wide can cause a short flame to burn in the equipment. Any major pressure drop in the gas supply line could also be responsible for the short flame. A defective regulator is always a possibility when you are looking for the cause of a short flame. And, a vent that is plugged in the regulator can

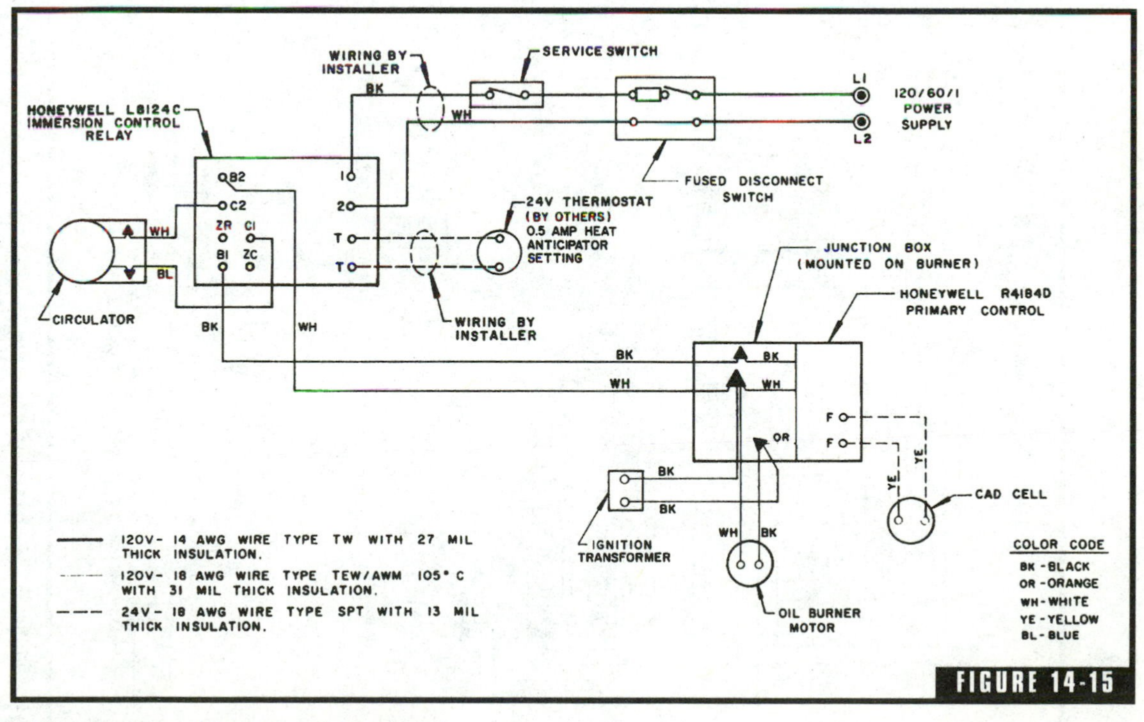

FIGURE 14-15

Wiring diagram. (*Courtesy of Burnham.*)

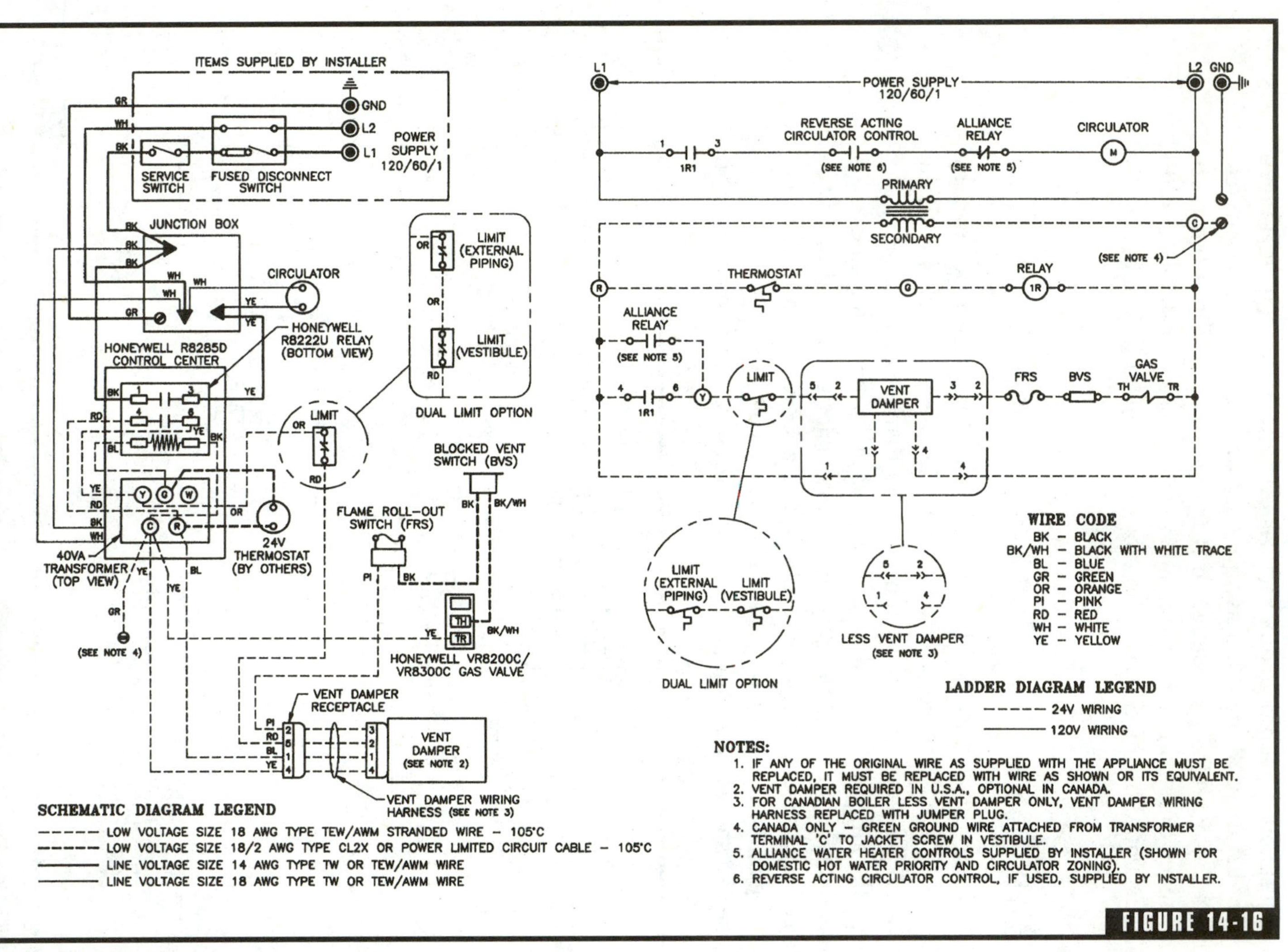

Wiring diagram. (*Courtesy of Burnham.*)

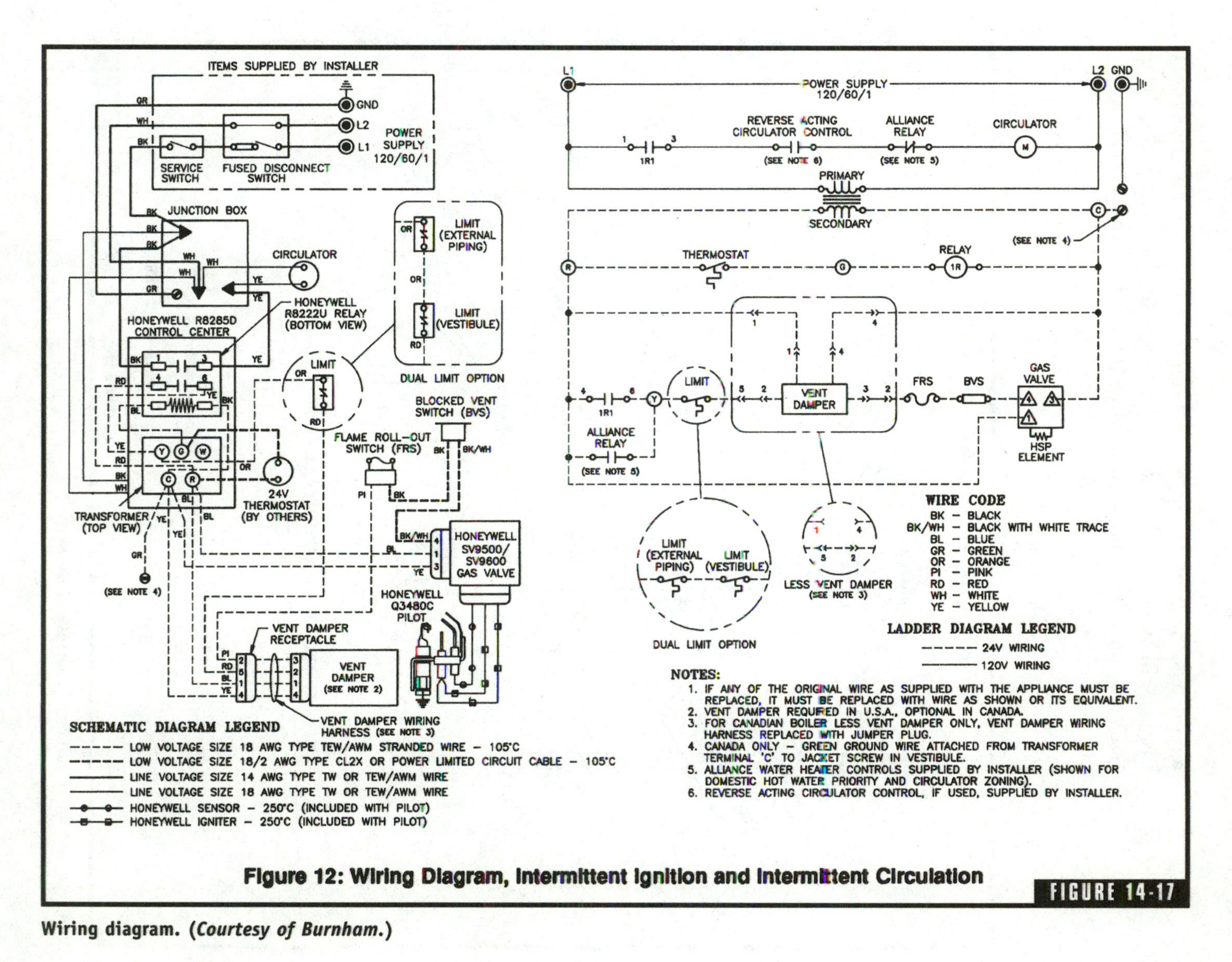

FIGURE 14-17

Wiring diagram. (*Courtesy of Burnham.*)

also cause the problem of a short flame. Some trial-and-error techniques will be needed, but one of the causes listed above is almost certain to be the reason for a short flame (Figs. 14-18 and 14-19).

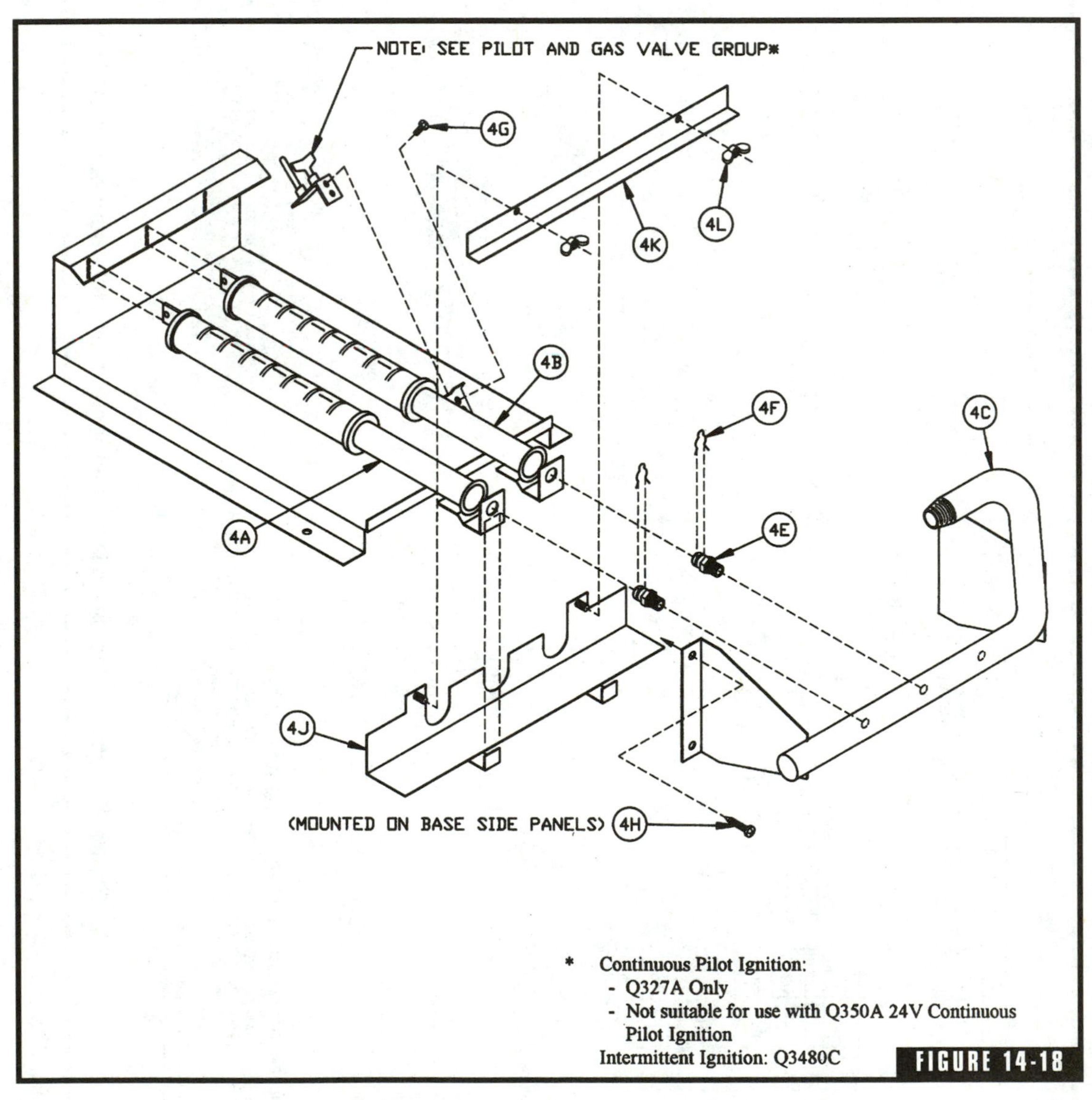

FIGURE 14-18

Manifold and main burners. (*Courtesy of Burnham.*)

Long Flames

Long flames in gas-fired equipment are usually limited to one of three causes. The first thing to check for is an air shutter that is not open far enough. When it is open too far, the flame will burn short, but if not open far enough, the flame will be long. The air shutter must be bal-

Key No.	Description	Part No.	Quantity									
			202	202X	203	204	205	206	207	208	209	210
4. Manifold and Main Burners (50 mm Main Burners Only)												
4A	Main Burner	8236091	—	—	1	2	3	4	5	6	7	8
4B	Main Burner with 45° Pilot Bracket	8236092	—	—	1	1	1	1	1	1	1	1
4C	Manifold	82260034	—	—	1	—	—	—	—	—	—	—
		82260044	—	—	—	1	—	—	—	—	—	—
		82260054	—	—	—	—	1	—	—	—	—	—
		82260064	—	—	—	—	—	1	—	—	—	—
		82260074	—	—	—	—	—	—	1	—	—	—
		82260084	—	—	—	—	—	—	—	1	—	—
		82260094	—	—	—	—	—	—	—	—	1	—
		82260104	—	—	—	—	—	—	—	—	—	1
4E	50 mm Main Burner Orifices, Natural Gas, High Altitude, U.S.A. and Canada, 2000-5000 Ft. Altitude											
	Main Burner Orifice, #38	822680	—	—	2	—	—	—	—	—	—	—
	Main Burner Orifice, #37	822601	—	—	—	3	4	5	6	7	8	9
	50 mm Main Burner Orifices, LP/Propane, High Altitude, U.S.A. and Canada, 2000-5000 Ft. Altitude											
	Main Burner Orifice, #53	8226002	—	—	2	—	—	—	—	—	—	—
	Main Burner Orifice, #52	822641	—	—	—	3	4	5	6	7	8	9
4F	External Hitch Pin Clip	822604	—	—	2	3	4	5	6	7	8	9
4G	Screw, Machine, Philips Head w/Captive Lockwasher, #10-32 x ¼"	80860674	—	—	1	1	1	1	1	1	1	1
4H	Screw, Self Tapping, Phillips Pan Head, ¼-20 x ½"	80860700	—	—	4	4	4	4	4	4	4	4
4J	Lower Injection Shield	618600301	—	—	1	—	—	—	—	—	—	—
		618600401	—	—	—	1	—	—	—	—	—	—
		618600501	—	—	—	—	1	—	—	—	—	—
		618600601	—	—	—	—	—	1	—	—	—	—
		618600701	—	—	—	—	—	—	1	—	—	—
		618600801	—	—	—	—	—	—	—	1	—	—
		618600901	—	—	—	—	—	—	—	—	1	—
		618601001	—	—	—	—	—	—	—	—	—	1
4K	Upper Injection Shield	718600303	—	—	1	—	—	—	—	—	—	—
		718600403	—	—	—	1	—	—	—	—	—	—
		718600503	—	—	—	—	1	—	—	—	—	—
		718600603	—	—	—	—	—	1	—	—	—	—
		718600703	—	—	—	—	—	—	1	—	—	—
		718600803	—	—	—	—	—	—	—	1	—	—
		718600903	—	—	—	—	—	—	—	—	1	—
		718601003	—	—	—	—	—	—	—	—	—	1
4L	Wing Nut, ¼-20, Size B	80860900	—	—	2	2	2	2	3	3	3	3

FIGURE 14-19

Parts list for manifold and main burners. (*Courtesy of Burnham.*)

anced to produce a flame of a desired length. If there are obstructions in air openings or at the blower wheel, the result can be a long flame. Too much input to the flame source will also produce a long flame. Since the causes for long flames are few in number and easy to check for, this problem is not very time consuming when it comes to troubleshooting it.

A Gas Leak at the Regulator Vent

A gas leak at the regulator vent is a sign of a damaged diaphragm. The diaphragm most likely has a hole in it. In any case, the damaged diaphragm must be replaced. Gas cannot be allowed to leak from any part of a system. In such a case, all attention should be focused on the regulator vent, since that is where the problem is (Fig. 14-20).

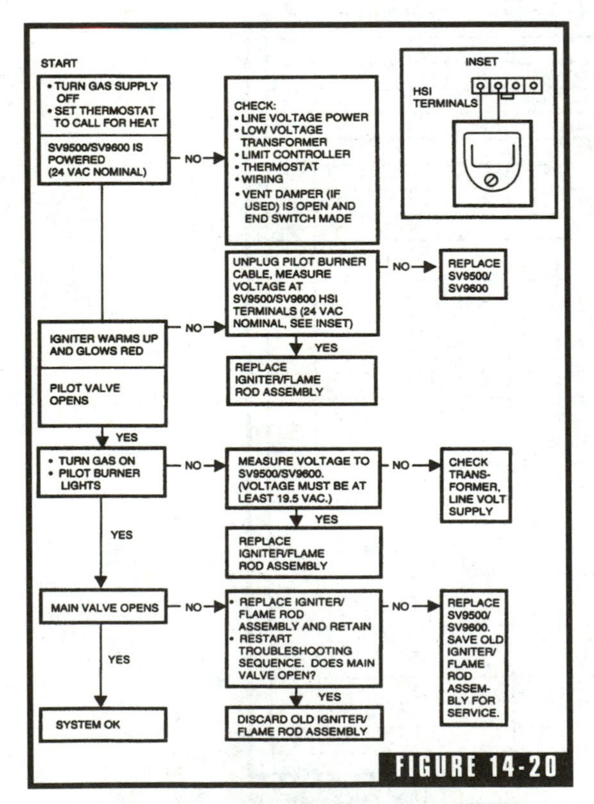

FIGURE 14-20

Troubleshooting guide for intermittent ignition. (*Courtesy of Burnham.*)

Valve Won't Close

Sometimes the main gas valve on a unit will not close when the blower stops. This is a problem that is usually caused by a defective valve or some type of obstruction on the valve seat. You can inspect the seat for damage or buildup, but be prepared to replace the entire valve. It is unsafe to have a gas valve that does not close properly when it should.

Be Careful

When you are working with gas, whether natural or manufactured, you must be careful. The fuel is volatile. It is said that familiarity breeds contemp. Don't allow this to be the case in your work. If you make a mistake with an oil-fired boiler, you might only get oil on a floor or incur some odors. A mistake with a gas system could result in a major explosion. Pay attention to what you are doing and stay focused on your task. Make sure anyone who might enter your work area is aware of what you are doing and what your needs are. For example, the last thing you need is for a casual observer to come in and fire up a cigarette lighter when you are bleeding off some gas. Stay sharp and stay safe.

CHAPTER

15

A Basic Overview of Forced Hot-Air Heat

Forced hot-air heat has been in use for years and years and years. It's no longer as popular as it once was. This is because heat pumps came along and offered both heating and cooling from one system with only one layout of duct work. But, forced hot-air furnaces are still made and installed. They are used in all climates, even extreme ones. I live in Maine, where the winters are very cold. A majority of homes in Maine rely on hot-water baseboard heat, but there are many houses where forced hot-air furnaces keep the homes warm all through the winter. In more temperate climates, hot-air furnaces are even more abundant.

Heat pumps did take a major bite out of the market share once enjoyed by hot-air furnaces. But, the additional cost of a heat pump can put it out of consideration for some buyers. Plus, if homeowners live in an area where air conditioning is not needed, such as where I live, the heat pump doesn't present so many advantages. Plus, heat pumps that are air-based units don't perform economically in extremely cold temperatures. This leaves plenty of room for hot-air furnaces to take advantage of.

As plentiful as hot-air furnaces are, there are people who don't like them. This can be the result of many factors. Dust is certainly one reason why hot-air furnaces are sometimes disliked. Since the units blow air into rooms to circulate heat, they also produce dust. The heating

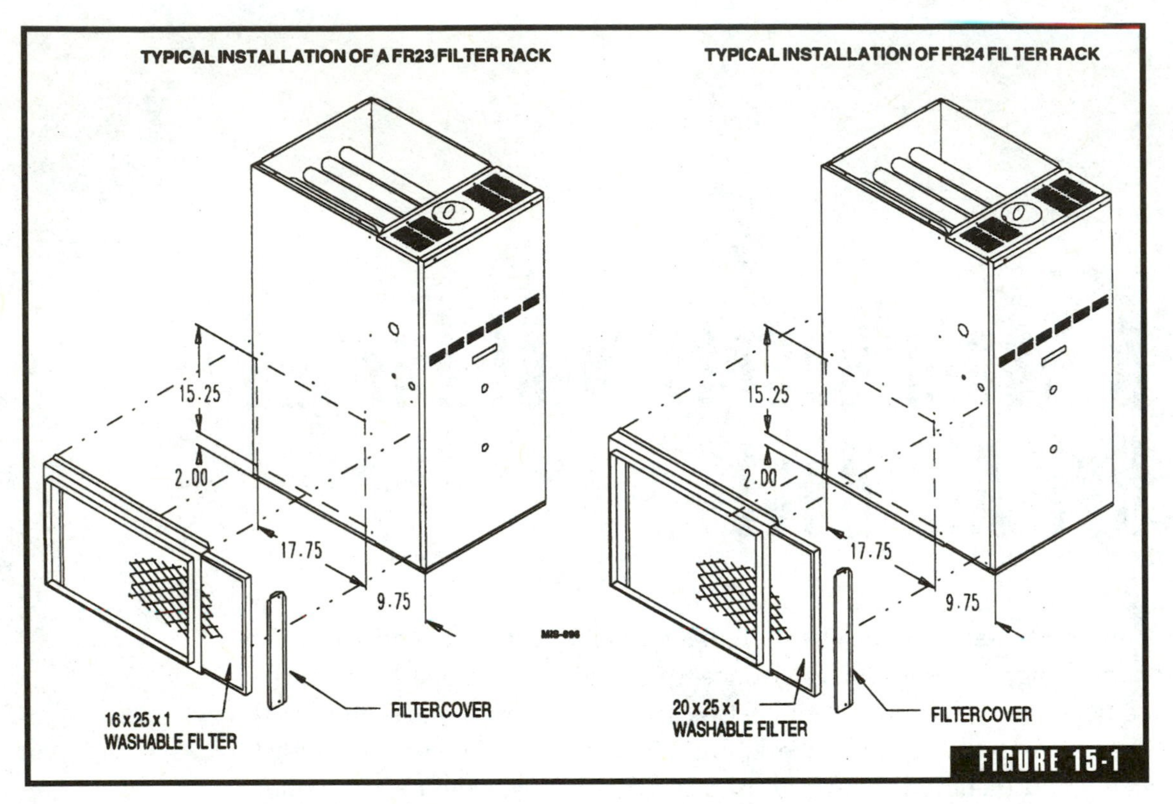

Filter locations. (*Courtesy of Bard.*)

system can also spread particles that people with allergies are sensitive to. A clean filter in the furnace helps to control the problem, but doesn't eliminate it. There are allergy filters available that capture more minute particles and result in cleaner homes (Fig. 15-1).

Duct work is another drawback to hot-air furnaces. The trunk lines and supply ducts require much more space for installation than some types of heating systems need. Basement ceilings can be cluttered with ducts that reduce headroom. Living space may have to endure boxed areas where ducts pass from one level of a home to another. These drawbacks are minor to some customers and a major issue to others. There is much to consider when thinking of offering and installing a forced-air system, so let's explore some of the topics that you should be prepared to answer questions about when you talk with potential customers.

Extended Plenum Systems

Extended plenum systems are just one type of heat distribution that is common when working with forced-air furnaces. These systems are popular and abundant. They are simple to install and perform very well. An extended plenum system consists largely of long runs of one size of rectangular duct work. This makes fabrication and fitting easy. There are few fittings needed, and a job can go together quickly. Rarely are any special fittings needed for a traditional extended plenum system.

If you were to conduct a survey of the types of air distribution systems used with forced-air furnaces, you would probably find that extended plenum systems are the type most often used. The system is usually one long line of what is called trunk duct. As the duct extends farther from the furnace, the trunk line is generally reduced in size. This improves air flow. Branch ducts are connected to the main trunk line to provide heat to various parts of a building.

Balancing air flow is essential for a forced-air heating system. The balancing of air flow with an extended plenum is easy, especially when the heating system is centrally located within a building. Branch ducts usually connect to the top or side of a trunk line. One disadvantage to this type of system is that the trunk line is large and requires substantial space. Since the trunk line is too large to conceal in average joist bays, the trunk duct is usually installed below floor or ceiling joists. When a trunk extension is installed in a basement, it can greatly reduce the headroom of the basement. It's common to design heating systems so that the main ducts run along beams. When needed, the trunk can be enclosed in a box to minimize the inconvenience of reduced headroom.

The versatility of an extended plenum system allows it to be used in almost any style of home. Single-story, two-story, and multilevel home designs can all be fitted with an extended plenum system. The duct work can be installed in a crawl space or an attic, or concealed in lowered ceilings. Branch ducts are typically small enough to be concealed between floor and ceiling joists. Vertical risers can be boxed in a closet or duct chase. All in all, an extended plenum system is most often very cost-effective to install and efficient to operate.

Design Considerations

There are some design considerations which must be addressed when deciding if an extended plenum system is suitable for a particular job. The maximum length of a trunk line is a major factor to consider when contemplating an extended plenum system. Ideally, a trunk line should not run for more than about 24 feet. A longer trunk line is possible, but air flow and air balancing can suffer with longer runs. This does not mean that a system is limited to houses of only 24 feet in length. To overcome the obstacle of length, you can place the furnace in the middle of a home and extend trunks from two sides for total coverage of about 48 feet, or so.

It's not always feasible to locate a furnace in a central point. For example, the furnace may need to be located near an end wall to coincide with exhaust venting or a flue location. This is not a big problem. But, suppose the house is 44 feet long and the furnace has to be placed near one of the far ends? You can't run the trunk more than 24 feet without the risk of poor air flow, so what can you do? The answer is simple. Use a reducing plenum system to overcome the distance. By decreasing the size of the plenum as it extends farther through a building, you can improve air flow and air balancing in runs that must exceed the recommended 24-foot length.

A Radial Duct System

A radial duct system can be an excellent choice for single-story homes. Next to extended plenum systems, radial systems rank high on the list of preferred air distribution plans. However, the use of a radial system may not be suitable for homes and buildings with multiple stories and levels. The appearance of a radial system is quite different from that of an extended plenum system. In fact, a drawing of a radial system might look like some type of robotic monster with numerous legs because a radial system doesn't use a trunk line. Each supply duct is connected directly to the plenum at the furnace, creating quite a network of small ducts.

The concept behind a radial system is to run each branch duct from the furnace plenum directly to the room to be served by the air distribution. When all the branch ducts are taken from the furnace plenum, the result is a uniform air flow, since all of the supplies originate from the same location. Many designers call for dampers to be installed in each branch duct, near where the branch attaches to the plenum. This allows the dampers to be serviced from a central location that offers suitable access. The dampers allow anyone to control air flow from room to room. Since branch ducts tend to be short, direct runs, the air flow works well.

If a customer is having a single-story home built and is looking for an inexpensive duct system, a radial system is a prime consideration. The system is simple and easy to install. Labor for the job is minimal, and material cost is quite reasonable. Since small ducts are used with radial systems, they can be concealed in typical joist bays. However, it's common for radial systems to be installed below living space with all of the duct work exposed. Sometimes the systems are installed above living space, in attics. It is common for the ducts to be in unheated areas, so the insulating of all ducts may be required to maintain energy efficiency.

If your customer is having a single-story home built on a concrete slab, you can offer a radial duct system that will be embedded in the slab. This not only conceals the duct work but if the ducts are left uninsulated, the home will benefit from a radiant-type of floor heating. However, to achieve radiant heating advantages, the perimeter of the slab area must be insulated well. This is done best with rigid, foam insulating boards.

Remember that radial systems are best suited for single-story homes. The systems can be used in different applications in some circumstances with the right design principles, but single-level buildings are the most feasible building types to install a radial system in. A central location for the furnace is important in a radial system. Long runs on individual branch ducts are not good for air flow. Return air ducts should be kept in a central location and close to the furnace.

Weaving a Web

Weaving a web of duct work can create a spider system. This type of heating layout is a variation of a typical radial system. This type of system pulls design characteristics from both radial systems and systems that utilize trunk ducts. It's something of a hybrid. These systems are growing in popularity and are edging out both standard radial systems and extended plenum systems. Given the proper design considerations, spider systems offer a number of advantages.

When a spider system is installed, the trunks and branches are run as directly as possible from the plenum to the point of air distribution. The ducts used for branches may be solid or flexible. If flexible ducts are used, they must be well supported, to avoid sagging and kinking that would affect air flow. Balancing dampers are difficult to install with some types of duct material, and this must be considered as a system is designed.

Hiding a spider system from view is not usually a concern. The systems are typically installed in crawl spaces, attics, or in concrete slabs. These locations are ideal for spider systems, since spider systems work best when the duct branches can be kept short. The ease of popping ducts up from below floor space or dropping them down from an attic makes the installation of spider systems easy. The simplicity of spider systems makes them cost-effective to install.

When it comes to energy efficiency, most spider systems rate very well. The duct runs are often made of insulated flexible duct and duct board mixing boxes. This adds to the efficiency. Since spider systems are usually installed in unheated areas, the insulated ducts are a must for maximum efficiency. Spider systems use perimeter loops and branches to keep air flowing properly. Return air usually comes from an outlet that is close to the heating unit.

It's not uncommon for a spider system to use junction boxes. The boxes allow air to flow into them and then to branch out to multiple duct branches. Junction boxes are made from duct board. Flexible ducts are normally used for branch ducts. They should be kept in straight runs and supported well. All flexible ducts should be cut precisely. Any leftover material may sag and reduce air flow. It's important to make sure that all flexible ducts are installed so as to avoid kinking. Any type of sagging, kinking, or obstruction can result in pressure drops.

Getting Return Air to the System

Getting return air to the system when a spider layout is used can be a bit more complex than other types of air-handling systems. If a spider system is to function properly, the return air is a critical factor. If not designed properly, the return air portion of a spider system can be inefficient and noisy and can produce a low air flow from supply ducts. The sizing of a return system and the locations of return grilles are significant factors in the successful installation of a spider system. Don't let this scare you off; just be aware of the need for close attention to details when it comes to return air with spider systems.

There are two types of return air systems which can be used with a spider system. An active return is one that is attached directly with duct work to the air handler in the system. Passive returns are not attached to the air handler. The two types of returns should normally be used in conjunction with each other. Some examples of passive returns include transfer grilles, short ducts that pass over walls, short ducts that pass through walls, and even undercuts in doorways.

Central return systems typically contain one, or more, large grilles that are placed in walls that are in close proximity to an air handler. In the case of multiple stories, return grilles are placed on each level of a home or building. It's standard procedure to undercut doors to rooms so that air can pass under them and find its way to return grilles. These systems are good in many respects. For example, a central return requires less duct work than a multiple-room return system. Friction loss in a central return system is less than that experienced with a multiple-room return system. By having less friction return on the system, the blower requirements for the system are reduced.

A central return system works well in homes where open designs are employed. The cost of installing a central system is lower than the expense incurred to install a multiple-room system, and filter grilles can be used with a central system to clean air before it returns to the duct work. As many advantages as central systems have, they also have their bad points.

Central return systems may create more noise than a multiple-room return. Since large ducts are typically needed with a central return, hiding the duct can be a problem. Big return grilles in walls might offend some homeowners, but they are necessary with a central system. Not everyone likes the idea of undercutting doorways, and central returns have been known to short-circuit air flow from supply registers that are close by. An alternative to a central system is a multiple-room return. The use of multiple-room returns does offer certain advantages over central systems.

A multiple-room return system tends to produce less noise than a central system. Air flow to individual rooms is usually better when a multiple-room system is used, even if the doors to the rooms are closed. There is no need to undercut doors. Since smaller return ducts can be used with multiple-room systems, the ducts can be hidden in walls and joists more easily than the larger ducts required with a central system. Wall grilles used with multiple-room systems are smaller and less obtrusive than the large ones used with central systems. There are, however, prices to be paid for the advantages provided with multiple-room systems.

In order to filter return air with a multiple-room system, a filter must be installed near the air handler. If the air unit is in a crawl space or attic, as it usually is, maintenance of the filter can be a pain in the neck, due to access. Extensive duct work is required for a multiple-room return system. The amount and layout of the duct system is similar to that of the supply ducts. This means considerably more money must be spent on both labor and materials to install a multiple-room system. Finding suitable routes for ducts in a multiple-room return system can be troublesome. Each job must be assessed individually to determine which type of return system is best suited to the job. In some cases, a combination of the two types of returns can be used to provide an ideal system solution.

Passive Return Applications

Passive return applications are used with central return systems. The purpose is to improve air flow when doors to rooms are closed. If you've ever lived in an old house where the upstairs was heated with the help of floor grilles that opened air space between the lower living space where a stove was placed and the upper level, you've experienced a form of passive usage. In forced-air systems, the grilles that allow air to pass from room to room are used for air circulation so that return air can be pulled back to the air handler through the main return grille. Without a multiple-room return system, it's essential to provide some form of free-flowing ventilation between heated spaces and the heating return grille. If doors are open, this is not a problem, but closed doors reduce air circulation. This is why doors are undercut in a passive return system.

Some people don't like passive return systems. Having a wall grille that is open from a bedroom or den to the living room, where the return grille is located, can reduce privacy. Voices, like air, will travel through the grilles. Another disadvantage to the grilles is their appearance. Having grilles, even expensive, decorative ones, on various walls is not considered to be attractive decorating by many people. Before you plan on installing a passive return system with a central return system, discuss the pros and cons with your customers. You may find that your customers would prefer to pay more for a multiple-room return system than to put up with the drawbacks of a passive system.

Boxed Cavities

Boxed cavities, like joist bays and wall space between studs, can serve as ducts for return air. This process is known as panning. The materials used to close off a cavity can be sheet metal, hardboard, drywall, or any other suitable material. Using this practice eliminates the need for some duct boxing that might interfere with headroom, and it is usually acceptable by local code officials. Since actual duct material is eliminated, the cost of panning can be less. Joists and stud bays that are panned are not extremely efficient, however.

Panning often results in air leakage. It's difficult, to say the least, to box in framed areas so that air is not lost around seams. The air leakage is a strong reason for not using panning principles. Many negative factors revolve around panning. For example, a panned space may leak air and spread pollutants that a sealed and filtered system would not. It's not practical to attempt complete sealing of panned areas, in most cases. Energy efficiency is usually not great with panned areas, and it's possible that some rooms may become depressurized.

Reducing Noise in Return Systems

Noise can present some level of displeasure in any type of heating system. This is especially true of duct systems when they are made with sheet metal. Noise coming from an air handler or other mechanical vibrations tend to be transmitted along the ducts. Multiple-return systems are generally less noisy than central return systems. There are a few tricks of the trade that help to reduce noise from a duct system.

To reduce noise, locate all mechanical equipment as far from living space as possible. Be especially sure to place heating equipment away from rooms that see the most use. If you are installing equipment that is likely to vibrate, install it on some type of shock-absorbing pad. Flexible connections to ducts will also help to reduce the transference of noise. When a blower compartment is lined with acoustical duct, it is quieter. And, putting a 90° elbow in the return duct will create an offset to reduce the direct path to equipment that can add to noise levels.

Some other pointers for reducing noise include sizing ducts properly. Ducts that are too small will be noisy. Fittings used in return lines should have curved or rounded throats, as opposed to angular throats. This, too, will reduce noise. Installing turning vanes, or even a less-expensive splitter in elbows, will produce a more even air flow that is not as noisy.

A little planning can go a long way in making a heating system enjoyable, instead of detestable. Invest time in the planning and design of your systems before you install them. It's much easier to make changes on paper than it is on a job site. Most problems associated with heating, ventilation, and air-conditioning (HVAC) systems can be avoided with the proper planning and professional installation procedures.

Forced Hot-Air Furnaces

Forced hot-air furnaces are responsible for heating many homes. These furnaces can be run with electricity or fueled by oil or gas. The choice of a fuel type is sometimes partially limited. For example, natural gas is not available in all locations. However, manufactured gas, electricity, and fuel oil are generally available in most populated areas. Heat pumps have cut into the forced-air furnace market. But, in areas where temperatures rarely require air conditioning, hot-air furnaces rule over heat pumps. Hot-air furnaces can be used in almost any heating application, so they are a competitive choice within the heating industry.

People who are having homes built have many choices when it comes to heating systems. Hot-air furnaces are one of the potential choices. There are, however, factors to consider when choosing a particular type of furnace. Deciding between gas-fired, oil-fired, and electrical furnaces can be tough. Picking the proper placement for a furnace that requires duct work may be a major decision in terms of the effectiveness of the heating system. All of the factors must be weighed before a buying decision is made (Figs. 16-1 to 16-3).

Air Flow Capacities

Model	Blower Size Dia. x Width	Motor H.P.	Motor Speed	CFM Inches H2O .20	.30	.50	Nominal Tons Cooling	*Rated Heating CFM
DCL065D36A	10 x 8	1/3	Low	840	815	710	2.0	1,100
			Med	1080	1045	975	2.5	
			High	1310	1260	1150	3.0	
DCL080D48A	10 x 10	1/2	Low	1125	1110	1025	2.5	1,212
			Med-Low	1350	1315	1200	3.0	
			Med-High	1550	1470	1350	3.5	
			High	1755	1670	1500	4.0	
DCL095D60A	11 x 10	3/4	Low	1305	1281	1220	3.0	1,600
			Med-Low	1470	1435	1360	3.5	
			Med-High	1705	1660	1600	4.0	
			High	2000	1950	1810	5.0	
DCL110D60A	Double 10 x 6	3/4	Low	1335	1330	1270	3.0	1,670
			Med-Low	1535	1520	1435	3.5	
			Med-High	1740	1685	1560	4.0	
			High	2005	1950	1800	5.0	

*Recommended design air flow for best operation and efficiency of the furnace.

Clearances — Installation and Service (Inches)

		Minimum Installation Clearances									Minimum Service Clearances		
		Furnace				Plenum							
Model	Type of Installation	Front	Back	Left Side	Right Side	Top	Sides	① Duct	Vent Pipe	Floor	Front	Back	Sides
DCL065D36A	Closet	4	0	0	0	1	1	1	0	C	24	24	18②
DCL080D48A	Closet	4	0	0	0	1	1	1	0	C	24	24	18②
DCL095D60A	Closet	4	0	0	0	1	1	1	0	C	24	24	18②
DCL110D60A	Closet	4	0	0	0	1	1	1	0	C	24	24	18②

① For the first three (3) feet from plenum. After the first three (3) feet, no clearance required.
② Required on one side for access to rear of furnace to service blower compartment and air filters.
C Floor may be combustible material.

Dimensions and Filter Sizes (Inches)

Model	A	B	Filter Size
DCL065D36A	19	18	2 — 9.5 x 20 x 1
DCL080D48A DCL095D60A	23	22	1 — 10 x 20 x 1 & 1 — 13 x 20 x 1
DCL110D60A	26	25	2 — 13 x 20 x 1

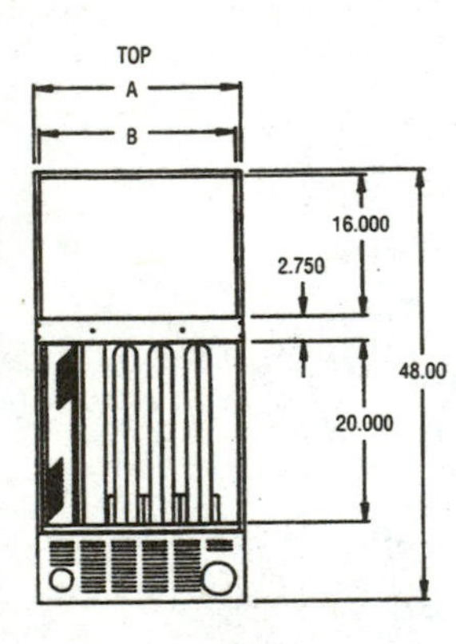

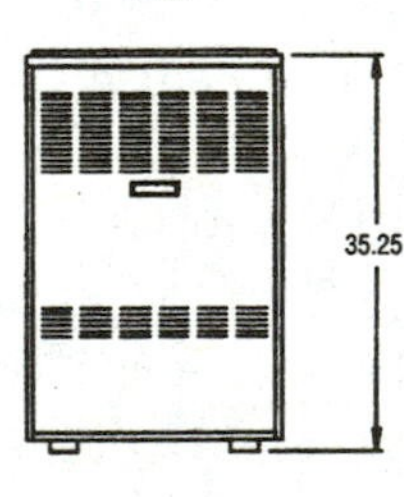

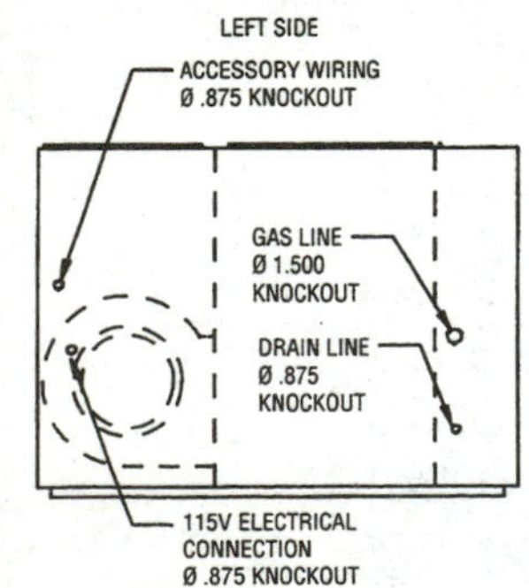

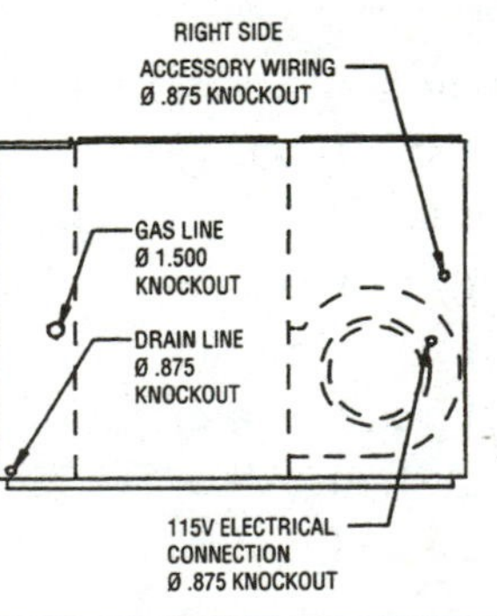

FIGURE 16-1

Gas-fired furnace. (*Courtesy of Bard.*)

Specifications

Model Number	Flue Location	① Nozzle Size	Input BTUH	Heating Capacity BTUH②	② AFUE	③ AFUE	Heating Blower Speed	Temp. Rise Range	① Nozzle Size	Input BTUH	Heating Capacity BTUH②	② AFUE	③ AFUE	Heating Blower Speed	Temp. Rise Range
	Factory Installed Standard								Field Installed Option						
FLF085D36C	Front	.75	105,000	85,000	80.0	84.7	Med	60-90	.65	91,000	74,000	80.5	84.2	Low	60-90
FLR085D36C	Rear	.75	105,000	86,000	81.0	85.0	Med	60-90	.65	91,000	75,000	81.5	85.3	Low	60-90
FLF110D48C	Front	1.00	140,000	113,000	80.0	84.5	Med-High	60-90	.85	119,000	97,000	80.5	85.7	Med-Low	60-90
FLR110D48C	Rear	1.00	140,000	114,000	81.0	84.2	Med-High	60-90	.85	119,000	98,000	81.5	84.8	Med-Low	60-90
FLR140D60C	Rear	1.25	175,000	143,000	80.0	84.2	Med-High	60-90	1.10	154,000	126,000	80.5	84.3	Med-Low	60-90

① 80° hollow cone spray pattern.

③ Annual fuel utilization efficiency with electro-mechanical flue damper.

② Annual fuel utilization efficiency and heating capacity based upon U.S. Government standard tests using D.O.E. isolated combustion rating procedure.

Electrical Ratings

Model Number	Volts-HZ-Phase	Total Unit Amps	Blower Motor HP	Blower Motor FLA	Burner Motor HP	Burner Motor FLA	Minimum Circuit Ampacity	Maximum Time Delay Fuse or HACR Circuit Breaker
FLF085D36C, FLR085D36C	115-60-1	8.1	1/3	5.6	1/7	2.5	15	15
FLF110D48C, FLR110D48C	115-60-1	13.0	1/2	10.5	1/7	2.5	16	20
FLR140D60C	115-60-1	15.0	3/4	12.5	1/7	2.5	19	20

Airflow Capacities

Model Number	Blower Size Dia. x Width	Motor H.P.	No. of Speeds	Speed	CFM In. H_2O .20	CFM In. H_2O .30	CFM In. H_2O .50	Nom. A/C Tons
FLF085D36C	10 x 9	1/3	3	Low	860	840	785	2
				Med	1,190	1,160	1,080	2-1/2
				High	1,405	1,365	1,235	3
FLR085D36C	10 x 9	1/3	3	Low	880	865	790	2
				Med	1,210	1,185	1,090	2-1/2
				High	1,425	1,380	1,275	3
FLF110D48C	10 x 10	1/2	4	Low	1,180	1,160	1,065	2-1/2
				Med-Low	1,335	1,315	1,235	3
				Med-High	1,580	1,540	1,415	3-1/2
				High	1,810	1,760	1,635	4
FLR110D48C	10 x 10	1/2	4	Low	1,040	1,030	1,000	2-1/2
				Med-Low	1,235	1,220	1,150	3
				Med-High	1,500	1,465	1,355	3-1/2
				High	1,760	1,685	1,525	4
FLR140D60C	11 x 10	3/4	4	Low	1,220	1,210	1,190	3
				Med-Low	1,425	1,420	1,400	3-1/2
				Med-High	1,690	1,675	1,640	4
				High	2,470	2,430	2,250	5

Dimensions (Inches)

Model Number	Cabinet A Width	Cabinet B Depth	Cabinet C Height	Plenum Openings D x E Supply	Plenum Openings D x F Return	Flue Connection Location	Flue Connection Dia. G	Flue Connection H	Flue Connection J	Air Filters① Size	Air Filters① No. Used
FLF085D36C	23	47-1/4	40-1/4	22 x 20	22 x 16	Front	6	—	5-1/4	11-1/2 x 17	2
FLR085D36C	23	47-1/4	40-1/4	22 x 20	22 x 16	Rear	6	34	—	11-1/2 x 17	2
FLF110D48C	23	47-1/4	44-1/4	22 x 20	22 x 16	Front	6	—	5-1/4	10 x 20/13 x 20	1 ea.
FLR110D48C	23	47-1/4	44-1/4	22 x 20	22 x 16	Rear	6	38	—	10 x 20/13 x 20	1 ea.
FLR140D60C	26	50	50	25 x 20	25 x 16	Rear	6	43-5/8	—	13 x 20	2

① Washable type filter 1" nom. thickness

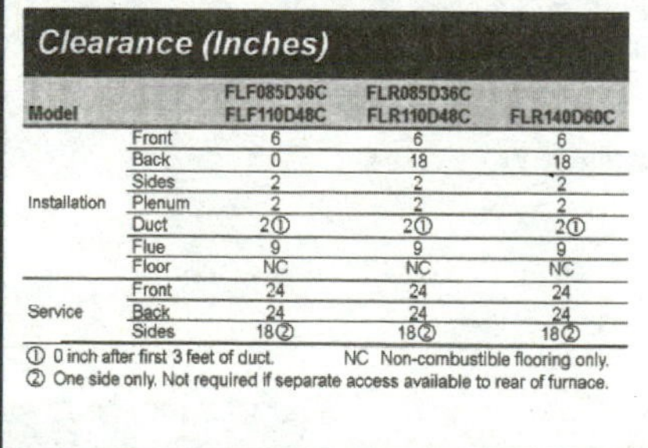

Clearance (Inches)

Model		FLF085D36C FLF110D48C	FLR085D36C FLR110D48C	FLR140D60C
Installation	Front	6	6	6
	Back	0	18	18
	Sides	2	2	2
	Plenum	2	2	2
	Duct	2①	2①	2①
	Flue	9	9	9
	Floor	NC	NC	NC
Service	Front	24	24	24
	Back	24	24	24
	Sides	18②	18②	18②

① 0 inch after first 3 feet of duct. NC Non-combustible flooring only.

② One side only. Not required if separate access available to rear of furnace.

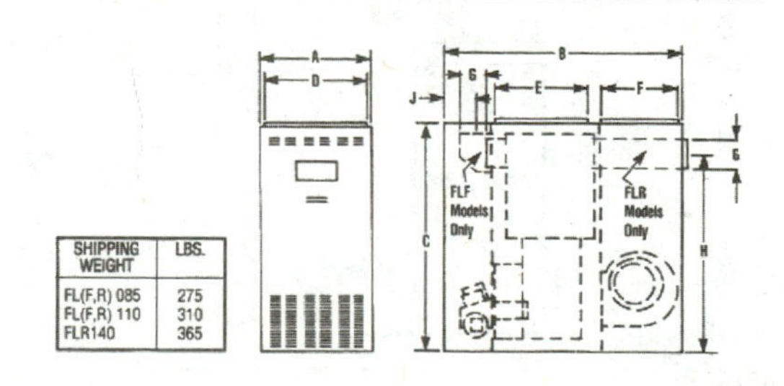

SHIPPING WEIGHT	LBS.
FL(F,R) 085	275
FL(F,R) 110	310
FLR140	365

FIGURE 16-2

Oil-fired furnace. (*Courtesy of Bard.*)

Choosing a Means of Operation

Choosing a means of operation for a forced-air furnace might be confusing for some people. Since there are several choices, people have to sift through the pros and cons, as they relate to the individual's personal taste. There is some statistical data that can make the process somewhat scientific, but many buyers will base their decisions on personal preferences. Availability of fuel types can be a consideration. Not all types of fuel are available in all areas. However, most of the fuels used to operate a furnace are typically available. Natural gas is one exception, as it is not always available in rural locations (Fig. 16-4).

Electricity

Electricity is a common form of power for heating systems. Since electricity is available in most residential locations, it's considered to be readily available. The cost of electricity varies from place to place, but it's usually a viable consideration. However, the cost of electricity is

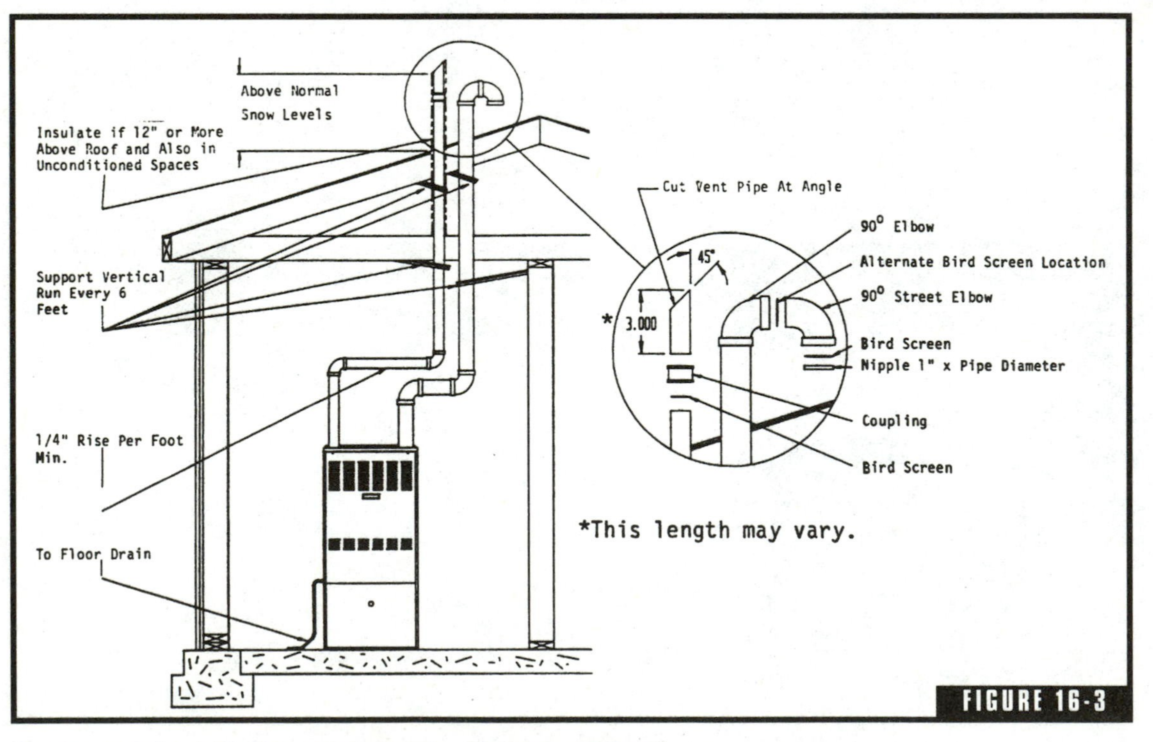

Recommended ventilation requirements. (*Courtesy of Bard.*)

Furnace Specifications

Model	Input BTUH	Heating Capacity BTUH	① AFUE	Temperature Rise Range	Vent ⑤ Piping Size	Vent ⑤ Piping Max Eq Ft	Combustion Air ⑤ Intake Piping Size	Combustion Air ⑤ Intake Piping Max Eq Ft	Gas Pipe Connection Size	Shipping Weight
DCH036D30C	36.000	33,000	90.5	20-50	2"	30'	2"	30'	1/2"	180
					2 1/2" or 3"②	65'	2 1/2 or 3"②	65'		
DCH050D30C	50,000	46,000	92.0	25-55	2"	30'	2"	30'	1/2"	210
					2 1/2" or 3"②	65'	2 1/2" or 3"②	65'		
DCH065D36C	65,000	59,000	91.0	35-65	2"	30'	2"	30'	1/2"	215
					2 1/2" or 3"②	65'	2 1/2" or 3"②	65'		
DCH080D48C	80,000	72,000	90.5	30-60	3" ②	65'	3" ④	65'	1/2"	245
DCH095D60C	95,000	86,000	90.0	35-65	3" ②	65'	3" ④	65'	1/2"	250
DCH110D60C	110,000	99,000	89.0	40-70	4" ③	65'	4" ④	65'	1/2"	300

① AFUE and capacity rating based upon D.O.E. Isolated Combustion Formula. ② 2" PVC pipe required inside cabinet. ③ 2" ABS pipe required inside cabinet only. ④ 3" PVC pipe required inside cabinet.
⑤ Vent pipe and combustion air pipe must be same size and approximately the same equivalent length

Electrical Ratings

Model	Volts	Hz	Phase	Total Unit AMPS	Blower Motor HP	Blower Motor FLA	Inducer Motor HP	Inducer Motor FLA	Minimum Circuit Ampacity	Maximum Time Delay Fuse on HACR Circuit Breaker
DCH036D30C	115	60	1	6.0	1/3	4.6	1/35	1.4	15	15
DCH050D30C	115	60	1	6.0	1/3	4.5	1/35	1.4	15	15
DCH065D36C	115	60	1	6.5	1/3	6.5	1/35	1.4	15	15
DCH080D48C	115	60	1	10.5	1/2	10.5	1/35	1.4	15	20
DCH095D60C	115	60	1	12.5	3/4	12.5	1/35	1.4	16	20
DCH110D60C	115	60	1	11.0	3/4	11.0	1/35	1.4	15	20

FIGURE 16-4

Direct vent furnace specifications. (*Courtesy of Bard.*)

much higher in some places than it is in others. I recently moved about 300 miles north of where I had been living. Both my present and previous home are in Maine. You might think that the cost of electricity within the state would be uniform. It is not. Electrical rates in northern Maine are substantially higher than what they are in the central, coastal part of the state.

Cost is clearly a factor when choosing a type of energy. People who live in remote areas, as I do, are often accustomed to paying more for basic services. The cost of a gallon of gasoline where I live is about 10 cents per gallon higher than it is some 300 miles south. Electricity is more expensive where I live than it is where I used to live. But, fuel oil costs about the same. There is no natural gas available in either area. Based on this, fuel oil should be considered the best type of fuel for me to use in my present home, and as it happens, it is the fuel I use in this location.

Every customer is likely to have a preference in a fuel type. Electricity is a strong contender for many reasons. Some of the best reasons are that essentially no effort is required to obtain it, it is clean, and it doesn't produce offensive odors. There is some argument that electricity is not good for winter heating, since electrical power is sometimes lost. In reality, many heat sources can't function properly without electricity. For example, an oil-fired burner is still dependent on the availability of electrical current. Most residential areas don't lose electrical power often or for long periods of time, so the concern of its constant availability is not usually a factor.

Electricity is a good form of power to use when installing a furnace. The power source is common, normally available in any residential location, safe to use, and produces no irritating odors. Cost can be a prohibitive factor, but electricity is a widely used choice for heating systems.

Oil

Fuel oil is typically hauled to a location by truck. This makes fuel oil a viable heating option in almost any location. Prices of fuel oil fluctuate from year to year and from season to season. Many astute homeowners buy a bulk of fuel oil during summer months, when prices are lower, to use in winter. Oil is not considered an explosive risk, so consumers sometimes feel safer with it being on the premises than they would with gas fuels. However, oil can produce fumes. The smell may be minimal, but it can still affect the senses of some people. There are consumers who are not willing to have any type of oil smells present in their homes. While the likelihood of actually noticing the smell of an oil burner, under normal conditions, is remote, this reason, alone, can turn some people off on the idea of an oil-fired furnace.

Gas

There are two types of gas (Fig. 16-5) commonly used with heating systems. Both natural gas and manufactured gas (LP gas as an example) are suitable fuels for a furnace. The cost of operating a furnace with gas as a fuel can be lower than other operating options. But, natural gas is not always available. It is a common utility in cities, but rural locations don't generally have access to natural gas. Still, people with homes in the country can turn to LP gas. Some people are afraid that the LP gas canisters needed to supply fuel to a furnace are dangerous. Under normal conditions, the LP gas containers are safe. But, there is a risk of explosion if gas leaks exist, and LP gas is a powerful explosive force if the circumstances allow for it.

Furnace Specifications

Model Number	Input Btuh	Heating Capacity Btuh	① AFUE	Temperature Rise Range	Vent Piping ③ Size	Vent Piping ③ Max. Eq. Ft.	Combustion Air Intake Piping ③ Size	Combustion Air Intake Piping ③ Max. Eq. Ft.	Gas Pipe Connection Size	Shipping Weight
DCC050D30A	50,000	46,000	91.0	30-60	2" 2 1/2" or 3" ②	30' 65'	2" 2 1/2" or 3" ②	30' 65'	1/2"	210
DCC065D36A	65,000	59,000	91.0	30-60	2" 2 1/2" or 3" ②	30' 65'	2" 2 1/2" or 3" ②	30' 65'	1/2"	215
DCC080D48A	80,000	72,000	90.5	40-70	3" ②	65'	3" ②	65'	1/2"	245
DCC095D48A	95,000	86,000	90.0	40-70	3" ②	65'	3" ②	65'	1/2"	250

① Annual fuel efficiency and capacity ratings based upon D. O. E. Isolated Combustion Formula.
② 2" pipe required to exit cabinet. 2" ABS required inside cabinet.
③ Vent pipe and combustion air pipe must be same size and approximately the same equivalent feet.

Electrical Ratings

Model Number	Volts	Hz	Phase	Total Unit Amps	Blower Motor HP	Blower Motor FLA	Inducer Motor HP	Inducer Motor FLA	Minimum Circuit Ampacity	Max. Time Delay Fuse or HACR Circuit Breaker
DCC050D30A	115	60	1	6.0	1/3	4.5	1/35	1.4	15	15
DCC065D36A	115	60	1	6.5	1/3	6.5	1/35	1.4	15	15
DCC080D48A	115	60	1	10.5	1/2	10.5	1/35	1.4	15	20
DCC095D48A	115	60	1	10.5	1/2	10.5	1/35	1.4	15	20

FIGURE 16-5

Direct vent counterflow furnace. (*Courtesy of Bard.*)

The metal containers used to store LP gas are easily transported by truck. This makes the fuel suitable for even remote locations. Generally speaking, LP gas is a safe fuel choice and it can certainly be a cost-effective fuel. Furnaces set up to work with natural gas can generally be converted to use with LP gas. It is, however, extremely important to make sure that the heating unit being served is set up for the type of fuel being supplied.

Regardless of the type of fuel being used, hot-air furnaces are all installed in a similar fashion. The connection of power or fuel is different, but the main furnace and the duct connections are essentially the same. Furnaces that use electricity are the easiest to connect in terms of operating power. Both types of gas-fired units are piped in a similar manner. Oil-fired burners are piped differently from gas units. With all of this said, it is still safe to say that the heating units themselves are installed in much the same way.

Picking a Place

Picking a place for a furnace is a major factor in the efficiency of a heating system. Typically, a furnace that uses ducts to distribute air flow should be placed in a central location. For ducts to carry air efficiently they must be sized properly. This usually means not extending a trunk line more than 24 feet in length. However, if a reducing line is used, the distance of the linear run can be greater. It is possible to install a furnace at one end of a home and still make it effective. The

Downflow Installations Minimum Installation Clearances									Minimum Service Clearance				Minimum Ventilation Openings for Confined Spaces Square Inches of Free Area ✡
	Furnace					Plenum							
Model	Front	Back	Left Side	Right Side	Top	Sides	Duct	Vent Pipe	Floor	Front	Back	Sides	
TDH042	1	0	0	2**	0	0	0	6*	C	24	0	0	100 (2 required)
TDH063	1	0	0	0	0	0	0	6*	C	24	0	0	100 (2 required)
TDH084	1	0	0	0	0	0	0	6*	C	24	0	0	100 (2 required)
TDH105	1	0	0	0	0	0	0	6*	C	24	0	0	100 (2 required)
TDH126	1	0	0	0	0	0	0	6*	C	24	0	0	100 (2 required)

✡ See Section 7 – Combustion & Ventilation Air for additional details
C Floor may be combustible material
* Clearance may be 1 inch when Type-B vent is used
** Clearance may be 0 inch when Type-B vent is used

FIGURE 16-6

Minimum clearances. (*Courtesy of Bard.*)

distance that ducts must extend is important, and this is something that any designer, builder, or remodeler must keep in mind.

To say that a furnace should be placed in a central location is not enough. Should the furnace be installed in a closet? Is it better to put the furnace in a basement or deep crawl space? Can a furnace be installed in an attic? The location of choice for a furnace would probably be a basement, but not all homes have basements. Fortunately, the versatility of forced-air furnaces allow them to be installed in crawl spaces and attics. This helps to keep the heating units out of living space and can reduce the noise sometimes associated with furnaces (Figs. 16-6 and 16-7).

Noise is a factor when thinking of a place to put a heating unit. Installing a furnace under bedrooms, for example, is not a good idea. Anytime that a heating unit is placed near living space, the risk of noise exists. This is not desirable. Vibration is another factor. How a heating unit is installed can affect the amount of vibration experienced. Design considerations are a serious issue of any heating system. Choosing where and how to install a heating unit has a lot to do with efficiency, noise reduction, and service. Remember, someone is going to have to get to the heating unit to service it. Just because you may be able to install it way back in a low crawl space doesn't mean that you should. People who have to change the air filters and perform other routine service may not appreciate your installation location (Fig. 16-8).

Every job can present different installation opportunities. As a contractor or installer, you must evaluate each job on an individual basis. There are, however, some principles that tend to prove fruitful.

	Horizontal Installations Minimum Installation Clearances								Minimum Service Clearances			Minimum Ventilation Openings for Confined Spaces Square Inches of Free Area ☼
	Furnace					Plenum						
Model	Front	Back	Left Side	Right Side	Top	Sides	Duct	Vent Pipe	Floor	Front	Back	
TDH042	1	0	1	0	0	0	0	6•	C	24	0	100 (2 required)
TDH063	1	0	0	0	0	0	0	6•	C	24	0	100 (2 required)
TDH084	1	0	0	0	0	0	0	6•	C	24	0	100 (2 required)
TDH105	1	0	0	0	0	0	0	6•	C	24	0	100 (2 required)
TDH126	1	0	0	0	0	0	0	6•	C	24	0	100 (2 required)

☼ See Section 7 – Combustion & Ventilation Air for additional details
C Floor may be combustible material
• Clearance may be 1 inch when Type-B vent is used

FIGURE 16-7

Minimum clearances for horizontal installations. (*Courtesy of Bard.*)

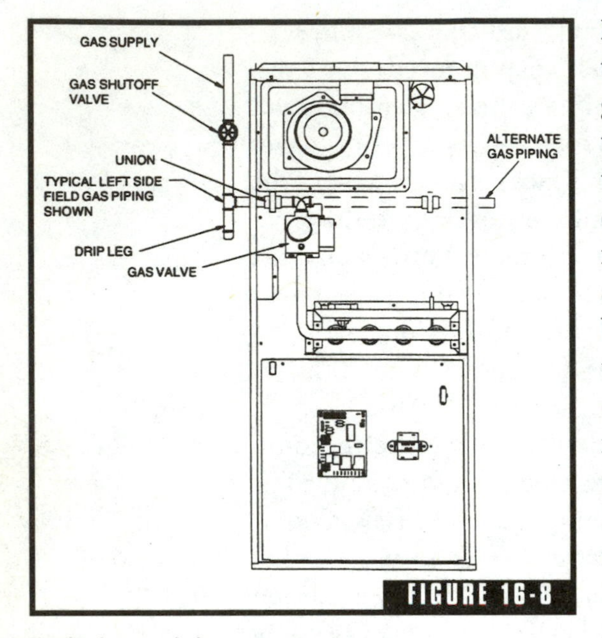

Typical gas piping arrangement. (*Courtesy of Bard.*)

First, find a central location for the furnace that you will be installing. If a basement is going to be present, try to find a location in the basement to install the furnace in. Put the furnace up on blocks to reduce the risk of damage to it if water seeps into the basement. Try to put the heating unit as far away from frequently used rooms as possible. For example, you may install a furnace under a foyer, a laundry room, or some other type of room that does not see much recreational living use. If you will be hanging a heating unit from wood members, some shock-absorbing devices on the mounting brackets can reduce vibrations. It's not difficult to overcome the problems of placing a hot-air furnace, but the process does require thought and planning.

The multiple installation locations possible with forced-air furnaces are part of what makes this type of system popular. Homes that are built with a single-story design can enjoy a complete heating system that is nearly invisible. Even multistory homes can benefit from the installation locations of a furnace. Heat pumps typically have an outside unit. This unit can be what some people consider an eyesore. Plus, most of the frequently used heat pumps require an inside unit, in addition to the outside unit. This reduces living space and can increase noise when heating a house. In over 20 years within the trade, I've never seen a boiler installed in an attic. Boilers are installed in crawl spaces, but this practice is relatively rare. Most boilers are installed in basements. Few heating systems are installed in attics, but hot-air furnaces can be.

Unheated Space

We've just discussed installing furnaces in unheated spaces, such as crawl spaces and attics. When this is done, the duct work for the system must be insulated if the system is to produce efficient heat. We will talk more about this in Chapter 17, but the subject is important enough to acknowledge here. Insulating the ducts is not a problem, but it is time consuming and does require additional materials. If a furnace can be placed in a heated environment, like a basement, the insulation may not be needed and installation costs are likely to be less.

Furnaces installed in attics should not have their output ducts in ceilings. The reason is simple. Warm air rises. It doesn't fall. Air-conditioning ducts are just the opposite. Since attic-mounted units require duct extensions so that distribution grilles can be placed close to floor level, they require more material than would be required for a heating unit installed in a basement or crawl space. This added material cost must be factored into the decision of where to install a heating unit. Attics do offer good installation opportunities for forced-air furnaces, but don't overlook the cost of insulating duct work and extending the ducts down partition walls.

When contemplating the installation location of furnaces, you should make sure that the surface that the heating unit will be installed on is level. If the unit is installed in an area where surface or seeping water may invade its space, place the unit on leveled blocks. Provide adequate air space for all heating units installed. The amount of air space needed may vary from manufacturer to manufacturer and from fuel type to fuel type. Refer to all manufacturer's recommendations before making final installation provisions. Be careful not to install furnaces in areas where freezing temperatures may be present. In most cases, the residual heat from a furnace and its ducts will prevent freezing temperatures, but there may be times when this is not the case. You can't afford for condensation which may be present with a heating unit to freeze.

Installing Gas Furnaces

Installing gas furnaces requires the need for some form of ventilation for exhaust fumes. Special piping is also needed (Figs. 16-9 to 16-11). Gas piping is usually done with black, iron pipe or copper tubing. Ventilation can be done in a number of ways, depending upon the type of system being used. Regular flues and chimneys are sometimes used. PVC pipe, like that used for plumbing systems, can be used as a vent

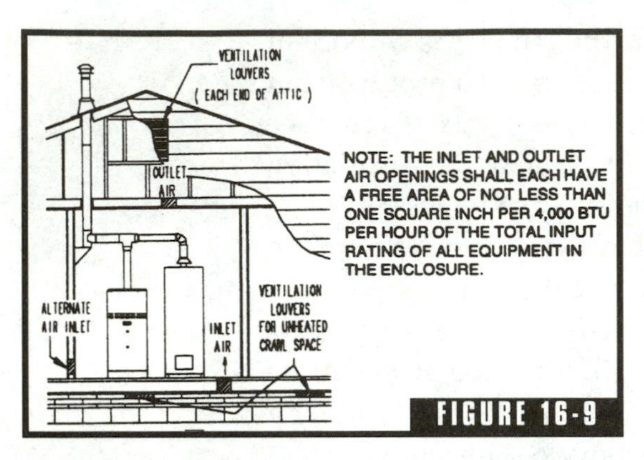

FIGURE 16-9

Detail of a system getting air from outside. (*Courtesy of Bard.*)

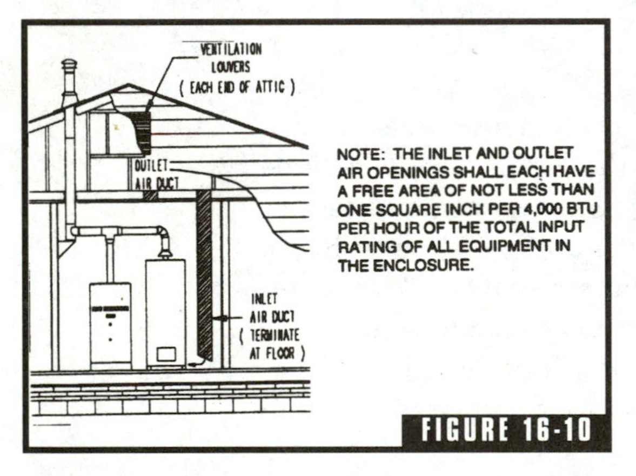

FIGURE 16-10

Detail of a system getting air from vertical ducts. (*Courtesy of Bard.*)

FIGURE 16-11

Detail of a system getting air from horizontal ducts. (*Courtesy of Bard.*)

for a gas appliance (Fig. 16-12). And, there are furnaces that use direct venting through walls with metal sleeves. The key is to use the type of venting recommended by the manufacturer of the heating unit being installed. We could spend pages talking about this, but some illustrations can make the process fast and easy to understand, so I will give you the information in the form of illustrations rather than text (Fig. 16-13).

Any piping (Fig. 16-14) for gas coming through a foundation wall or a concrete floor should be sleeved. The sleeve should be at least two pipe sizes larger than the pipe passing through it. Screw connections are used with iron and steel pipe. Flare fittings are used with copper tubing. Compressions should not be used, since they may vibrate loose or be hit and jarred loose. All copper tubing should be secured and protected from punctures or other damage. Copper tubing should never be installed in a way that it may rub against an abrasive material, such as a concrete wall. If copper tubing is installed close to concrete or some other abrasive material, install some foam pipe insulation around the tubing to prevent the risk of holes being rubbed in the copper.

There was a time when it was allowable for the exhaust vent from a furnace to share a flue or chimney with some other heating device, such as a wood stove. This is no longer the case. Every furnace should have its own, private ventilation source. Do not install more than one furnace on a flue or chimney. Never combine a furnace with wood stove on a common flue. When multiple furnaces are installed on a common flue, either or both of the furnaces can malfunction (Fig. 16-15).

If you are installing a PVC exhaust vent for a heating system, you must clean the pipe properly and make sure that the solvent (glue) used to make the joints is suitable. A primer should be applied to both pipe and fittings before joints are made. It's a good idea to rough up the edges of the pipe with sandpaper before making a joint, but this is not a requirement. Make sure that the solvent being used is rated for the working temperature at which the glue is being applied. By this, I mean that if you are working in cold temperatures, use a solvent that is rated for the cold temperatures.

Pipe solvents are highly flammable and produce severe odors. Use them only in well-ventilated areas and don't allow any open flames around the substances. It's not a good idea to smoke around the fumes or the solvents. Both skin and eye contact with any pipe solvent should be avoided. If the contents of a can of solvent are lumpy, thick, or otherwise unusual, don't use them. Glue that gets cold can turn lumpy and may not make good joints. Don't take any chances. Use only top-notch materials.

Once an installation is complete, you have to check for gas leaks. This is done best with a soapy-water solution. Never use an open flame to look for leaks. When a system is fully connected, turn on the gas from the gas main to the unit shutoff valve and paint each pipe or tubing connection with

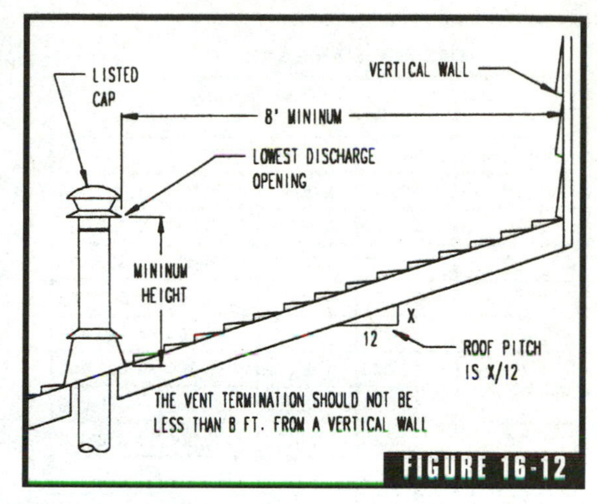

Vent termination requirements. (*Courtesy of Bard.*)

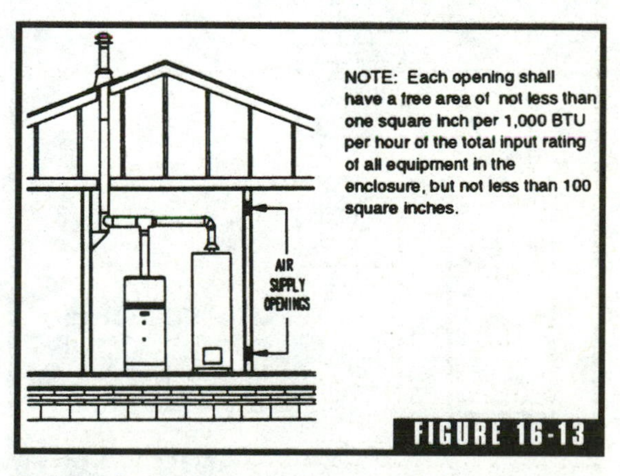

Detail of a system getting all air from inside a building. (*Courtesy of Bard.*)

GAS PIPE SIZES - NATURAL GAS

Length of Pipe Feet	Pipe Capacity BTU Per Hour Input Pipe Size			
	1/2"	3/4"	1"	1-1/4"
10	132,000	278,000	520,000	1,050,000
20	92,000	190,000	350,000	730,000
30	73,000	152,000	285,000	590,000
40	63,000	130,000	254,000	500,000
50	56,000	115,000	215,000	440,000
60	50,000	105,000	195,000	400,000
70	46,000	96,000	180,000	370,000
80	43,000	90,000	170,000	350,000
100	38,000	79,000	150,000	305,000

FIGURE 16-14

Pipe sizes for natural gas. (*Courtesy of Bard.*)

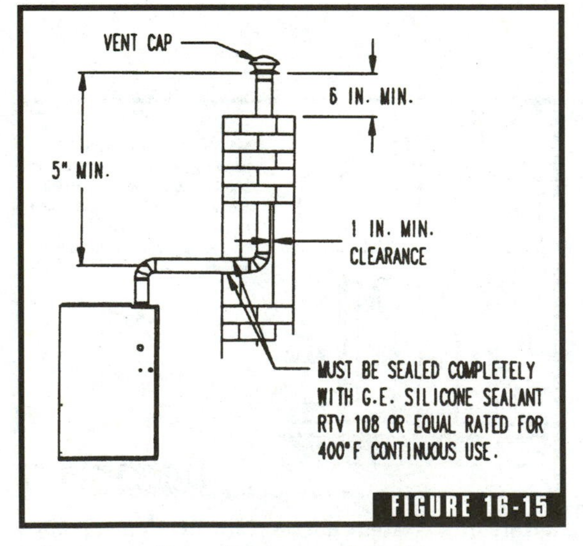

Type B vent liner. (*Courtesy of Bard.*)

soapy water. If bubbles are produced at any locations, they indicate a leak. Assuming that the piping to the unit cutoff checks out, turn on the unit cutoff and check the joints between it and the heating unit. If there are no leaks, you can fire the system and set it up for proper operation.

Electric Furnaces

Electric furnaces don't require a lot of discussion in terms of how to hook them up to their power source. When you are installing a furnace that uses electrical power, you only need a dedicated circuit, a disconnect box, and some wire and wire nuts. It's a simple hookup and there is no need to look for leaks. In most jurisdictions, electrical connections to new equipment are required to be made by licensed electricians. Based on this fact, there is not a lot for us to discuss on the situation here. The basics are that a single circuit is required for an electrically operated furnace. It is common for a disconnect box to be required. The box needs to be in close proximity of the furnace. This allows the electrical power to the unit to be cut off without searching out the electrical service panel, and it's a good idea even if it is not required. Aside from the separate circuit and the probable requirement of a disconnect box, the electrical connection to a furnace is nothing more than standard electrical procedure. The wire is run to the unit, through a stress-relief fitting, and connected to the wires of the furnace by being

DIMENSIONS (Inches) LO-BOY MODELS

Model Number	Cabinet: A Width	Cabinet: B Depth	Cabinet: C Height	Plenum Openings: DxE Supply	Plenum Openings: DxF Return	Flue Connection: Location	Flue Connection: G Dia	Flue Connection: H	Flue Connection: J	Air Filters (1): Size	Air Filters (1): No. Used
FLF085D36C	23	47-1/4	40-1/4	22x20	22x16	Front	6	--	5-1/4	11-1/2x17	2
FLR085D36C	23	47-1/4	40-1/4	22x20	22x16	Rear	6	34	--	11-1/2x17	2
FLF110D48C	23	47-1/4	44-1/4	22x20	22x16	Front	6	--	5-1/4	10x20/13x20	1
FLR110D48C	23	47-1/4	44-1/4	22x20	22x16	Rear	6	38	--	10x20/13x20	1
FLR140D60C	26	50	50	25x20	25x16	Rear	6	43-5/8	--	13x20	2

(1) Permanent washable type filter 1" nom. thickness.

FLF Models Only
FLR Models Only

DIMENSIONS (Inches) HI-BOY MODELS

Model Number	Cabinet: A Width	Cabinet: B Depth	Cabinet: C Height	Plenum Opening: DxE Supply	Plenum Opening: FxH (2) Return	G Flue Diameter	(1) Filter Size
FH085D36C	23	31-1/2	56	22x20	23x14	6	16x25
FH110D48C	23	31-1/2	60	22x20	23x14	6	16x25
FH110D60C	23	31-1/2	60	22x20	23x14	6	20x25

(1) Permanent washable type filter.
(2) Left or right side return air option. Must be cut-in by installer.

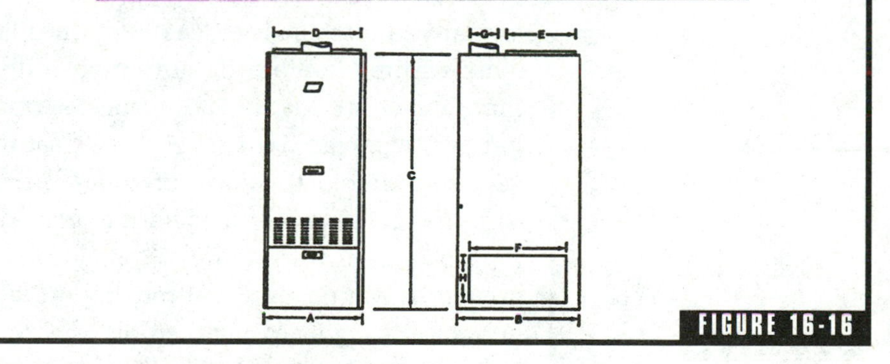

FIGURE 16-16

Furnace dimensions (*Courtesy of Bard.*)

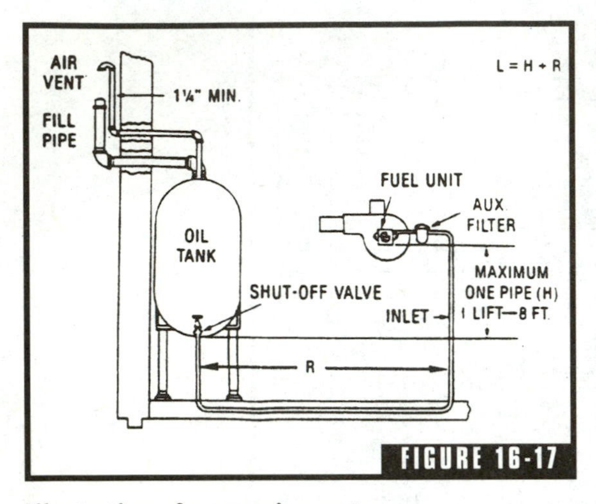

FIGURE 16-17

Illustration of a one-pipe system. (*Courtesy of Bard.*)

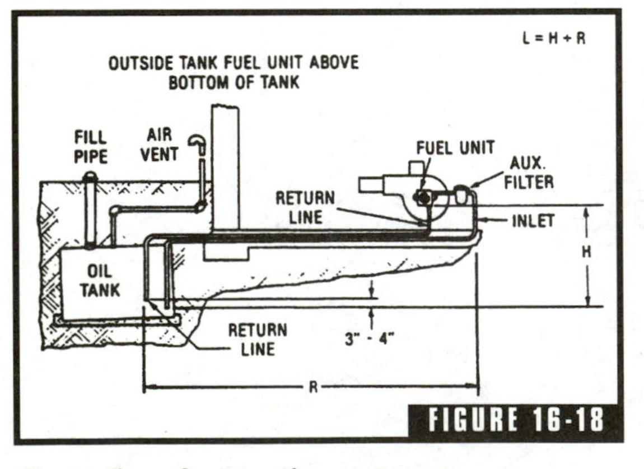

FIGURE 16-18

Illustration of a two-pipe system. (*Courtesy of Bard.*)

twisted under wire nuts. It's no big deal, but the work is usually required to be done by a licensed electrician.

Oil-Fired Furnaces

Oil-fired furnaces (Fig. 16-16) are popular in some areas when natural gas is not available. This seems to be especially true in colder climates. For example, every hot-air furnace I've seen in Maine has been an oil-fired unit. In comparison, I don't recall ever seeing an oil-fired furnace during my many years of working in Virginia. The furnaces that I worked with in Virginia were either natural-gas units or depended upon electricity to operate. Why is this? I don't have any hard statistics to support my theory, but I believe it has to do with local customs and the cost of operation.

Furnaces in Virginia are not needed for as many months as furnaces in Maine. The temperatures in Virginia are much milder than those in Maine, so furnaces don't have to work as hard or as long. Natural gas is available to many of the cities in Virginia, but natural gas is hard to come by in most areas of Maine. A majority of homes in Maine use oil-fired boilers and hot-water baseboard heat, so oil is a common fuel in the state.

The basic installation of an oil-fired furnace is the same as that for a gas-fired unit or an electric unit, except for the fuel considerations. When an oil-fired unit is used, it requires an oil storage tank (Fig. 16-17). Copper tubing, fittings, and components are needed to get oil from the tank to the furnace. The piping of the oil tank and supply line is done as it is for an oil-fired boiler, which we have already discussed. Due to the need for an oil tank, there is more cost in both labor and material when installing an oil-fired boiler. This disadvantage may be outweighed when the operating cost of the furnace is calculated over several years (Fig. 16-18).

The installation of either type of furnace is usually considered to be the easy part of a heating job. It is the installation of the duct work that most installers consider the difficult part. A furnace creates heat, but it is the duct work that balances and delivers the heat. If there are problems in an installation, it is likely that they are caused by some flaw in the design or installation of the ducts. With this in mind, let's turn to the next chapter and explore the installation procedures for duct work.

Installation of Duct Work

The installation of duct work is a job that many people dislike. Even though the work is sometimes easier with the new materials available to work with, it can still be somewhat of a chore. When sheet metal is used for ducts, and it frequently is, the risk of getting cut on the sharp edges of the metal is great. When I entered the trade, just about all of the ducts were made with sheet metal. The sharp edges and burrs were constant risks during the work day. With today's flexible ducts and duct boards the use of sheet metal is not as intense, but there are still plenty of duct systems being fabricated out of sheet metal.

Available space is often the biggest problem that HVAC installers face. Finding enough room to install ducts without dropping a lot of ceilings or building a lot of boxed chases requires planning. The location of plumbing is another consideration which must be factored into the decision of where to place ducts. In the old days, where I worked, some serious moments broke out among HVAC installers and plumbers over joist territory. It can be difficult to find enough space to keep everyone happy.

Assuming a heating design has been done and that the routing of ducts has been established, the actual installation process is fairly simple. The theory is very simple, and the execution of a detailed heating plan usually works pretty well. There are, however, frequent on-site adjustments needed to work around unplanned obstacles. Hanging ducts in a full basement isn't bad work, but working with trunk lines in a crawl space can be quite another story. Unlike electrical wiring, where wires can be turned easily and run in small places, ducts are not so forgiving. HVAC mechanics need a good plan, and they need to work their plan.

System Components

The system components of duct work include the ducts used, fittings, dampers, registers, grilles, and so forth. Not all components are needed in all systems. The three primary types of ducts used in residential applications are plenum ducts, trunk ducts, and branch ducts. A system will have both a supply and a return plenum. All systems will have registers and grilles. Register boots, the fittings used to turn ducts either up in a floor or out of a wall, are needed in all systems. Any number of fittings might be used, but their use depends on the system design and site conditions. Transition fittings are commonly used whenever a change of direction is needed. This is done to minimize air friction and turbulence within the system.

Plenum Ducts

Plenum ducts are generally rectangular and are used on the supply side of an air handler. They may also be used on the return side of the heating unit. The plenum is a box, usually made of sheet metal, that either distributes air to (on the supply side) or collects it from (on the return side) major ducts. Plenums are custom made, often on site, to fit the air handler snugly. It's not uncommon for the plenum to be insulated for better efficiency.

Trunk Ducts

Trunk ducts are the big ducts that make long runs and feed branch ducts. The main supply duct is called the trunk duct, whether on the supply side or the return side. These ducts connect directly to the plenum that they receive their air from. As the trunk ducts extend the length of a job, branch ducts are connected to them to feed supply registers or return grilles. Most trunk ducts are fabricated into a rectangular shape.

Trunk ducts for residential use start, typically, with minimum dimensions of 8 inches high and 8 inches wide. When size is increased, it is usually done in 2-inch increments. Lengths of trunk ducts tend to range from 4 to 8 feet, but they can be made in different sizes. The assembly of a trunk duct is usually done on site. When fabricated into two L-shaped halves, as is the normal procedure, trunk ducts can be put together quickly on a job site. Sheet metal is the material most often used to make trunk ducts, but the ducts can be made from fiberglass duct board or flex duct.

Branch Ducts

Branch ducts are the feeder ducts. They originate from either a plenum or from a trunk duct and deliver air to specific registers. They are also used to return air to the air handler. Branch ducts are much smaller than trunk ducts. Unlike trunk ducts, which are rectangular, branch ducts are round. They are often fabricated from sheet metal, but flexible duct material can also be used for branch ducts. It is also possible to obtain round fiberglass ducts. Since the ducts are round, they provide good air-flow characteristics and generally require fewer types of fittings.

Oval branch ducts are frequently used in vertical walls. The oval shape has a higher pressure loss than a round or square duct, but the oval shape allows the duct to be concealed in a standard stud wall.

Duct Fittings

Duct fittings are numerous. They range from elbows to dampers to flexible connectors. There is usually a fitting design for any need, and when there isn't, a creative mechanic can fabricate a fitting. Since ducts can be large and cumbersome, having a wide array of fittings to work with makes installations easier. The basic fittings used in residential systems include the following:

- Elbows
- Starting Collars
- Reducers
- Takeoff Fittings
- Register Boots
- Dampers
- Flexible Connections
- End Caps
- Stackheads

A starting collar should be installed between the plenum and the trunk duct. Some installers omit starter collars to reduce costs, but the elimination of the collar increases the resistance to air flow, which is not desirable. Another optional, but desirable fitting, is a flexible connector. This is a flexible material that is bonded to metal at each end. The flexible connector should be installed between the starting collar and the trunk duct. Noise is greatly reduced when the flexible connector is used. Another reason for using a flex connector is to provide a little give and take in the alignment of trunk ducts.

Dampers are not always needed or installed, but they are used in both trunk and branch ducts to balance airflow. The best use of dampers is achieved in supply trunk ducts. When dampers are installed in trunks, they can control air flow with less hardware, and any noise caused by the damper will be less noticeable than it would be in a branch duct.

Elbows are used to facilitate 90° changes in direction of ducts. It is also possible to use elbows for 45° turns, and customized elbows can be made when conditions demand it. Reducers are used to make a gentle transition from one size of duct to another. Regular reducers are designed to reduce from one specific duct size to another. An alternative to regular reducers is a reducing adapter. This fitting has a broad operating range and can be used for reduction in different sizes, often with a range of up to 4 inches in change of size. End caps are just what they sound like. They are placed at the end of trunk runs to terminate the duct. Most mechanics make their end caps on the job site.

Takeoff fittings are used to connect branch ducts to trunk ducts. It's common for a takeoff fitting to go from a rectangular duct to a round duct. There are top takeoffs and side takeoffs so that branch ducts can originate on either the side or the top of a trunk duct. Installers who are looking to cut corners on cost sometimes avoid the use of takeoff fittings. To do this, they connect branch ducts directly to trunk ducts. This can be done, but the risk of air leakage and reduced air flow exists when a takeoff fitting is not used.

When ducts are going to terminate in a vertical wall, stackheads are used to end the run of the branch duct. These fittings are equipped with "ears" so that they can be secured to wall studs by driving nails through the ears. They can be used with both supply and return ducts. A stackhead can be made to work with either rectangular or oval ducts.

Register boots are needed to provide a transition from branch ducts to register grilles. The boots can be made to fit either round or rectangular branch ducts. Boots are used for both floor and ceiling registers. Some boots have an integral balancing damper built into them. Even though the built-in ducts are available, many professionals avoid them. Most mechanics agree that a damper in a branch duct that is close to the trunk duct is a more effective and quieter way to control air flow.

Registers, Grilles, and Diffusers

Registers, grilles, and diffusers are used as finish trim for air outlets and inlets. Grilles are most often installed on return lines. Registers and diffusers are more often used on supply ducts. There are many types of registers, grilles, and diffusers to choose from. Most are fairly plain in their design and color, but decorative units are made for the more discriminating consumer. Supply registers have a built-in damper. The damper may be a single-blade design or an opposed-blade design. The opposed-blade design gives more uniform air flow. All supply registers can be used in floors, ceilings, walls, or baseboard installations.

Diffusers are a more specialized means of diverting air flow. They are most often used in ceilings. These applications are most often reserved for the delivery of cool air, rather than warm air. But, there are certain types of ceiling diffusers that are intended for use with heating systems. Available in either round or rectangular shapes, diffusers use wide spacing of defection louvers to provide maximum free area for air flow. The diffuser sends air out in a flat, blanketing pattern. When a diffuser has a curved-blade system, it can be used with satisfactory results for heat distribution. There are many types and styles of diffusers available for a broad range of usage.

Return grilles cover return ducts and are not equipped with integral dampers. Grilles can be used with supply ducts, but there is not local way of regulating air flow when a grille is installed. However, a damper in the supply duct can rectify this situation. Better grille designs include a built-in filter attachment. This allows a filter to be installed at the grille location so that air passing through the filter is cleaned before it enters the duct system on a return line. A hinged cover on the grille allows access to the filter. This type of grille makes service much easier, especially when the air handler is located in a crawl space or attic.

Duct Materials

Duct materials come in a variety of types. Sheet metal has been and still is a prime choice for the fabrication of ducts. As we have already discussed, most trunk ducts are made in L-shaped halves. The sections are put together with drive straps. These are flat pieces of metal with grooves in them that can be driven over raised edges on trunk ducks. Sheet metal offers a number of advantages. The fact that sheet metal is inexpensive, rugged, durable, and available in many sizes and thicknesses (gauges) is just some of what makes it an appealing choice for duct material. In addition, sheet metal can be made into both rectangular and round shapes, can be formed into transitional fittings, and can pass through fire barriers, which means you've got a winner. As an added benefit, the smooth surface of a metal duct offers low resistance to air flow.

Any material with a lot of advantages is likely to have some disadvantages. Sheet metal ducts are no exception. Due to the connection methods used with sheet metal ducts, there is a likelihood of some air leakage. Metal conducts sound, so the transmission of sound along a metal duct can be a problem. Flexible routing of a metal duct is not easy. When insulation is required for metal ducts, it must be applied either in the field or in a shop setting. Additionally, the cost of labor involved with metal ducts can be higher than the labor needed for other types of duct materials.

Fiberglass Duct Boards

Fiberglass duct boards are an excellent choice for duct material when an insulated duct is needed. Trunk runs of duct board are available in standard lengths of 4 and 8 feet. Standard rectangular ducts are normally connected by using shiplap joints and pressure-activated tape. The corner joints for duct boards are either shiplapped or V-grooved and taped. The joints are tested and allow minimum air leakage. Fiberglass duct boards are light in weight, have no sharp edges, are easy to install, dampen sound travel, are already insulated, and come with a vapor barrier included as a part of the construction.

There are not many disadvantages to duct board, but there are a few. The cost of the material is more than that of sheet metal. But, if the duct is being installed in an unconditioned space, where insulation is needed, the final cost of labor and material may be about the same as other duct options. One risk that has to be taken into consideration is the fact that the prefabricated ducts can be crushed on a job site or while in transit. A final potential drawback is the longevity of the material. So long as the ducts are closed properly, the life span is good.

Flexible Branch Ducts

Flexible branch ducts are generally made with blanket insulation that is covered with a flexible vapor barrier on the outside and supported by a helix wire coil on the inside. Metal collars allow the flexible ducts to be connected to trunk ducts. A standard length for this type of duct material is 25 feet. As an insulated duct, the inside diameter of a flexible duct is likely to range from 3 to 14 inches. Some specialty ducts that are flexible and insulated can go up to 20 inches in diameter. Uninsulated versions of the duct span diameter measurements ranging from 4 to 20 inches.

Flexible ducts are favored for many reasons. Some of them include the fact that flexible duct can be purchased with factory-installed insulation, has a low material cost, can be used in unconditioned spaces, requires fewer connections and joints than some other types of duct material, and is simple and inexpensive to install. On the downside, the ducts can be damaged easily by being torn, crimped, or crushed. Damage on the inside of a duct might go unnoticed. And, due to its design, flexible ducts are responsible for higher resistance to air flow.

Duct Insulation

Duct insulation is a primary concern when ducts are installed in unconditioned space. Fiberglass duct board is already insulated, and flexible ducts can be purchased with insulation already installed. But, sheet metal systems require manual insulation. Insulation can be installed either inside or outside a duct. Duct material that is installed in an attic or crawl space is most likely to be wrapped on the outside with insulation. The use of an insulation liner is more often used in line return register boxes and supply trunks that will be difficult to reach once installed.

Fiberglass duct liner is available in a number of thicknesses, which range from 1/2 inch to 2 inches. The material is treated for use in lining the inside of rectangular metal duct work. A typical thickness found in metal ducts is 1 inch. This insulation is used for thermal protection. If the only reason for insulating a duct is to dampen noise, a 1/2-inch-thick material is common.

When fiberglass insulation is used to wrap a metal duct, the thickness of the insulation may range from 1 1/2 to 3 inches. When thermal protection is needed, duct wrapping is more effective than duct lining. However, wrapping a duct with insulation does very little to reduce noise in a heating system. When a duct system is wrapped, the insulation is held in place by taping joints where the insulation meets. The material used to wrap a duct may be faced or unfaced.

Special Ducts

Special ducts are used when the ducts are to be installed in concrete slabs. Either plastic or plastic-coated metal ducts are used below concrete. The use of these special ducts makes it possible to prevent groundwater from entering the duct system. Putting ducts in concrete slabs is good in some cases, since the duct is hidden from view, but there are disadvantages associated with the procedure.

When ducts are embedded in concrete, they are very expensive to access and repair if the need ever arises. High groundwater may find a void in which to invade the underground duct system. Ducts must be installed on a grade or slope, similar to drainage pipes in a plumbing system, to avoid a buildup of condensation. And, installing the ducts must be done before concrete is poured, which creates an opportunity for the ducts to be damaged during construction.

Don't Get Sloppy

When installing duct work, don't get sloppy. A few seemingly minor slipups during installation can have a negative effect on a duct system for years to come. For example, failing to make and seal tight joints can allow air leakage and promote poorer performance for a heating system. The improper placement of dampers can create unneeded noise. Failure to use starting collars and flexible connectors can add to the noise associated with a heating system. A poor job of insulating ducts can result is lower efficiency ratings. Extending a trunk duct too far, usually more than 24 feet, without reducing its size can reduce airflow performance. Any single defect can amount to trouble, and a combination of defects can result in many headaches. Take the time to plan your jobs carefully. Then, make sure that the system is installed according to your plan. If you do this, you should have happier customers and fewer complaints.

CHAPTER 18

Maintaining and Troubleshooting Oil-Fired Furnaces

Proper maintenance of a forced-air furnace is essential if you want to get the most efficiency from the heating unit. Something as simple as cleaning or replacing a clogged filter can have a dramatic effect on how well a furnace functions. Even with regular maintenance, there may be times when a furnace fails to function properly. When this happens, troubleshooting skills are needed to locate the cause of the problem. This chapter is going to address both the maintenance and troubleshooting of oil-fired hot-air furnaces. We will talk about gas-fired units in Chapter 19.

Maintaining Oil-Fired Furnaces

Maintaining an oil-fired furnace is a task that can be done mostly on an annual basis. However, filters in all forced-air systems should be check, cleaned, or replaced more frequently. The amount of time that is allowed to pass between cleaning or replacing filters varies from system to system. For example, a homeowner who has pets that shed a lot of hair will need to change filters more often than one without pets. If a home is subjected to more dust than usual, such as when local construction is creating unusually high amounts of dust, filters are more likely to need cleaning or replacing more often. Most home-

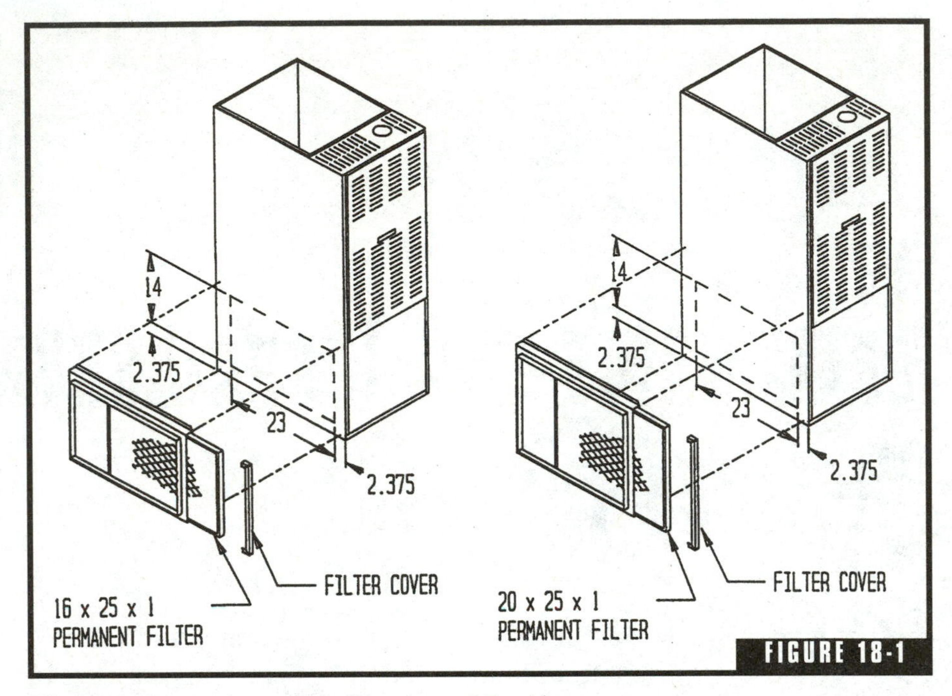

Filter location and removal. (*Courtesy of Bard.*)

owners are capable of maintaining their own furnace filters, so long as they are shown where the filters are, how to access and remove them, and how to reinstall them. Whenever you install a new system, make sure that you take a few minutes to explain filter maintenance to your customers (Fig. 18-1).

Many furnaces are sold with filters that are intended to be cleaned rather than replaced. Advise your customers what types of filters their units have and what the proper maintenance procedure is. Assuming that a home is average in its air condition, filters should be cleaned or replaced annually and at least twice during the heating season. But, it is a good idea to check filter conditions more frequently, to avoid overworking a heating system.

The bearings in the motor of an oil burner should typically be lubricated at least twice during a heating system. A few drops of a high-grade, SAE-20 motor oil will do the trick. It's important that only a few drops, say between 2 and 4 drops are used. Adding too much oil is not good for the equipment. Some equipment parts, such as blower

TABLE 11 TEMPERATURE RISE RANGES, LIMIT CONTROL SETTINGS, AND HEATING BLOWER SPEEDS

MODEL	NOZZLE (1)	RISE RANGES	HEATING BLOWER SPEED	LIMIT SETTING	ON	OFF
FH085D36C	.65	70 - 100	Low	170	110	90
	.75	60 - 90	Med	170	110	90
FH110D48C	.85	70 - 100	Low	170	110	90
	1.00	60 - 90	Med Low	170	110	90
FH110D60C	.85	60 - 90	Low	170	110	90
	1.00	60 - 90	Med Low	170	110	90
FLF085D36C	.65	60 - 90	Low	230	140	110
	.75	60 - 90	Med	230	140	110
FLF110D48C	.85	60 - 90	Med Low	230	140	110
	1.00	60 - 90	Med High	230	140	110
FLR085D36C	.65	60 - 90	Low	250	140	110
	.75	60 - 90	Med	250	140	110
FLR110D48C	.85	60 - 90	Med Low	240	140	110
	1.00	60 - 90	Med High	240	140	110
FLR140D60C	1.10	60 - 90	Med Low2	220	140	110
	1.25	60 - 90	Med High	220	140	110
FC085D36C	.65	70 - 100	Low	180	110	90
	.75	60 - 90	Med	180	110	90

(1) 80 degree hollow cone spray pattern

FIGURE 18-2

Temperature riser ratings. (*Courtesy of Bard.*)

motors, may be permanently lubricated and require no oiling. Check the manufacturer-provided paperwork that comes with a furnace to determine what needs to be oiled and how often it should be serviced.

An annual inspection of all oil-fired furnaces should be conducted. These inspections are performed best by professionals. Some homeowners have the skills to perform at least some, if not all, of the tests, but it is generally recommended that professional service technicians be called in for the annual inspections.

Routine Annual Inspections

There are several types of tests which should be performed on routine annual inspections. One of the first steps of a maintenance inspection is to look for oil stains on equipment and flooring. You should inspect all parts of the oil storage and circulating system for leaks. Be especially sure to check the nozzle of the burner to see that no oil is escaping from it when the burner is not operating (Fig. 18-2).

Both the suction line and the pump on the equipment should be checked for air. All air should be out of the system. The reason for this is to ensure that the burner has instant oil shutoff at the nozzle and that there will be no noise from the pump that is associated with air. Any inspection should include checking the burner flame. It should be clear, quiet, and free of odor. If there is anything out of the ordinary with the flame, check the nozzle to make sure that it is the right size.

It may seem silly, but it's a good idea to check the thermostat performance. This is an easy test to perform, and it can turn up a problem that would be uncomfortable on a cold day. To conduct the test, turn the thermostat way up. It should be set several degrees above room temperature. When this is done, the burner should cut on. Then, turn the thermostat to its lowest setting. When the thermostat is put at a setting that is well below room temperature, the burner should cut off.

Another recommended test is one involving the primary control. This should be done in compliance with the recommendations provided by the manufacturer of the equipment that you are working on. Here are some examples of how the test might be done. You can simulate a flame failure by shutting off the oil supply with a valve. This should be done while the burner is on and running. Within 15 seconds the safety switch should lock out. The ignition should stop, the motor should stop, and the oil valve should close. If the safety switch doesn't lock out, there is a problem with the control.

To test for ignition failure you shut off the oil while the burner is turned off. Then, put the burner through a typical starting sequence. The safety switch should lock out when the lack of a flame is

Model	Minimum Ventilation Opening-Square Inch	Recommended Opening 2 Required	
		Size	Sq. In.
FH085D36C	240	8 x 16	128
FH110D48C	280	9 x 18	162
FH110D60C	280	9 x 18	162
FLF085D36C	290	8 x 19	152
FLR085D36C	290	8 x 19	152
FLF110D48C	340	9 x 19	171
FLR110D48C	340	9 x 19	171
FLR140D60C	360	9.5 x 19	180
FC085D36C	240	8 x 16	128

FIGURE 18-3

Ventilation requirements. (*Courtesy of Bard.*)

determined by the control. If the safety doesn't function, you have a problem with the control (Fig. 18-3).

Sometimes the electrical power to a burner goes off. This is a common problem in some areas during winter ice storms. To test a burner for a power failure, turn off the power to the burner while it is running. Once the burner has had time to go out on its own, turn the electricity back on. The burner should restart. If it doesn't, the primary control is malfunctioning (Figs. 18-4 to 18-6).

When the operating tests are not satisfactory, you should check the wiring and installation of the control first. If this inspection doesn't produce the problem, check the control circuit. You may have to replace the flame detector or the burner-mounted relay.

Other simple things to check during a routine inspection include the amount of oil in the oil supply tank. All filters should be checked and cleaned or replaced, if necessary. It's also a good idea to check all air openings on both the supply and return system for obstructions. There shouldn't be any, but it's best to find them before a heating season if there are any.

Troubleshooting Oil Burners

Noise is not an uncommon complaint with oil-fired burners. When a noise is noticed, it is a sign that something is wrong. The problem can be one of any number of things. The type of noise heard can be an indicator of what to look for. There are three typical types of noise associated with oil burners. Some give off pulsing sounds. Thumping noises are also common. Rumbling is another type of noise that you might experience with an oil burner. The nozzle in the burner is always a good place to start your investigation (Fig. 18-7).

Start by replacing the existing nozzle with one that has a wider spray angle. Try the unit and see if it is still producing excess noise. When a nozzle with a wider spray angle will not quiet the noise made by a burner, try a nozzle that has a smaller opening. The installation of a new nozzle that is one size smaller than the one previously installed could be all it will take to quiet the boiler. It is necessary to make adjustments and test after each one to see if the problem is corrected. Another course of action, especially if the oil burner is producing pul-

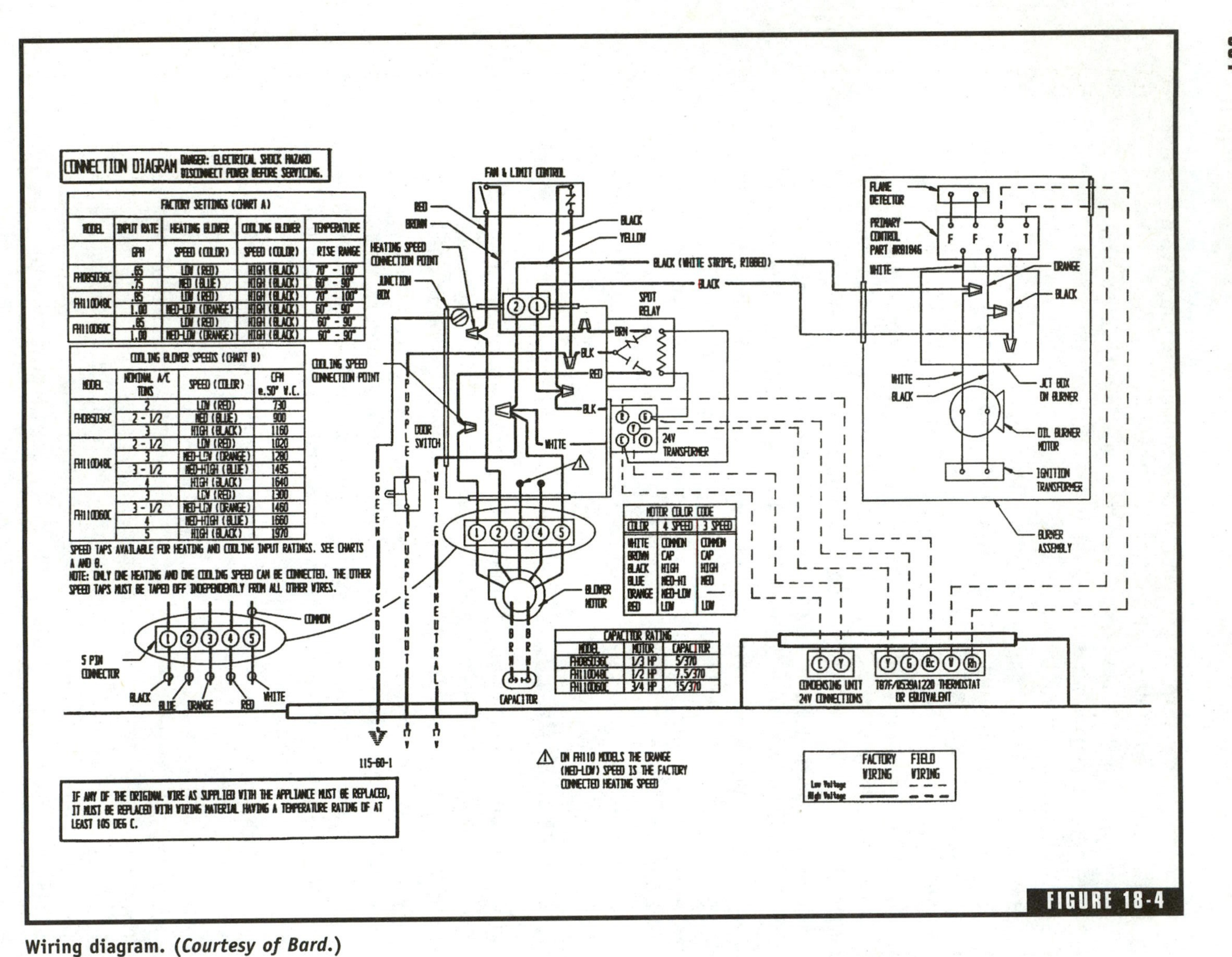

Wiring diagram. (*Courtesy of Bard.*)

FIGURE 18-5

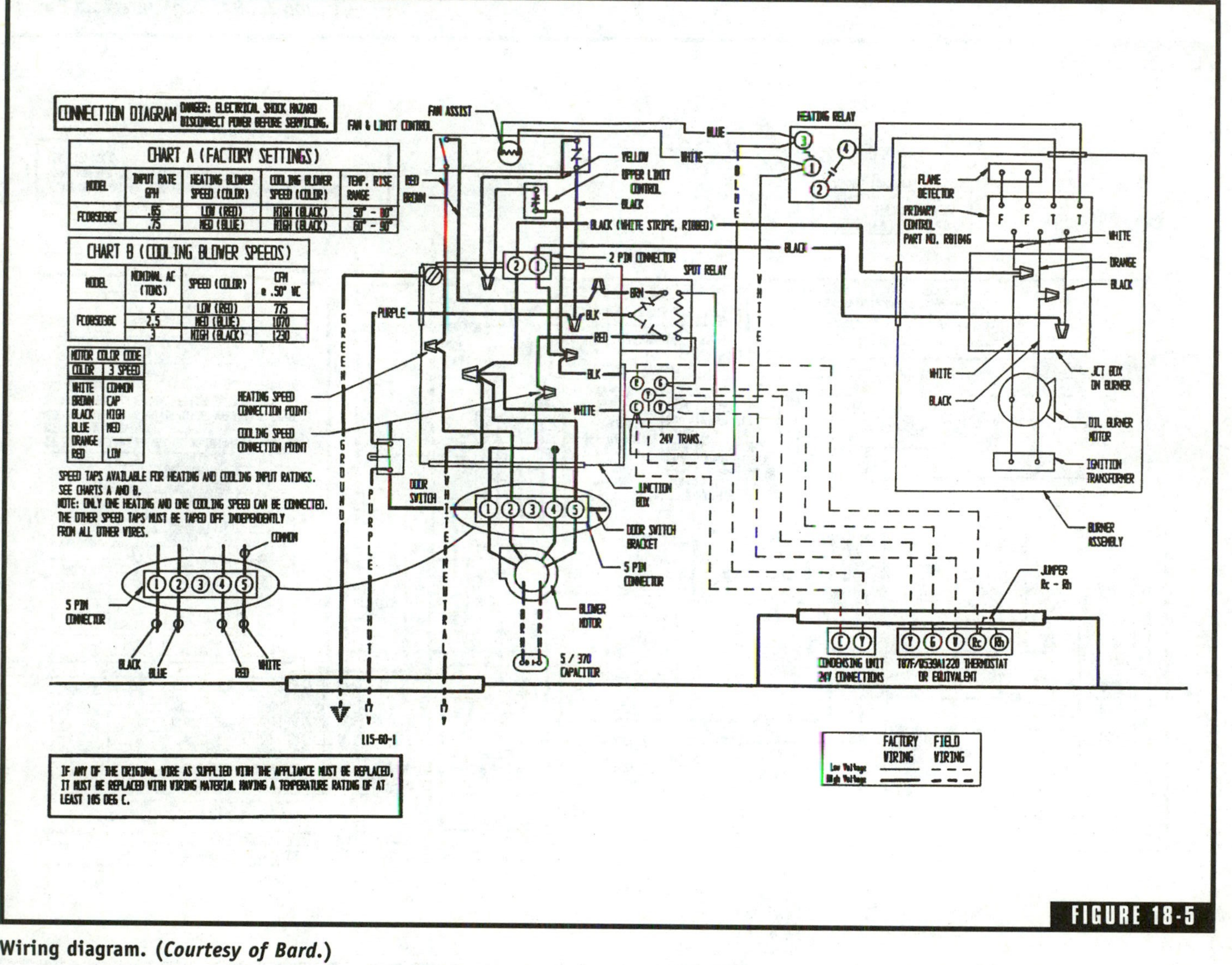

Wiring diagram. (*Courtesy of Bard.*)

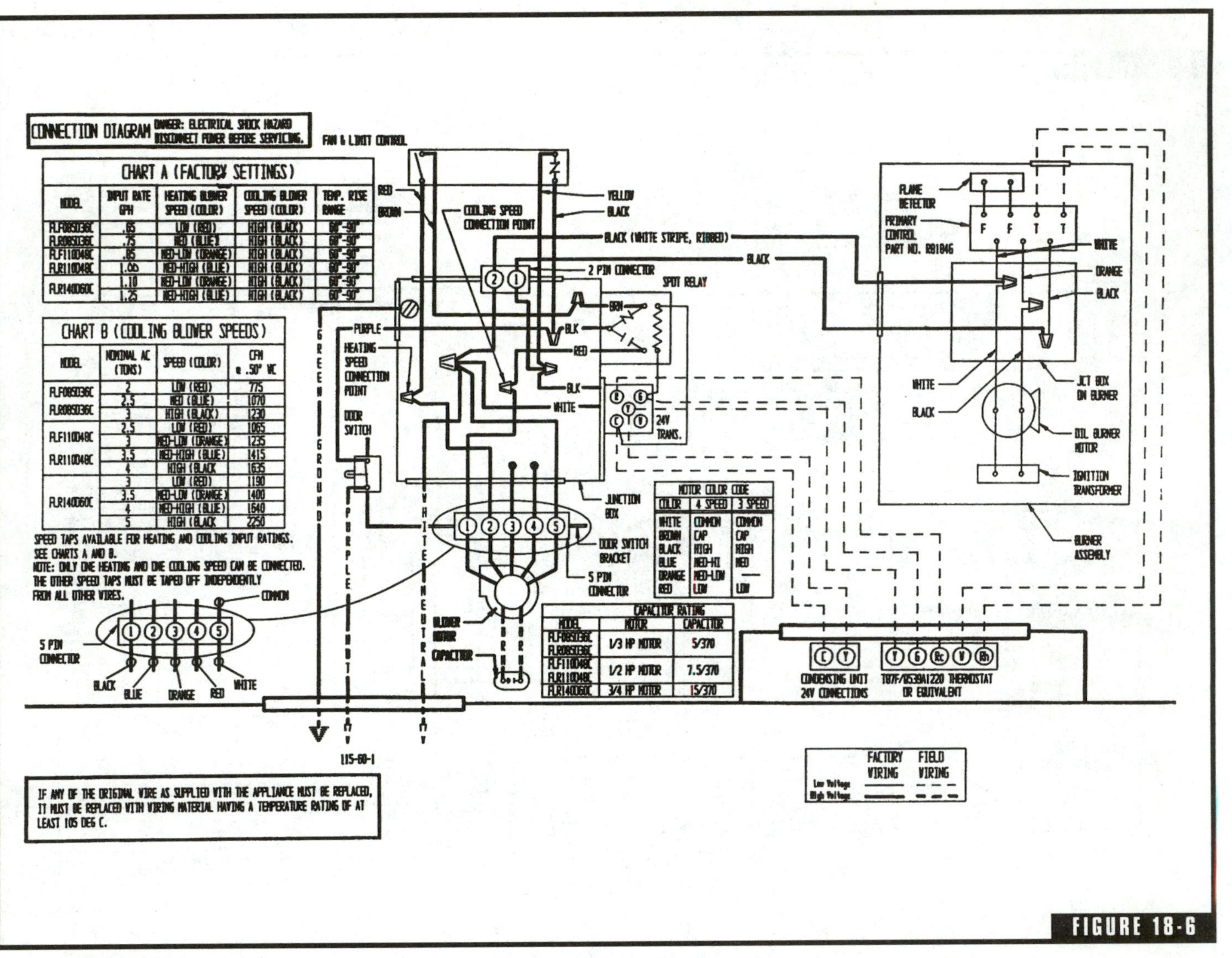

CHART A (FACTORY SETTINGS)

MODEL	INPUT RATE GPH	HEATING BLOWER SPEED (COLOR)	COOLING BLOWER SPEED (COLOR)	TEMP. RISE RANGE
FLF085036C	.65	LOW (RED)	HIGH (BLACK)	60°-90°
FLR085036C	.75	MED (BLUE)	HIGH (BLACK)	60°-90°
FLF110048C	.85	MED-LOW (ORANGE)	HIGH (BLACK)	60°-90°
FLR110048C	1.00	MED-HIGH (BLUE)	HIGH (BLACK)	60°-90°
FLR140060C	1.10	MED-LOW (ORANGE)	HIGH (BLACK)	60°-90°
	1.25	MED-HIGH (BLUE)	HIGH (BLACK)	60°-90°

CHART B (COOLING BLOWER SPEEDS)

MODEL	NOMINAL AC (TONS)	SPEED (COLOR)	CFM @ .50" WC
FLF085036C FLR085036C	2	LOW (RED)	775
	2.5	MED (BLUE)	1070
	3	HIGH (BLACK)	1230
FLF110048C FLR110048C	2.5	LOW (RED)	1065
	3	MED-LOW (ORANGE)	1235
	3.5	MED-HIGH (BLUE)	1415
	4	HIGH (BLACK)	1635
FLR140060C	3	LOW (RED)	1190
	3.5	MED-LOW (ORANGE)	1400
	4	MED-HIGH (BLUE)	1640
	5	HIGH (BLACK)	2250

MOTOR COLOR CODE

COLOR	4 SPEED	3 SPEED
WHITE	COMMON	COMMON
BROWN	CAP	CAP
BLACK	HIGH	HIGH
BLUE	MED-HI	MED
ORANGE	MED-LOW	—
RED	LOW	LOW

CAPACITOR RATING

MODEL	MOTOR	CAPACITOR
FLF085036C FLR085036C	1/3 HP MOTOR	5/370
FLF110048C FLR110048C	1/2 HP MOTOR	7.5/370
FLR140060C	3/4 HP MOTOR	15/370

FIGURE 18-6

Wiring diagram. (*Courtesy of Bard.*)

sating noises, is to install a delayed-opening solenoid on the nozzle line.

There is another consideration. If you are getting a noisy fire and the nozzle replacements don't fix the trouble, check to see where the oil supply tank is located. Assuming that nozzle replacement and a delayed-opening solenoid have not satisfied you, the problem might be cold oil. When an oil tank is installed outside, the oil can get cold enough to create a noisy burner. To remedy this, pump the fuel oil through a nozzle that is the next size smaller at a pressure of about 125 psi. One of these four courses of action should put the burner into good operating condition.

Excess Smoke

Excess smoke can come from a defective nozzle or something as major as a damaged combustion chamber. When you are called to troubleshoot a smoking burner, check first for dirt. If the air-handling parts of the oil burner are dirty, smoke output can be present. The dirt prohibits the burner for functioning properly. A good cleaning may be all that it will take to solve the problem. Assuming that dirt is not the problem, turn your attention to the nozzle in the burner. If it's the wrong size, the nozzle could be the cause of the smoke.

The air-handling parts to check for dirt include the fan blades, air intake, and air vanes in the combustion head. When these components

MINIMUM CLEARANCES--INCHES

	Minimum Installation Clearances								Minimum Service Clearances		
	Furnace			Plenum		(1)	Flue				
Model	Front	Back	Sides	Top	Sides	Duct	Pipe	Floor	Front	Back	Sides
FH085D36C	4	1	2	2	2	2	9	C	24	--	--
FH110D48C	4	1	2	2	2	2	9	C	24	--	--
FH110D60C	4	1	2	2	2	2	9	C	24	--	--
FLF085D36C	6	0	2	2	2	2	9	NC	24	24	18*
FLR085D36C	6	18	2	2	2	2	9	NC	24	24	18*
FLF110D48C	6	0	2	2	2	2	9	NC	24	24	18*
FLR110D48C	6	18	2	2	2	2	9	NC	24	24	18*
FLR140D60C	6	18	2	2	2	2	9	NC	24	24	18*
FC085D36C	6	1	2	2	2	2	9	NC**	24	--	--

(1) For the first three feet from plenum. After 3 feet, no clearance required.
C - combustible flooring NC - non-combustible floor
* Maintained on one side or the other to achieve filter access and/or blower service.
** Floor must be non-combustible. Can be installed on combustible flooring only when installed on special base part No. CFB7 available from factory.

FIGURE 18-7

Minimum clearances. (*Courtesy of Bard.*)

are clean, they allow the system to work much more efficiently. Nozzle problems are not difficult to fix. In the case of a smoking burner, the replacement of an existing nozzle with a smaller one can be all it takes to stop the smoking. Another tactic with the nozzle is to install one that has a narrower spray angle. If the cause of the smoke is a cracked combustion chamber, it's likely that the boiler will need to be replaced. It may be possible to repair the combustion chamber, but if repairs are made, you must make sure that there is no smoke or gas escaping the combustion chamber (Fig. 18-8).

Fuel Odors

Fuel odors can occur with oil burners. There are three basic reasons why oil burners produce odors. Chimney obstructions are one of the first things to consider when you are faced with a burner that is stinking up a mechanical room. If the chimney checks out, the next logical step is to check the ignition time on the burner. If it is delaying, that can account for the buildup of excess fumes. The third possibility is that there is too much air going through the burner.

The draft in a chimney or flue over a fire from a boiler should not be less than .02 to .04. If you do a draft test and find the draft to be below suitable levels, you will have to do a physical inspection of the chimney or flue. Maybe a bird has started building a nest in the flue. It's possible that leaves or other obstructions are blocking the draft. There is a good chance that odors being produced from oil burners will be associated with a poor draft.

Once you know that the chimney or flue is clear, look for delayed ignition. This is a sure cause of odor buildup. There are many factors which may contribute to delayed ignition. A simple test may prove that the electrode setting is not correct. If there are insulator cracks, delayed ignition can result. Dirt, such as soot or oil, can foul an electrode and cause it to fire improperly. When a pump is set with an incorrect pressure setting, you can experience delayed ignition. A bad spray pattern for a nozzle can also be responsible for delayed ignition. If a nozzle becomes clogged, it can cause slow ignition. Another potential cause of delayed ignition is an air shutter that is opened too far.

Electrode settings and nozzles are the main issues to consider. Sometimes replacing an existing nozzle with one that has a solid spray pattern will correct the problem. More likely than not, a simple nozzle

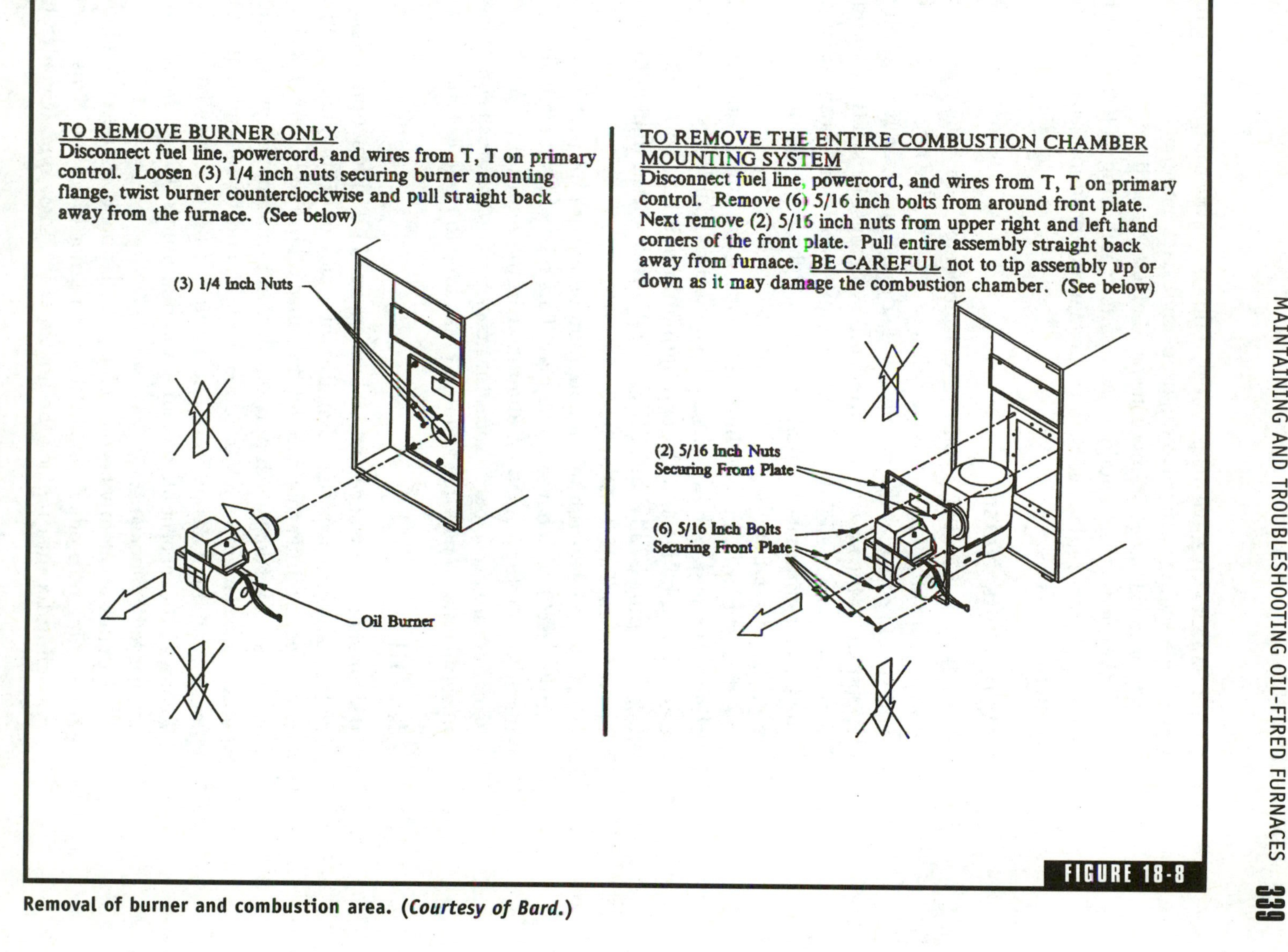

FIGURE 18-8

Removal of burner and combustion area. (*Courtesy of Bard.*)

replacement will solve the problem. If this is not the case, resetting the electrode settings can solve the problem. Check manufacturer suggestions for nozzle angles, flow rates, and electrode settings. If you have a burner that is using a hollow spray pattern and firing 2.00 gallons per hour (gph), try replacing it with a nozzle that produces a solid spray pattern.

More about Noise

We talked earlier about some of the most common causes of noisy operation in an oil burner. In addition to what you've already learned, you should consider the possibility that there may be a bad coupling alignment in the system. To correct this, you must loosen the fuel-unit mounting screws slightly and shift the unit in different positions until the noise is eliminated. If the noise stops, tighten the mounting screws and pat yourself on the back.

If air gets into an inlet line, noise can occur during operation. To eliminate this possibility, check all connections for leaks. All connections should be made with good flare fittings. It may be worth checking the flaring of all tubing. If you find a leak, either tighten the connection or replace it with a new one.

Have you ever run into a tank hum on a two-pipe system with an inside tank? If you haven't, you may. This type of problem is fixed by installing a return-line hum eliminator in the return line. As you can see, there are not a lot of causes for noisy burners, so this is one aspect of service and repair that shouldn't take long to troubleshoot.

No Oil

It's common knowledge that oil burners need oil to function. As basic as this is, sometimes the possibility of a lack of oil is ignored. When you've got a burner that will not fire, check the supply tank to see that it contains an adequate supply of oil. Don't trust the oil gauge; it may be stuck. Use a dip stick to probe the tank and to determine the fuel level. If the tank has oil in it, check the fuel filter. It may be clogged. Remove and inspect all fuel filters and strainers. Clean them if they appear to be blocked.

Nozzles frequently become clogged. Obviously, an obstructed nozzle is going to produce less oil for the burner to fire on. This is easy enough to fix. Just replace the nozzle with a new one. If you've followed all of these steps and still have a problem, look for a leak in the intake line. If you find one, try tightening all fittings where you suspect the leak. Check the unused intake port plug to make sure that it is tight and not leaking. Air can also invade a system around a faulty filter cover or gasket, so check these components to make sure that they are in good shape and are not leaking. When you are checking the intake line for leaks, make sure that there are no kinks in the line which may be obstructing it.

Occasionally, a two-pipe system will become airbound. If this happens, you must check the bypass plug. Insert one if necessary and make sure that the return line is below the oil level in the supply tank. When you have an airbound single-pipe system, you should loosen the port plug or easy-flow valve and bleed oil for about 15 seconds after all foam is gone from the bleed hose. Then check all connections and plugs to make sure that they are tight. Getting the air out is not hard, and it can be all it takes to get a system back up and running.

There are a couple of other things to check if you are having trouble getting oil to the burner. For example, check to see that there are no slipping or broken couplings. If there are, they must be tightened or replaced. It may sound weird, but you may find a situation where the rotation of a motor and fuel unit is not the same as indicated by the arrow on the pad at the top of the unit. Should this happen, install the fuel unit so that the rotation is correct. The last thing to consider is a frozen pump shaft. If you find this to be the problem, the unit must be sent out for approved servicing or factory repair. Before you install a replacement unit, check for water and dirt in the oil tank.

Leaks

Leaks happen in oil systems. The vibration of equipment can create small leaks. There are many reasons why a system may suffer from leakage, but the leaks should not be left unattended. Sometimes removing a plug, applying a new coat of thread sealant, and reinstalling the plug will take care of a leak. When leaks show up at nozzle plugs or near pressure adjustment screws, the washer or O-ring at the connection may need to be replaced. Time can take its toll on both washers and O-rings. The covers on housings sometimes leak. When this happens, you may get by with just tightening the cover screws, but be prepared to replace a defective gasket. If a seal is leaking oil, the fuel unit must be replaced.

Blown seals can occur in both single- and two-pipe systems. A blown seal in a single-pipe system may mean that a bypass plug has been left out. Check it and see. You may have to install one. You may find that you will have to replace the whole fuel unit. Two-pipe systems with blown seals could be affected by kinked tubing or obstructions in the return tubing. If this is not the case, plan on replacing the fuel unit.

Bad Oil Pressure

Bad or low oil pressure might be as simple as a defective gauge. It's worth checking. If the gauge proves to be defective, replace it and you are home free. Assuming that the gauge is reading true, look for a nozzle where the capacity is greater than the fuel unit capacity. Should you find this to be the case, replace the fuel unit with one that will offer an adequate capacity.

Pressure Pulsation

Pressure pulsation may be caused by a partially clogged strainer or filter. When this is the problem, all you have to do is clean the strainer and the filter element. Air leaks are another possible cause for pulsating pressure. If you are experiencing an air leak in an intake line, you should tighten all fittings where leaks may be present. Another place to look for air leaks is the cover of the unit. If you suspect leaks on a cover, tighten the cover screws first and then check for leaks. If leaks persist, remove the cover and install a new cover gasket. Then, of course, replace the cover and tighten the cover screws.

Unwanted Cutoffs

Unwanted cutoffs can be caused by several factors. If you have a burner that is cutting off when it shouldn't be, start by testing the nozzle port with a pressure gauge. Put the gauge in the nozzle port of the fuel unit and run the system for about 1 minute. Then, shut the burner down. If the pressure drops from normal operating pressure and stabilizes, the fuel unit is working properly. This means that air is the culprit of your troubles. But, if the pressure drops to zero, the fuel unit has given up the ghost and needs to be replaced.

Filter leaks can cause unwanted cutoffs. Check the face of the cover and the cover gasket to see if there are any leaks. If there are, replace the gasket and secure the cover tightly. Another thing to check is the cover strainer. This can be fixed by tightening the cover screws of the strainer. Sometimes an air pocket builds up between the cutoff valve and the nozzle. If this happens, run the oil burner while stopping and starting it until all smoke and afterfire disappears.

An unwanted cutoff could be the result of an air leak in the intake line. Check all fittings for possible leaks. Tighten any suspicious connections. Check the unused intake port and return plug to make sure that the plug is tight. If you are still having trouble, look for a partially clogged nozzle strainer. Clean obstructed strainers or replace the nozzle. If there is a leak at the nozzle adapter, change the nozzle and the adapter.

Maintaining and Troubleshooting Gas-Fired Furnaces

Routine maintenance of a gas-fired furnace is important. Any time when gas and flames are in close proximity there is a need for safety precautions. In the case of gas-fired burners, this translates into maintenance. Some of the maintenance required for a gas burner is similar to that required for an oil burner, but there are also differences. Some of the maintenance procedures, such as checking air, can be done by homeowners. Much of the routine maintenance, however, should be performed only by trained professionals. With this in mind, let's run down some of the recommended maintenance for gas burners.

Frequent Inspections

Frequent inspections should be performed on certain parts of a gas heating system. Air filters are top priority. They should be checked every month when the heating system is being used. If the filters are dirty, they should be cleaned or replaced. Many furnaces use filters that are intended to be cleaned, rather than replaced. If permanent filters are meant to be used in a system, do not replace them with disposable fiberglass filters.

Most blower motors are permanently lubricated and need no oiling. A visual inspection of the drain line that is used with condensing furnaces should be checked monthly. If the drain becomes blocked, the furnace will not operate properly. Any obstructions in the drain line should be removed. A visual inspection should also be made of the vent and air intake for the heating system. If any sags, dips, or leaks are found, they must be repaired. The vent and intake should be inspected both inside and outside of the building. If the screens are blocked, clean them.

Annual and Semiannual Inspections

It's a good idea to perform a thorough inspection of gas heating systems at least once a year. Inspecting the system before the heating season and then again once during the heating season is a wise move. Before an inspection is made, you should turn all thermostats off and turn off the gas to the heating system. Wait about 5 minutes to proceed with your inspection. You should also turn off all the electrical power to the heating system.

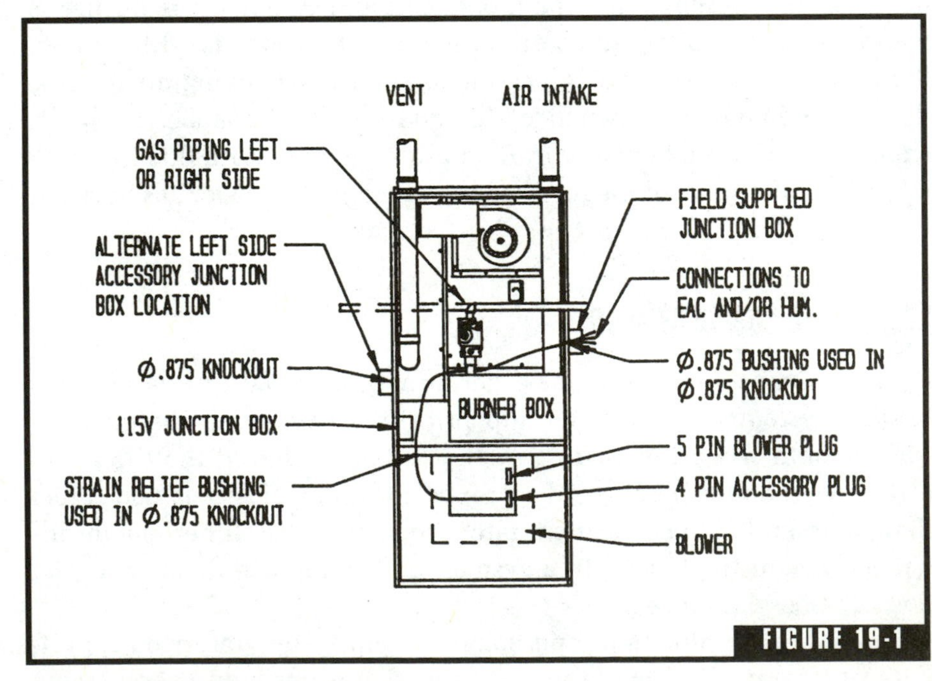

Furnace detail. (*Courtesy of Bard.*)

Air Blower Assembly

The first step in checking out the circulating air blower assembly is to remove the blower service panel. Then loosen the retainers for the blower base. You may have to disconnect pin connectors or wiring from the blower in order to remove it from its housing. Once the blower is removed, you can inspect it and then reinstall it.

Hot Surface Ignitor

The hot surface ignitor should be checked during an annual or semi-annual inspection. Always refer to suggested manufacturer's recommendations whenever possible. The inspection is done with an ohmmeter. You will be checking for resistance. For example, one model of furnace recommends a resistance of 45 to 250 ohms. Anything over 250 ohms on this unit may be indicating a fracture or hairline crack. If this is the case, the ignitor should be replaced.

The Burner

The burner (Fig. 19-1) of a gas-fired unit should always be checked during an inspection. Look for any scaling, sooting, or blockage of the ports. If any of these conditions exist, the burner should be removed and cleaned with a wire brush. Remember to remove the hot surface ignitor before removing the burner, to avoid damage to the ignitor. The grommet that seals a gas manifold vertical riser to the top panel of a burner box is a critical seal. If this type of seal is removed, the reinstallation of it is crucial. If the seal is not reinstalled just as it was removed, the heating system may fail and could cause property damage or bodily injury.

Heat Exchanger

The heat exchanger and flue gas passageways also need to be inspected on at least an annual basis. The heat exchanger should not require any cleaning. The exception to this would be a heat exchanger in a system where sooting is occurring. To look for sooting, you can inspect the heat exchanger with a light and a mirror

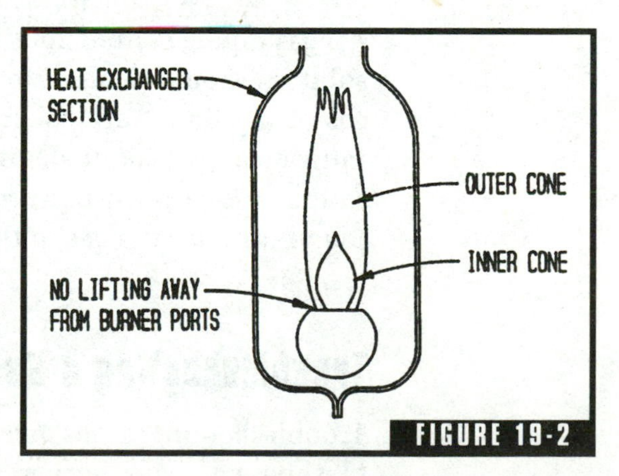

FIGURE 19-2

Example of a proper flame. (*Courtesy of Bard.*)

that is attached to an extension handle. Should soot or scale be found, the heat exchanger must be cleaned. In most cases, cleaning will not be necessary.

Check the Flame

When conducting a routine inspection, you should check the flame in the burner (Fig. 19-2). A strong blue flame should be present. It's okay if the flame has a little orange around its edges, but there should not be any yellow in or around the flame. All of the flame should be in the center of the heat exchanger compartment. The flame should never be impinging on the heat exchanger surfaces. You should watch the flame until the blower starts. Expect a short delay while waiting for the burner to start. There should be no change in the size or shape of the flame when the burner starts. If you notice the flame wavering or blowing in any way, you may have discovered a heat exchanger that is leaking.

The flame sensor should run at a current of a minimum of 0.20 microamp for the furnaces that I'm most familiar with. Of course, you should always rely on detailed information provided by the manufacturer of equipment on which you are working. In the case of the furnaces that I use most, the flame sensor should be checked if the burner drops out at the end of a 9-second flame proving period. It's possible that the sensor rod is coated with some type of airborne foreign substance, such as silicone.

Gas Piping

All gas piping connections should be checked for leaks. A soapy water solution should be used to test for leaks. Apply the soapy water to all gas connections and look for bubbles. Any bubbles in the test solution indicate a gas leak. It should go without saying, but obviously the gas to the system must be turned on for this type of testing. If a leak is found, turn off the gas to the system and repair the leak as needed.

Troubleshooting a Gas Burner

Troubleshooting a gas burner for a furnace is very similar to troubleshooting a gas burner that is used with a boiler. We've already talked about troubleshooting for boilers. Much of that information can be applied to furnaces. Since the procedures are so similar, the material that follows is essentially the same as what was provided for boilers. I'm repeating it here to make this chapter a standalone resource for troubleshooting gas furnaces (Fig. 19-3).

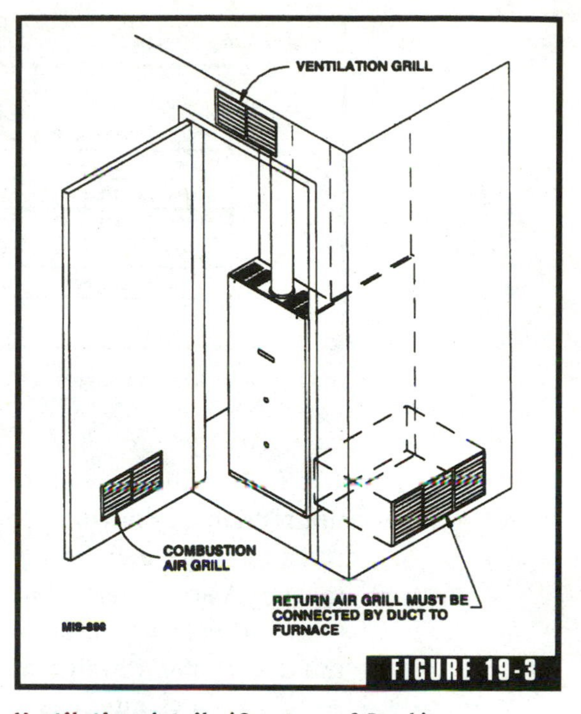

FIGURE 19-3

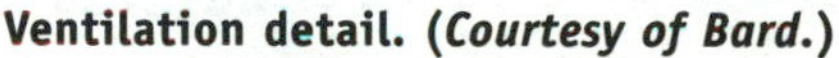
Ventilation detail. (*Courtesy of Bard.*)

Standard Safety Measures

There are some standard safety measures to take into consideration whenever you are working with a gas-fired heating system. The first rule is to pay attention to any odors. If you smell any appreciable amount of gas or other strange odors, ventilate the area before working in it. Avoid electrical devices, if possible. Open windows when you can. If gas accumulates heavily, it can explode. The quicker you can rid the area of fumes, the safer you will be.

Always shut off the gas supply to any device you are planning to work on. Cut the gas valve off and wait several minutes before proceeding with your work. If there are multiple cutoff valves, close all of them. Wait at least 5 minutes to give residual gas time to vacate the system. You may already know, but if you don't, be advised that LP gas is heavier than air and will not dissipate upwardly in a natural fashion. LP gas lingers below air, gathering in low areas and creating a possibility for an explosion risk. In minute quantities, this is not a problem, but it is a fact that any service technician should be aware of.

If you will be working with or near electrical wiring, you should turn off the power supply. This may mean removing a fuse, but in most homes it is as simple as flipping a circuit breaker. There may also be an individual disconnect box near the heating unit. If this is the case, you can turn off the power by throwing the switch on the disconnect box (Figs. 19-4 to 19-9).

TABLE 8—AMPACITY RATING

MODEL	VOLTS/HZ/PH	TOTAL AMPS	BLOWER MOTOR		INDUCER MOTOR		MINIMUM CIRCUIT AMPACITY	MAXIMUM TIME DELAY FUSE OR HACR CIRCUIT BREAKER
			HP	FLA	HP	FLA		
TU042	115/60/1	9.25	1/3	4.5	1/40	1.0	15	15 AMP
TU063	115/60/1	11.3	1/3	6.5	1/40	1.0	15	15 AMP
TU084	115/60/1	15.3	1/2	10.5	1/40	1.0	15	15 AMP
TU105	115/60/1	17.3	3/4	12.5	1/40	1.0	16	20 AMP
TU126	115/60/1	17.3	3/4	12.5	1/40	1.0	16	20 AMP

FIGURE 19-4

Ampacity ratings. (*Courtesy of Bard.*)

If you will be working with electrical matters, don't attempt to jump or short the valve coil terminals on a 24-volt control. Doing this can short out the valve coil or burn out the heat anticipator in the thermostat. Also, never connect millivoltage controls to line voltage or to a transformer. If you do, you may be burning out the valve operator or the thermostat anticipator.

If you have any reason to suspect a gas leak, check all connections where leaks may be possible. Don't use any type of flame to test for leaks. I've known plumbers who tested joints on natural-gas piping with their torch flames, but don't do this, and, certainly, never do it with LP gas. Natural gas usually just burns in a leak situation, but LP gas is likely to explode. I say again, don't search for leaks with any type of flame. Soapy water is the right substance to use when looking for leaks in gas piping. Obviously, to find a gas leak the gas supply must be turned on.

When in search of a leak, ventilate the area by opening windows and other ventilating openings. With the gas on, apply a generous quantity of soapy water to all gas connections. When the water is applied to a connection where a leak exists, the escaping gas will create bubbles in the soapy solution. Take note of the leak and continue testing the rest of the gas connections. Once you have checked all connections, you can turn off the gas supply and fix the leaks. When fixing leaks, don't take shortcuts. Treat each leak with respect. Loosen all leaking joints, apply thread sealant as required, and tighten the joints. Sometimes just tightening a fitting is enough. Regardless of what you have to do to correct a leak, always test the connection carefully after it is repaired. Test the joint just the way you did to find the leak in the first place.

Never bend the pilot tubing at the control once the compression nut has been tightened. If the tubing is bent after the compression nut is tightened, a leak can result. Compression nuts don't take well to vibration and manipulation. Any substantial movement of a compression fitting after it is tightened can create a leak. Compression fittings are fine when they are installed and treated properly.

The Troubleshooting Process

There are only about nine types of failures that are typically encountered with gas-fired burners. While there are not a lot of potential problems, there are several possible solutions. But, using a methodical plan during the troubleshooting process makes the job easier. Experience is the best teacher, in many ways, but a good checklist of what to do and in what order to do it is the next best thing. And, that's what I'm about to give you. So, with the safety issues covered, let's move into the cause-and-repair section to give you some guidance in finding and fixing problems with gas-fired burners.

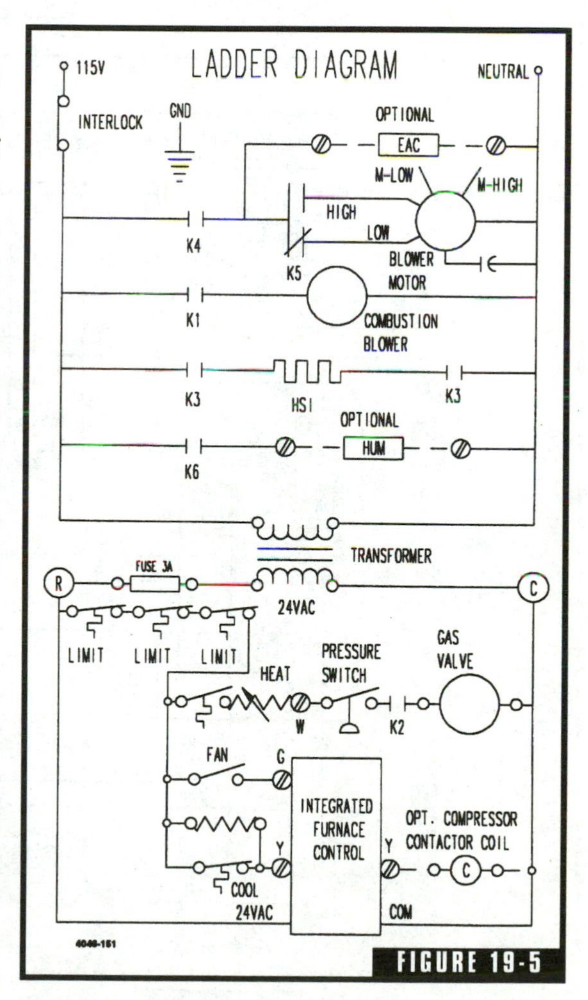

FIGURE 19-5

Wiring diagram. (*Courtesy of Bard.*)

No Pilot Light

When a gas-fired burner fails to operate, it could be due to the lack of a functioning pilot light. There are four basic reasons why this problem might exist. First, check to see if there might be air in the gas line. Open the connections and purge any air that may be trapped in the tubing or pip-

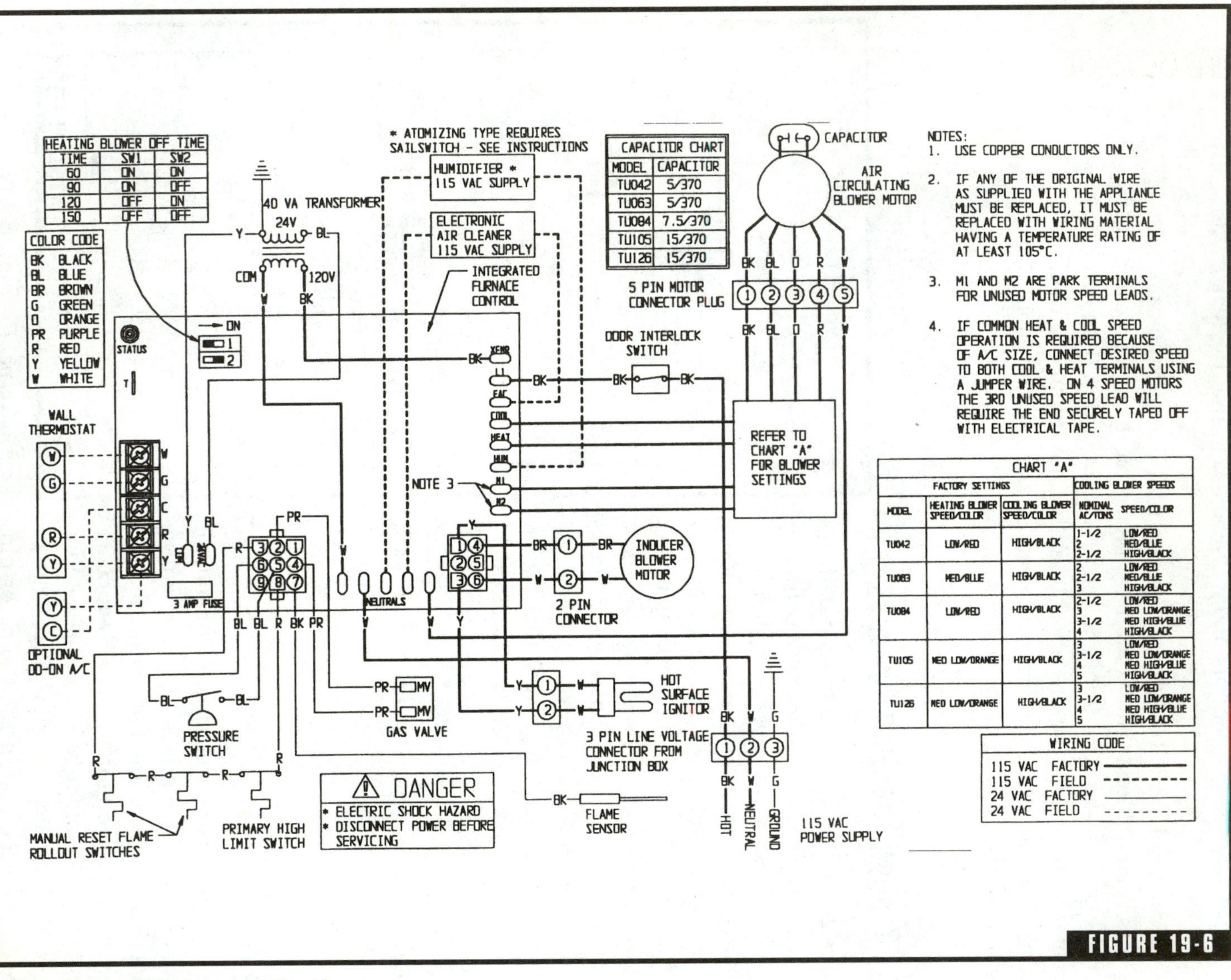

FIGURE 19-6

Wiring diagram. (*Courtesy of Bard.*)

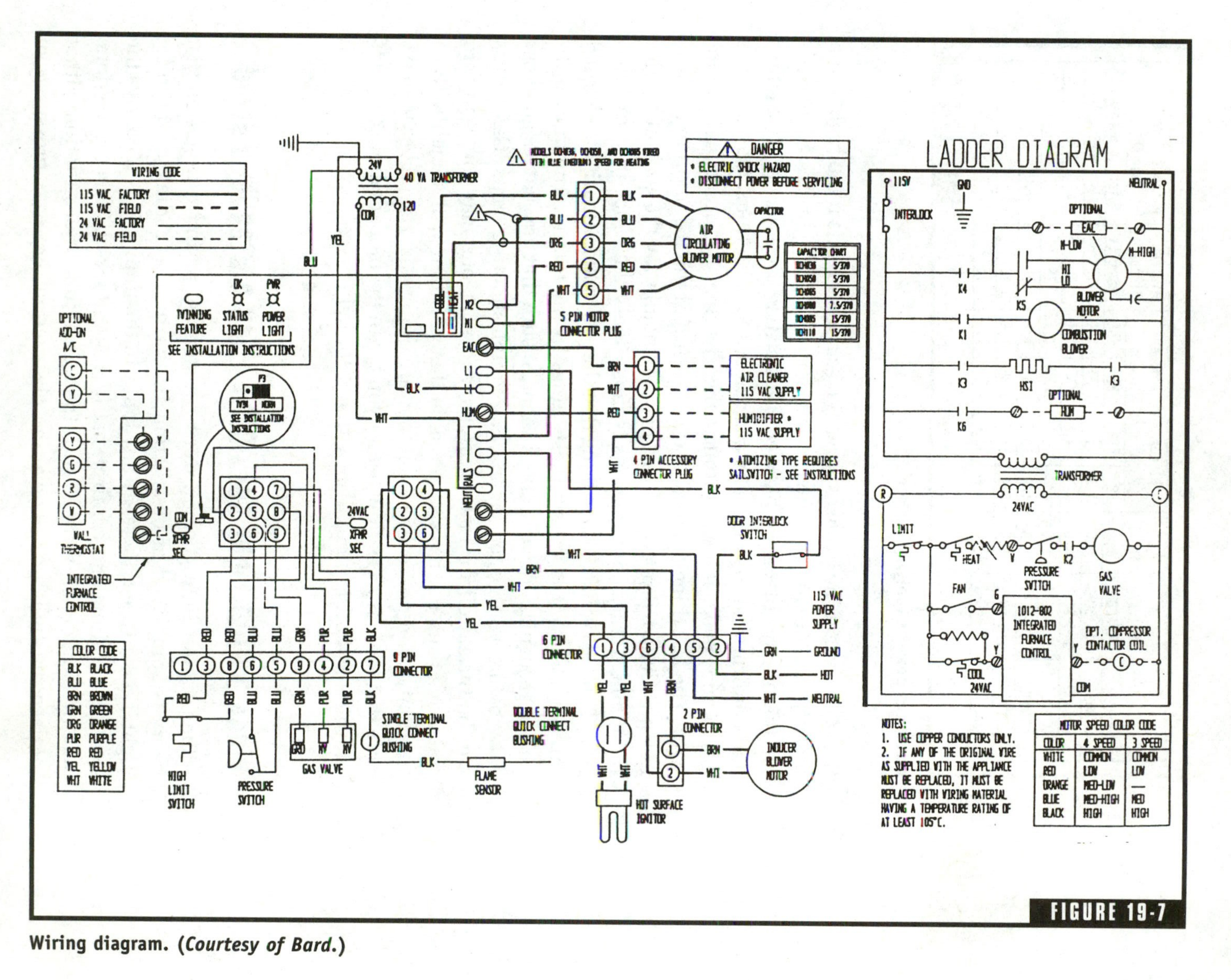

Wiring diagram. (*Courtesy of Bard.*)

Model	Volts/ HZ/PH	Total Amps	Blower Motor		Inducer Motor		Minimum Circuit Ampacity	Minimum Time Delay Fuse OR HACR Circuit Breaker
			HP	FLA	HP	FLA		
DCH036	115-60-1	6.0	1/3	4.5	1/35	1.4	15	15
DCH050	115-60-1	6.0	1/3	4.5	1/35	1.4	15	15
DCH065	115-60-1	6.5	1/3	6.5	1/35	1.4	15	15
DCH080	115-60-1	10.5	1/2	10.5	1/35	1.4	15	20
DCH095	115-60-1	12.5	3/4	12.5	1/35	1.4	16	20
DCH110	115-60-1	11.0	3/4	11.0	1/35	1.4	15	20

FIGURE 19-8

Electrical power supply. (*Courtesy of Bard.*)

ing. A second thing to check for is the gas pressure. Is it too high or too low? Either situation can result in a failing pilot light. It's also possible that the problem is a blocked pilot orifice. Do a visual inspection and confirm that the orifice is clear of any obstructions. Obviously, if gas cannot make its way through the opening, it cannot fuel a pilot light. The last thing to check is the position of the flame runner. If it is not positioned properly, it can be responsible for the lack of a pilot light. One of these four reasons is most likely the cause of your problem. This, of course, is assuming that the gas to the pilot light is turned on. Sometimes it is the simplest things in life that are the most difficult to recognize. Always check to make sure that you have a good gas flow before you go too far in your troubleshooting steps. Gas valves can be turned off for many reasons, and someone may forget to turn the valve back on. It's also possible that children will somehow turn off a valve and not realize what they are doing. Just remember to make sure that there is a gas flow before you begin the more complex aspects of troubleshooting a problem with a pilot light.

When a Pilot Light Goes Off during a Standby Period

When a pilot light goes off during a standby period it might be that there is some type of obstruction in the tubing leading to the pilot light. Look to see if there are any kinks in the tubing. If not, remove the section of tubing and confirm that it is not blocked. This can be done just by blowing air through it. If there is a kink or obstruction, either replace the tubing or remove the obstruction. Low gas pressure can also cause a pilot light to go off during standby. Check the gas pressure and confirm that it is within acceptable levels for the equipment that you are working on.

You already know that a blocked pilot orifice can prevent a pilot light from burning. But, did you know that a blocked orifice can also be responsible for a pilot light that goes off during standby? It can, so check the orifice if your problem persists. If there is a loose thermocouple on 100 percent shutoff, a pilot light can go off during a standby period. It's also possible that the thermocouple is defective. When a pilot safety is not working properly, it can cause a pilot light to go off during standby. Another consideration to check is the draft condition. If there is a poor draft, a pilot light may not stay lit. A final possibility is a draft tube that is set into or flush with the inner wall of the combustion chamber. These same seven reasons for failure can apply to a safety switch that needs frequent resetting.

If a Pilot Light Goes Off When the Motor Starts

If a pilot light goes off when the motor of a unit starts, there are only three probable causes. The first one to look into is the possibility of some type of restriction or obstruction in the tubing to the pilot light. You need to loosen the connection on the tubing and confirm gas flow. If the tubing is blocked or kinked and cannot be repaired, replace the tubing. Once again, gas pressure could be at fault. If the pressure is too high or too low, the pilot light function can be affected. Confirm that the gas pressure is within suitable limits for the equipment that you are working with. A final consideration is the possibility that there may be a substantial pressure drop in the gas piping when the main gas valve opens. Putting the equipment through cycles until you can observe what happens as the main gas valve opens might be necessary.

Motor Won't Run

The first thing to check when you encounter a burner with a motor that will not run is the electrical circuit. If there is a local disconnect box, check it first to see that it has not been turned off. Next, go to the

MODEL DCH080	Orifice Size Chart									
Fuel Gas Type	Gas Heat Value Btu/Cu. Ft. *	0 to 2000 Feet	2001 to 3000 Feet	3001 to 4000 Feet	4001 to 5000 Feet	5001 to 6000 Feet	6001 to 7000 Feet	7001 to 8000 Feet	8001 to 9000 Feet	9001 to 10,000 Feet
Natural	800 - 849	2.75mm	#37	2.60mm	#38	#39	2.50mm	2.45mm	2.40mm	2.35mm
	850 - 899	#37	#39	#40	2.45mm	2.40mm	#42	2.35mm	2.30mm	2.25mm
	900 - 949	2.60mm	2.50mm	2.45mm	#41	2.40mm	2.35mm	2.30mm	#43	2.20mm
	950 - 999	#39	#41	2.40mm	#42	2.30mm	2.30mm	2.25mm	2.20mm	2.15mm
	1000 - 1049**	2.45mm	#42	2.35mm	2.30mm	#43	2.20mm	#44	2.10mm	#45
	1050 - 1100	2.40mm	2.30mm	#43	2.25mm	2.20mm	#44	2.10mm	2.10mm	2.05mm
Propane (LP)	2500***	#53	1.45mm	1.45mm	1.40mm	1.40mm	1.35mm	1.35mm	1.30mm	1.30mm

MODEL DCH095	Orifice Size Chart									
Fuel Gas Type	Gas Heat Value Btu/Cu. Ft. *	0 to 2000 Feet	2001 to 3000 Feet	3001 to 4000 Feet	4001 to 5000 Feet	5001 to 6000 Feet	6001 to 7000 Feet	7001 to 8000 Feet	8001 to 9000 Feet	9001 to 10,000 Feet
Natural	800 - 849	3.00mm	#32	2.90mm	#34	2.80mm	2.75mm	#36	#37	#38
	850 - 899	2.90mm	2.80mm	7/64"	2.75mm	2.70mm	#37	#38	#39	#40
	900 - 949	#34	2.75mm	2.70mm	#37	2.60mm	#38	#39	#40	#41
	950 - 999	2.75mm	#37	2.60mm	#38	#39	2.50mm	2.45mm	2.40mm	2.35mm
	1000 - 1049**	2.70mm	2.60mm	#38	#39	#40	#41	2.40mm	2.35mm	2.30mm
	1050 - 1100	2.60mm	2.50mm	2.45mm	#41	2.40mm	2.35mm	2.30mm	#43	2.20mm
Propane (LP)	2500***	1.65mm	1/16"	1.55mm	1.55mm	#53	1.50mm	1.45mm	1.45mm	1.40mm

MODEL DCH110	Orifice Size Chart									
Fuel Gas Type	Gas Heat Value Btu/Cu. Ft. *	0 to 2000 Feet	2001 to 3000 Feet	3001 to 4000 Feet	4001 to 5000 Feet	5001 to 6000 Feet	6001 to 7000 Feet	7001 to 8000 Feet	8001 to 9000 Feet	9001 to 10,000 Feet
Natural	800 - 849	#35	2.70mm	#37	2.60mm	#38	#39	#40	#41	#42
	850 - 899	#36	2.60mm	#38	#39	2.50mm	2.45mm	2.40mm	2.35mm	2.30mm
	900 - 949	#37	#38	2.50mm	#40	#41	2.40mm	2.35mm	2.30mm	2.25mm
	950 - 999	#38	#40	2.45mm	2.40mm	#42	2.35mm	2.30mm	2.25mm	2.20mm
	1000 - 1049**	2.50mm	2.40mm	#42	2.35mm	2.30mm	#43	2.25mm	#44	2.15mm
	1050 - 1100	2.45mm	#42	2.35mm	2.30mm	#43	2.20mm	#44	2.10mm	#45
Propane (LP)	2500***	#53	1.45mm	1.45mm	1.40mm	1.40mm	1.35mm	1.35mm	1.30mm	1.30mm

* At standard conditions: Sea level pressure and 60°F temperature.

** Standard factory supplied orifice size.

*** BTU/cu. ft. at 60°F temperature.

ORIFICE DRILL SIZE DECIMAL EQUIVALENTS

Drill No.	3.50mm	#29	3.40mm	3.30mm	#30	3.25mm	3.20mm	1/8"	3.10mm	#31	3.00mm	#32	2.90mm	#33
Decimal	.1378	.1360	.1339	.1299	.1285	.1279	.1260	.1250	.1221	.1200	.1181	.1160	.1142	.1130
Drill No.	#34	2.80mm	#35	7/64"	2.75mm	#36	2.70mm	#37	2.60mm	#38	#39	2.50mm	#40	2.45mm
Decimal	.1110	.1102	.1100	.1094	.1082	.1065	.1063	.1040	.1024	.1015	.0995	.0984	.0980	.0964
Drill No.	#41	2.40mm	3/32"	#42	2.35mm	2.30mm	#43	2.25mm	2.20mm	#44	2.15mm	2.10mm	#45	#46
Decimal	.0960	.0945	.0938	.0935	.0925	.0906	.0890	.0885	.0866	.0860	.0846	.0827	.0820	.0810
Drill No.	2.05mm	2.00mm	#47	5/64"	1.95mm	#48	1.90mm	#49	1.85mm	1.80mm	#50	1.75mm	#51	1.70mm
Decimal	.0807	.0787	.0785	.0781	.0767	.0760	.0748	.0730	.0728	.0709	.0700	.0688	.0670	.0669
Drill No.	1.65mm	#52	1.60mm	1/16"	1.55mm	#53	1.50mm	1.45mm	1.40mm	#54	1.35mm	#55	1.30mm	1.25mm
Decimal	.0649	.0635	.0630	.0625	.0610	.0595	.0590	.0570	.0551	.0550	.0531	.0520	.0512	.0492

FIGURE 19-9

Orifice sizing chart. (*Courtesy of Bard.*)

electrical panel for the home or building and check all fuses or circuit breakers. If all appears to be in order, use your electrical meter to confirm that power is reaching the motor. Also make sure that the wiring is connected to the controls properly. Before you get too involved, move around the building and check the thermostats. Make sure that they are turned up to a sufficient temperature to be calling for heat. A mechanic can feel quite foolish after spending considerable time performing major troubleshooting steps and then finding out that the thermostats were turned down for some reason. Assuming that the thermostats are calling for heat, you must check to see that the thermostats and limit controls are not defective or calibrated improperly.

Sometimes the bearings in a motor will seize up, due to a lack of lubrication. See if this might be the case on your job. Try turning the equipment. If it will not turn freely, you have probably identified your problem. In a worst-case scenario, you are dealing with a motor that has burned out, which will, of course, require replacement.

It Runs But Doesn't Heat

If you have a burner that runs but doesn't heat, check to see if the pilot light is burning. If the pilot is out, relight it. You may discover that the root of the problem lies within a defective pilot safety control. This could be as simple as resetting the safety or it could require replacement of the control. Another possible cause for the problem of a motor that runs with no flame is that the thermocouple might not be generating enough voltage for the system. The causes are limited, so the troubleshooting should go quickly. Gas pressure has a large effect on a gas-fired system. Check the pressure to make sure that it is within recommended working parameters. If it is, and you still have the problem, check to see if maybe the motor is running slower than it should be. This could be the cause of the motor running without a flame.

A Short Flame

A short flame in a gas-fired burner is not good. It might be caused by any one of about five possible problems. The place to start is with the pressure regulator. If it is set too low, the flame will be short. Also, check the air shutter. An air shutter that is open too wide can cause a short flame to burn in the equipment. Any major pressure drop in the gas supply line could also be responsible for the short flame. A defective regulator is always a possibility when you are looking for the cause of a short flame. And, a vent that is plugged in the regulator can also cause the problem of a short flame. Some trial-and-error techniques will be needed, but one of the causes listed above is almost certain to be the reason for a short flame.

Long Flames

Long flames in gas-fired equipment are usually limited to one of three causes. The first thing to check for is an air shutter that is not open far enough. When it is open too far, the flame will burn short, but if not open far enough, the flame will be long. The air shutter must be balanced to produce a flame of a desired length. If there are obstructions in air openings or at the blower wheel, the result can be a long flame. Too much input to the flame source will also produce a long flame. Since the causes for long flames are few in number and easy to check for, this problem is not very time consuming when it comes to troubleshooting it.

A Gas Leak at the Regulator Vent

A gas leak at the regulator vent is a sign of a damaged diaphragm. The diaphragm most likely has a hole in it. In any case, the damaged diaphragm must be replaced. Gas cannot be allowed to leak from any part of a system. In such a case, all attention should be focused on the regulator vent, since that is where the problem is.

Won't Close

Sometimes the main gas valve on a unit will not close when the blower stops. This is a problem that is usually caused by a defective valve or some type of obstruction on the valve seat. You can inspect the seat for damage or buildup, but be prepared to replace the entire valve. It is unsafe to have a gas valve that does not close properly when it should.

Be Careful

When you are working with gas, whether natural or manufactured, you must be careful. The fuel is volatile. It is said that familiarity breeds contempt. Don't allow this to be the case in your work. If you make a mistake with an oil-fired burner you might only get oil on a floor or incur some odors. A mistake with a gas system could result in a major explosion. Pay attention to what you are doing and stay focused on your task. Make sure anyone who might enter your work area is aware of what you are doing and what your needs are. For example, the last thing you need is for a casual observer to come in and fire up a cigarette lighter when you are bleeding off some gas. Stay sharp and stay safe.

CHAPTER 20

A Basic Overview of Solar Heating Systems

Solar heating systems are a great concept, and they can work very effectively. However, the cost of the systems and the regional climatic conditions often outweigh the advantages of heating systems powered by the sun. A number of considerations must be taken into account when thinking of a solar heating system. For example, some subdivisions prohibit the use of solar panels, due to their unconventional appearance. Cost is a big factor to consider, and so is the amount of time during winter days when a solar system can effectively collect energy. Simply put, solar systems are not for everyone, but they can be very good under the proper conditions.

There are many types of solar heating systems available. When they were first introduced in mass marketing, the systems gained a good bit of attention. However, over the past 10 years, or so, interest in solar heating systems seems to have fallen off. I suspect part of the reason for this is the high cost of creating complex systems. Another reason for the decreased interest in solar heating may be lower costs for fuel oil or the difficulty of maintaining quality heat with a solar heating systems in some parts of the country.

The late 1970s and the 1980s were the solar heating system's spotlight. This is when people were turning to solar systems to overcome highly priced fuel oil that was not guaranteed to remain available. As the oil situation leveled out, so did the interest in solar heating systems. During the high times of active solar heating systems much was learned about design issues, effectiveness, and so forth. A great number of system designs failed, were too expensive, or presented extreme servicing problems. Many systems were very complex in their design. Through more than a decade of evolution solar heating systems have had their ups and downs.

The cost of creating a solar heating system has been a problem that still exists. It just isn't cheap to build a working solar heating system. And, with today's lower cost of fuel oil, recovering the high installation cost of a solar heating system takes a long time. This situation influences builders and customers to opt for more traditional heating systems. Through all of the good and bad experience gained with solar heating systems, one type of solar system remains favored; it is the flat plate solar collector.

Flat Plate Solar Collectors

Flat plate solar collectors are one remaining type of active solar heating system that continues to have its followers. Throughout most of the United States the flat plate solar collector has proved to be one of the most cost-effective types of solar heating system. Availability of these collectors is also good. When these collectors are used, they can be grouped together as components to make almost any type of collector arrangement. The collectors are most often installed on the roof of a home and face in a southerly direction. However, the collectors can be mounted at ground level and still function.

The positioning of solar collector panels is crucial to the effectiveness of a heating system. Typically, the collectors should have a slope angle that is equal to the local latitude, plus an additional 10 to 15°. This type of setting generally maximizes the winter heat output. Roofs are a preferred mounting place for solar collectors. There are a couple of reasons for this. First, it is generally less expensive to mount the panels on a roof, since independent racks or holders are not needed, as they are with ground-mounted systems. Protection from damage is another reason why roofs are a preferred location. When collector panels are set at ground level, they are more likely to fall victim to vandalism or simple accidents which may break the panels.

Storing solar energy that is collected during the day is a major aspect of a successful solar heating system. Since more heat is needed at night, when there is no sunlight available for collection, the heat collected during the day must be stored for use at night. There are different methods used for this purpose. Some homes are designed with mass or heat storage elements that will absorb heat and then radiate it for several hours after the sun is no longer available. Concrete floors covered with thick tiles are one way of creating a storage mass for radiant heat. This, however, is not the same as storing energy to produce more even heat throughout a home.

To accomplish energy storage in larger amounts, water-filled thermal storage tanks are used. The tanks absorb energy during the day and retain it for use during the night. Circulating water through collectors and storing it is a common means of producing heat storage. In some cases, antifreeze solutions are circulated between the collectors and a heat exchanger that is coupled to the storage tank.

Installation Methods

Just as there are many types of solar heating systems available, so are there different ways to install various systems. There are four primary methods of installation. One type of system is the direct circulation system. Another type is a draindown system. Or, you might find that a gravity drainback system is more appropriate. The fourth type of system is the closed-loop system. To further our discussion on these four main systems, let's look at them individually.

Direct Circulation System

A direct circulation system offers many advantages. When a direct circulation system is used, there is no need for a heat exchanger between the solar collector and the storage tank. Since a direct circulation system is moving, it does not require an antifreeze solution. This also saves money. Other elements that are required with a closed-loop system that are eliminated with a direct circulation system include an expansion tank, a pressure relief valve, and air purging devices. The elimination of so many components naturally makes this type of system less expensive to install. When you eliminate system components, you are not only saving money on the cost of an installation, you are reducing the number of potential problems that might occur after installation.

When a heat exchanger is used between a storage tank and a collector, some heat is lost. This is not the case with a direct circulation system. The amount of heat, in degrees, that may be lost in a closed-loop system can be 10 to 15° more than what is lost with a direct circulation system. This is a significant amount of heat. Heat loss from collectors working with a closed-loop system reduce the effectiveness of a heating system. The facts that direct circulation systems cost less to install, work more efficiently, and reduce the components of a system which may require service all join together to make direct circulation systems a wise choice.

There is one drawback to a direct circulation system. Since these systems don't incorporate the use of antifreeze solutions, they must be protected from freezing temperatures. Any part of the circulation system that is not in heated space must be otherwise protected from freezing. To do this, pipes in the system are drained of all water prior to freezing conditions. The need for freeze protection is the only real negative aspect of a direct circulation system.

Draindown Systems

Draindown systems work automatically to protect circulation systems from freezing. Electronic controls are installed to facilitate the draining process. When sensors associated with the electronic controls sense temperatures near the freezing point, a command is sent to motorized valves to open them. When the valves open, the system is drained in the sections where drainage is required to prevent freezing.

When collectors and piping are installed with a draindown system, the piping and the collectors must be installed with a minimum grade of 1/4 inch to the foot of downslope for drainage. While a draindown system may seem convenient, it is not without its risk. Any failure of operation within the system can result in frozen pipes and collectors, which will shut down the system and may require costly replacements of damaged parts that were swollen or ruptured during the freezing.

Since draindown systems use fresh water when they refill after a draindown, the systems must be piped as open-loop systems. Since fresh, oxygenated water is allowed into the system after a draindown, it is not practical to use cast-iron or steel components with the system. Rust created from the oxygen-filled water will damage such components.

Gravity Drainback Systems

Some systems rely on gravity for drainage. In a gravity system, the pipes and collectors are drained automatically, by natural gravity, when the circulating pump stops. This type of system eliminates the needs for motorized valves and sensors that may fail to function properly. By eliminating the valves and electronics, the cost of a gravity drainback system is less than that of a draindown system. Just as is required with a draindown system, a drainback system must have all piping installed with a minimum of 1/4-inch-per-foot drop on the pipes for proper drainage.

A drainback system does not introduce new water into the system. Existing water is reused each time the circulator runs. When designing this type of system, you must be sure that the circulating pump is capable of pumping water to the highest point of the system. Gravity drainback systems are fairly simple and quite effective.

The concept of a gravity drainback system requires water to drain down whenever the circulating pump stops. Air enters the system to replace the water that is drained down. The air enters the system through an open tee that is located above the water level in the tank. It's possible to size the tank so that it works as an expansion tank for the collector loop and the distribution system.

Return piping from collectors to a tank should be sized to accept a minimum flow velocity of 2 feet per second. When this is the case, water returning from the collectors can entrain air bubbles and return them to the storage tank. The result is a pipe that becomes something of a siphon. When the siphon is created, it minimizes gurgling noises in the piping and increases the rate of water flow. By increasing the speed with which water is flowing through a collector, the efficiency of the collector is improved.

Gravity drainback systems are liked because of their simplicity. By not relying on electronics and motorized valves, a gravity system is typically more dependable and produces fewer service problems. The ability of this type of system to work almost flawlessly is advantageous in preventing frozen pipes and collectors. Another big advantage that a drainback system has over a draindown system is the fact that new water is not introduced into the system. This means that less-expensive components, such as cast-iron circulators, can be used. A direct circulation system with a gravity drainback system is generally considered the best type of active solar heating system to install.

Closed-Loop Antifreeze Systems

Closed-loop antifreeze systems require an external heat exchanger. The heat exchanger links the collectors to the storage tank. Some of this type of system is exposed to freezing temperatures at times and, therefore, must be protected from freezing. This is done by filling the system with antifreeze. A glycol-based antifreeze solution is used most often. The antifreeze solution is what transports heat from the collectors to the heat exchanger. Water is moved between a storage tank and the heat exchanger with a circulating pump. When solar energy is being collected, two circulating pumps are running. Since all pipes and collectors are filled with antifreeze, they can be installed in unheated areas and do not have to be installed so that they can be drained down effectively, although it is still preferable to install the piping so that it can be drained down.

Planning a Solar Heating System

Some people think of solar collectors as running at high temperatures. In reality, this is exactly opposite of what is desirable. Ideally, collectors should operate at temperatures that are as low as possible. When collectors operate at low temperatures, they are more efficient. The lower operating temperature allows a system to collect more energy.

It is rare for an active solar heating system to provide complete heating for a home or building. While the solar heating system may be the primary system, it is usually backed up by some other type of system. Without a backup system, it would be very difficult to have regular, dependable heat output. Radiant heating systems that are installed in floors, especially concrete floors, usually are the most effective way to distribute heat from an active solar heating system. Since radiant heat in a floor can perform well at lower water temperatures, it makes itself ideal for use with solar collectors. Hydronic fan-coil units that are sized to work with low water temperatures are also a viable means of distributing heat from an active solar heating system.

Backup heating systems used in conjunction with solar heating systems should be sized to work independently in heating the required space without assistance from the solar heating system. However, the auxiliary heating system must not maintain the thermal storage tank at temperatures suitable for use by the distribution system during nonsolar periods. This is a mistake that some people make, so retain the fact in your mind.

When the solar heating system is in use, the storage tanks must be allowed to cool down to room temperature. This is how the tank transfers its heat. Assuming that the tank is located in heated space, the heat from the tank cools off and offsets a portion of the heating load in the structure. Solar collectors will reestablish the higher temperature of the tank and allow the process to repeat itself.

The Cost Factor

The cost factor is probably the largest contributor in deciding when a solar heating system is sensible. Since a full-scale backup heating system of a traditional design should be installed in conjunction with a solar heating system, the cost of installing a solar system is substantially higher than the cost of installing a typical heating system. Customers are paying for a traditional heating system and then paying extra, usually thousands of dollars in extra money, for the benefit of heating with energy from the sun. This added cost is not always practical.

The decision on whether or not to go solar revolves around cost and energy conservation. Certainly, solar heating systems do reduce the need for fuel oil or electricity when running a heating system. However, the cost savings may not be enough to justify the installation of a solar heating system. When money is a motivating factor, and it usually is, a customer has to determine how long is too long to wait for a payback from the installation cost of a solar system. With present prices of heating fuels, it can take many years to recover the cost of installing a solar system. And, if the installation is being done in a region where extremely cold temperatures are encountered, the auxiliary heating system may be carrying the majority of the heating load. A lot of thought and research is required to install a solar heating system that will pay for itself in a reasonable period of time.

Depending on an active solar heating system to function properly with some types of heating, such as hot-water baseboard heating units, can be asking too much of the solar system. As noted earlier, radiant heat pipes embedded in concrete floors are the most sensible means of heat distribution when the heat source is the sun. Wall-mounted hot-water convectors require a much higher water temperature to operate well than in-floor radiant heat does.

Before any active solar heating system is installed, experts in the design and function of the system should be consulted. It will be expensive to have a professional design a complex solar heating system, but the money will have been well spent. If such a system is designed by someone with limited solar experience, an expensive system may be installed that simply will not perform adequately.

Solar energy definitely has its advantages, but it also has disadvantages. Each job has to be considered on an individual basis. The topography and location of a building lot may make a solar system impractical. Geographic regions vary in how well they will allow a solar heating system to work. The cost of installing a solar system can add thousands of dollars to the cost of building a home. Maintenance of a solar system is an additional burden that is not required when only a traditional heating system is used. To fully plan a system and to make a wise buying decision, each potential buyer must have expert advice that is directed to a particular case study. If you are considering the installation of a solar heating system, talk with experienced professionals and spend enough time in the planning stages to avoid the displeasure so many people have suffered from poorly planned systems.

CHAPTER

21

A Basic Overview of Wood-Fired Heating Systems

Wood has served as a heating fuel from generation to generation. Whether placed in a campfire, a fireplace, a wood stove, or a boiler, wood is a dependable source of heat. And, it is a renewable resource. A great number of people rely on wood for some or all of their home heating needs. Wood stoves are very popular in rural areas, and wood-fired boilers can be found in houses all across cold climates. When a home is equipped with an efficient wood-burning heating unit, the cost of fuel can be quite low. Some people cut their own wood and others buy it. There are, however, drawbacks to burning wood.

Anyone who has never used wood as a primary heat source may not realize what a major time commitment the process is. During different phases of my life I have heated my homes entirely with wood stoves. I enjoy the smell of wood burning, and the heat from a wood stove is unlike any other type of heater. When I was younger, cutting, splitting, stacking, and hauling wood was no big deal. In fact, it was good exercise and a great stress reliever. But there is no mistaking the fact that burning wood is a lot of work, and the dust created from wood stoves is a bother.

A friend of mine has a wood-fired boiler. He cuts his own wood and heats his home exclusively with the wood-fired boiler and hot-water baseboard heating elements. Like a wood stove, a wood-fired boiler requires a lot of effort to keep it running. The physical requirements of moving and loading wood can be too much for some people. Time, however, is one of the biggest problems most people encounter with wood-burning heaters. Unless someone is home most of the time during the heating season, getting a wood-burning heater to keep a house warm is difficult.

Air-tight wood stoves are capable of burning for several hours without reloading them with new wood. Wood-fired boilers have large fireboxes so that a lot of wood can be loaded at one time, and they too will burn for hours. Some wood-fired boilers are made as combination boilers. This means that they can burn more than one type of fuel. For example, a boiler that will burn wood may also be capable of burning fuel oil. Almost any boiler that will burn wood can also be set up to burn coal. Having a combination boiler that will burn wood or oil is a very good idea.

The benefits of combination boilers include being able to leave the boiler unattended for days at a time, when it will be running from the burning of fuel oil. If a home is equipped with only a wood-fired boiler, leaving the home during freezing temperatures for an extended stay somewhere else is not practical, as the plumbing in the home will freeze. A combination boiler eliminates this problem.

Wood heat became very popular in the mid-1970s. This is about the same time that solar power was becoming popular. The oil crisis and high prices of fuel oil had people looking for alternative heating options. Both solar and wood gained popularity, but wood maintained its following more successfully. Solar heating became less fashionable over time, but wood heating remains popular. During the growth process of wood heat a number of new boilers and wood stoves were designed. Today's modern wood-fired boilers are the result of many years of research.

Many Choices

People in the market for a wood-burning boiler have many choices available to them. Some wood-fired boilers rely on the natural draft created by a standard chimney. Others incorporate the use of a small blower to force air into the combustion chamber. Boilers that are called solid fuel boilers are designed to burn only wood or coal. Most solid fuel boilers can be set up to burn either wood or coal. These boilers partially control the heat output of a fire with the use of a thermostatically controlled air damper. Combustion chambers for solid fuel boilers are quite large to accommodate large loads of wood. There is an ash cleanout drawer located at the bottom of these boilers that must be maintained periodically.

Other choices include combination boilers that can use solid fuel or fuel oil. These boilers are more expensive, but well worth the additional cost. Having a combination boiler makes it possible to burn wood when it's convenient and oil when it's not suitable to burn wood. Some combination boilers have a single combustion chamber for both types of fuel. This type of set up is not ideal. When a single combustion chamber is used for both wood and oil, it is not possible for the boiler to operate as efficiently. A compromise is required to set the combustion chamber up for the two types of fuel. To make combination boilers more efficient, some manufacturers build boilers with two combustion chambers. This allows each fuel type to have an independent combustion chamber and results in a higher efficiency rating.

Problems can arise when a combination oil- and wood-burner is used. It's not uncommon for the soot and fly ash from the wood-fired operation to coat the head of the oil burner. This happens when the flame head of the oil burner is located in the same combustion chamber that is used for burning wood. Under such circumstances, it is necessary to clean the flame head frequently. This problem does not occur with boilers that are equipped with separate combustion chambers. Some manufacturers offer a double-door system on their single-chamber combination boilers. When this is the case, a user chooses which door to close, depending upon the type of fuel being used. With this type of design, the oil burner's flame head is not in the combustion chamber when wood is burning, so there is far less soot and fly ash to contend with. The disadvantage to the system is that the boiler has to be in one mode or the other and cannot have the oil fuel kick in automatically if the wood fire goes out.

When wood-fired boilers are used, they must basically burn out the wood that is in them. Unlike other fuel types which can be turned on and off, wood cannot be turned off. It is possible to reduce the intensity of a wood flame by closing an air damper, but this is not something that should be done too often or for long periods of time.

When a damper is closed, the fire in the combustion chamber burns slower and colder, creating creosote. This means that experience is required to regulate a wood-fired boiler so that it will not burn too hot or too cold.

Efficiency

The efficiency of a wood-fired boiler is difficult to predict. Heat output from a wood-fired boiler is highly dependent on the quality of the firewood being burned. If the wood is unseasoned, has a high moisture content, or is a type of wood that gives off low Btus, the boiler cannot produce as well as it would with seasoned hardwood as a fuel. Efficiency estimates on wood-fired boilers range from a low of about 50 percent to a high of about 80 percent. However, you must keep in mind that any efficiency rating with a wood-fired boiler is subject to the quality of the wood being burned.

In order to have high efficiency from a wood-fired boiler, the boiler must reach a high combustion temperature. For example, a wood-fired boiler that is equipped with a forced draft might hit a combustion temperature of 2,000°, which is plenty for good efficiency. At this temperature, the pyrolytic gases released from the firewood ignite and burn hot. This reduces the buildup of creosote in the chimney. However, if the combustion temperature doesn't rise to an excess of approximately 1,100°, the pyrolytic gases will not ignite and they will coat the chimney with a creosote buildup.

Transfer Storage Tanks

Wood-fired boilers lose efficiency when they are forced to burn out at low temperatures and creosote is created. One way to overcome this problem is the installation of a transfer storage tank. Instead of cutting a boiler back to a low burning temperature, the boiler is allowed to burn at a high temperature, with the unwanted heat being transferred to the storage tank. Then, when the heat is wanted from the storage tank, it can be delivered through a hydronic distribution system. Another advantage to installing a storage tank is that the firing time for the boiler is reduced to only a few hours a day. This makes it easier for people who cannot be around their home for hours at a time to keep the boiler burning wood.

If a heat storage tank is going to be used, the boiler must have a much higher heat output rating than what would be needed without the storage tank. The reason for this is that the boiler has to produce 24 hours worth of heat in only a few hours. It is generally recommended to use a forced-draft type of boiler when a storage tank is installed. This is because the forced-draft boiler can produce a higher firing rate. Don't expect a system like this to be cheap. The larger boiler and the insulated storage tank run up the cost of installation.

Another Option

Another option when a wood-fired boiler is desired but another fuel type is also wanted is to install two different systems. To obtain optimum performance with minimum problems, both a wood-fired boiler and an oil-fired boiler should be installed. This, of course, is expensive and is only worthwhile when wood is available at a reasonable cost and will be used steadily for heating purposes. There are different ways to pipe the two boilers to work together. One way is to set one boiler up as a primary boiler and have the other boiler piped as a secondary heat source. It's also possible to pipe the two boilers in a parallel or a series fashion.

When boilers are installed in a series piping design, with the wood boiler being the primary boiler, the wood boiler burns first. When the wood-fired boiler drops to a colder temperature, a temperature control brings on the conventional boiler to supplement and eventually take over for the wood-fired boiler. While this type of hookup is simple to install, it does not offer the most efficiency.

Using a parallel piping system allows each boiler to have its own circulator and check valve. This improves efficiency. The small cost of the extra circulator and check valve is returned soon in higher efficiency. Controls installed for the boilers allow the wood-fired boiler to run until it reaches a colder temperature and then the conventional boiler kicks in. The controls should be set to avoid having both boilers running at the same time. If they are both running simultaneously, the flow rate on the system will be increased and may become noisy.

Choosing between a combination boiler and the use of two separate boilers is a personal decision. Cost is clearly a factor. The highest efficiency is achieved with the use of two independent boilers, but the cost may be more than a customer is willing to pay for. All factors must be taken into consideration when choosing a wood-fired system. For some people, wood-fired boilers are an excellent choice, but they are not for everyone.

CHAPTER

22

Working with Coal-Fired Heating Systems

Coal-fired heating systems are not often used in modern homes. Some people like coal-fired stoves and continue to use them, but coal-fired boilers are rare in modern construction. However, most boilers that will burn wood can also be set up to burn coal. These are called solid fuel boilers. Coal is a good fuel, but it's dirty to handle and requires a substantial amount of space for storage. If a boiler has to be fed coal by hand on a regular basis, like loading wood into a boiler, the time commitment and effort can be more than an average person might want to deal with. Many older homes and buildings do still have coal-fired heating systems, and there are probably some new homes being built where coal will be responsible for the heating system's operation.

People who shovel coal into their heating system are faced with the routine of making sure that the fire doesn't go out. Of course, if they are using a combination boiler, this is not a major issue. But, the opening of the door to the combustion chamber to fuel the fire lets colder air in that cools the fire's temperature. It is also a fact that after loading a coal fire there is a period of smoking that takes place prior to normal combustion conditions, and this reduces efficiency. As with any solid fuel, the fuel tends to burn down to a level where it is not at peak efficiency. All of these are drawbacks to hand-loading coal. The alternative is an automatic loader that is most often called a stoker.

Coal and wood are both solid fuels and usually both can be used in the same boiler, but not under the same conditions. Wood has different characteristics than coal, and a boiler that is capable of using either fuel must be set up for one type or the other. In the case of coal, the draft requirements depend on several factors. The size of the coal grate, the size and type of the coal to be burned, the thickness of the fuel bed, and the boiler pass resistance all affect the draft requirements. Perhaps that most significant factor is the degree of resistance offered by the boiler. If the gases can't pass through the boiler fast enough, they will back up in the combustion chamber and take away oxygen that is needed to keep the fire burning hot.

Types of Coal

There are different types of coal that can be burned. Three of the main types are anthracite, bituminous, and semibituminous coal. Of the three, anthracite coal is the one used most often in residential applications. This type of coal burns a cleaner flame and produces a steadier heat than the other types of coal. It also burns longer than the other types of coal. While anthracite coal holds many advantages, it does cost more than other types of coal and it does require a higher temperature for combustion. Still, it is the coal of choice in residential applications.

Since coal should be purchased in a size that will work well with the fire grate and heating unit being fueled, it is common for coal to come in different sizes. Anthracite coal, for example, is available in many sizes, some of which are listed below:

- Buckwheat
- Stove
- Pea
- Egg
- Chestnut

Buckwheat-size coal comes in four sizes or grades. When using buckwheat coal, you should always maintain an even, low fire. A smaller mesh grate or a domestic stoker should be used. Make sure that new coal is exposed to hot coal so that combustion will not be delayed. And, keep the heating system warm at all times. If a system is allowed to cool down, the amount of time and coal required to bring the system back up to temperature is excessive.

Egg-sized coal is typically used in large firepots with 24-inch grates or even larger grates. To obtain maximum performance this type of coal should be placed in a bed that is at least 16 inches deep. When stove-sized coal it used, it should be used with a grate that is at least 16 inches in size and the bed of coal should be a minimum of 12 inches deep. Chestnut-size coal is used in firepots that are up to 20 inches in diameter, and the recommended bed depth ranges from 10 to 15 inches. Pea-sized coal can be used on a standard grate and the bed depth should be added to gradually until it reaches the sill of the fire door. A strong draft is required when burning pea-sized coal.

Bituminous Coal

Bituminous coal ignites and burns easily. The flame from this type of coal is long and produces excessive smoke and soot when it is not fired properly. To avoid smoking and soot buildup a side banking method of coal placement is recommended for bituminous coal. Basically, this means moving hot coals to one side of the grate and placing new coal on the opposite side. Side banking allows for a slower and more even release of volatile gases.

When feeding a fire with bituminous coal, you should add small quantities of coal at a time. This allows for better combustion. You should never fire bituminous coal over the entire fuel bed at a single loading. Doing so will result in a very smoky fire. A stocking bar can be used to break up a fresh charge of coking coal about 20 to 60 minutes after firing the coal. The stoking bar should not be brought up to the surface of the coal. When the stoking bar breaks up the fired coal, ashes should fall through the grate without any need for being shaken. Due to its tendency for smoking and sooting, bituminous coal is not an ideal choice for residential applications.

Semibituminous Coal

Semibituminous coal burns with a great deal less smoke than bituminous coal. However, semibituminous coal lights with greater difficulty than bituminous coal. A central cone method of loading is recommended for semibituminous coal. The coal is heaped into a pile to form a cone shape. Large lumps of coal fall to the sides and the fine cones remain in the center. Poking the coal in a cone formation should be done so that the cone is not stirred on knocked over. Regardless of what type of coal is used, the question of how to get it into the firebox is one that must be answered, so let's look at this question now.

Stokers

Many a ton of coal has been shoveled into a fire, but few people today opt for the manual labor associated with feeding a coal fire. Stokers are normally used to deliver coal to a firebox. There are four basic classes of stokers normally used with different types of heating systems. A Class 1 stoker delivers between 10 to 100 pounds of coal per hour and is the type most often associated with residential applications. These stokers are usually an underfeed type that is designed to burn anthracite, bituminous, semibituminous, lignite coal, and coke. Ash deposits can be removed automatically or manually. Anthracite coal is the type most often used in domestic heating systems. Underfeed stokers feed the coal upward from underneath the boiler. The action of a worm or screw carries the coal back through a retort from which it passes upward as the fuel above is being consumed. Ash from the fire is left on dead plates on either side of the retort.

Class 2 stokers deliver between 100 and 300 pounds of coal per hour. Larger heating systems can be served by Class 3 stokers at a rate of 300 to 1,200 pounds per hour or a Class 4 stoker that will produce over 1,200 pounds of coal per hour. Some systems are designed with stokers that work with hoppers. The stokers work well, but the hopper must be filled routinely with coal. If a stoker works with a bin, instead of a hopper, the coal doesn't have to be loaded by hand. When a coal delivery is made to the bin, that is the end of the handling process; the stoker does the rest.

Start-Up Procedures

Start-up procedures for coal-fired heating systems vary. There are, however, some general procedures which usually apply. Room thermostats should be set above room temperature so that they will call for heat. Both the coal feed and air settings must be set to their proper rate.

When the line switch is in the on position, the stoker can start. If the stoker has a hopper, fill it with coal and set the overfire air door about one-quarter to one-half the way open. Once the stoker has filled the retort with coal, the fire must be started. This is often done by putting paper, kindling wood, an a little coal on top of the retort pile and lighting it.

The draft for a coal-fired heating unit should be as little as possible without causing smoking from the fire door. As a fuel bed builds up, an adjustment of the air damper may be needed. The goal is to get a fire that is burning yellow with hardly any smoke. Air control is very important in maintaining a consistent burn rate. Automatic air controls don't normally require any manual adjustment. Motors and transmissions on stokers should be equipped with overload protection. Reset buttons should be present on both the motor and transmission that can be pushed to restart the equipment after the cause of an overload is eliminated.

The general maintenance for a coal-fired heating system is not too intense. Motors should usually be lubricated before the heating season and at least twice during the heating season. Any high grade of motor oil should be fine, but always read and comply with manufacturer recommendations. Transmissions normally require about 1 pint of good engine oil for lubrication. Some people change transmission oil once every 2 years and others change it annually.

Troubleshooting

Troubleshooting coal-fired systems is unlike working with an oil-fired boiler. If there is a problem where noise is the complaint, you should check to see if any pulleys or belts are loose. If they are, either tighten or replace them. Noise may come from dry motor bearings; oiling the bearings will correct this problem. Worn gears and gears that need to be oiled can make noise. Once you check the gears, you should either oil or replace them, subject to their condition.

There are usually three reasons why a motor will not start. If there are hard clinkers over or on the retort, a motor may not start. To remedy this problem you must remove the clinkers. If something gets caught in the feed screw of a stoker, it can cause the motor to not start.

Removing whatever is blocking the feed screw will solve this problem. Sometimes the end of the feed screw becomes worn to a point where it causes coal to be packed in a retort. This can keep a motor from starting. Removing the coal from the retort and replacing the worn part will solve the problem.

When a stoker won't shut off, it may mean that the controls are out of adjustment, that there is a dirty fire, no fire, or a dirty heating unit. If the fire is dirty, it must be rebuilt to burn cleanly. A fire that is out must be restarted. Clean the heating unit if it is dirty and correct control adjustments if needed.

If a heating system becomes filled with unburned coal, you should check for clinkers that may be clogging the retort. Remove any clinkers that are in the way. Sometimes the coal feed is set too high and causes an overstock of coal. When this happens, adjust the feed setting. A lack of air in the firebox can cause a buildup of coal. Open the manual damper first. If the damper being open doesn't solve the problem, check the air ports to see if they are clogged and clear them if they are. Also, check the wind box; it may be full of siftings that should be removed.

When you are faced with a stoker that will not run, you should look for a problem with the limit control. It may have shut the system down due to high temperatures. Let the limit control cool off and then see if the stoker will run. Another possibility for a stoker not running is that a boiler has shut down due to low water. Check the boiler to see that it has an adequate supply of water. A less likely cause of the problem is a gear case that has been submerged in water. If this has happened, you should drain and flush the gear case immediately, with the stoker turned off, and then refill the gear case with oil.

If there is smoke backing up into a hopper, the hopper may be low on coal and may need to be refilled. There could be a clinker blocking the retort that must be removed, or there could be a clog in the smokeback connection that must be cleared. It's also possible that there is a fire burning down in the retort. If this is the case, check the air supply to see that the fire is not getting too much air. You should also check to see that the coal feed is not too low.

When there is no fire in the combustion chamber, you should check to see that enough coal is being delivered to burn properly. Remove any clinkers that are blocking the retort and check to see if the switch to the unit may have been turned off accidentally. Check the fuse box or circuit breaker box to make sure that the failure is not due to a tripped circuit or blown fuse. The last likely cause will be a failure in the electric controls. Refer to the control specifications to confirm that they are functioning properly.

Coal is not a common fuel for residential heating purposes any longer, but there are still coal-fired heating systems in use. Very few new homes are built with coal-fired heating systems. It is unlikely that the average heating company serving residential customers will have much contact with coal, but if you do, you now have a pretty good idea of how the systems work and what to look for when they are not operating properly.

Index